QUALITATIVE METHODS IN
NONLINEAR DYNAMICS

PURE AND APPLIED MATHEMATICS

A Program of Monographs, Textbooks, and Lecture Notes

MONOGRAPHS AND TEXTBOOKS IN
PURE AND APPLIED MATHEMATICS

113. *D. L. Stancl and M. L. Stancl*, Real Analysis with Point-Set Topology (1987)
114. *T. C. Gard*, Introduction to Stochastic Differential Equations (1988)
115. *S. S. Abhyankar*, Enumerative Combinatorics of Young Tableaux (1988)
116. *H. Strade and R. Farnsteiner*, Modular Lie Algebras and Their Representations (1988)
117. *J. A. Huckaba*, Commutative Rings with Zero Divisors (1988)
118. *W. D. Wallis*, Combinatorial Designs (1988)
119. *W. Wiesław*, Topological Fields (1988)
120. *G. Karpilovsky*, Field Theory (1988)
121. *S. Caenepeel and F. Van Oystaeyen*, Brauer Groups and the Cohomology of Graded Rings (1989)
122. *W. Kozlowski*, Modular Function Spaces (1988)
123. *E. Lowen-Colebunders*, Function Classes of Cauchy Continuous Maps (1989)
124. *M. Pavel*, Fundamentals of Pattern Recognition (1989)
125. *V. Lakshmikantham et al.*, Stability Analysis of Nonlinear Systems (1989)
126. *R. Sivaramakrishnan*, The Classical Theory of Arithmetic Functions (1989)
127. *N. A. Watson*, Parabolic Equations on an Infinite Strip (1989)
128. *K. J. Hastings*, Introduction to the Mathematics of Operations Research (1989)
129. *B. Fine*, Algebraic Theory of the Bianchi Groups (1989)
130. *D. N. Dikranjan et al.*, Topological Groups (1989)
131. *J. C. Morgan II*, Point Set Theory (1990)
132. *P. Biler and A. Witkowski*, Problems in Mathematical Analysis (1990)
133. *H. J. Sussmann*, Nonlinear Controllability and Optimal Control (1990)
134. *J.-P. Florens et al.*, Elements of Bayesian Statistics (1990)
135. *N. Shell*, Topological Fields and Near Valuations (1990)
136. *B. F. Doolin and C. F. Martin*, Introduction to Differential Geometry for Engineers (1990)
137. *S. S. Holland, Jr.*, Applied Analysis by the Hilbert Space Method (1990)
138. *J. Oknínski*, Semigroup Algebras (1990)
139. *K. Zhu*, Operator Theory in Function Spaces (1990)
140. *G. B. Price*, An Introduction to Multicomplex Spaces and Functions (1991)
141. *R. B. Darst*, Introduction to Linear Programming (1991)
142. *P. L. Sachdev*, Nonlinear Ordinary Differential Equations and Their Applications (1991)
143. *T. Husain*, Orthogonal Schauder Bases (1991)
144. *J. Foran*, Fundamentals of Real Analysis (1991)
145. *W. C. Brown*, Matrices and Vector Spaces (1991)
146. *M. M. Rao and Z. D. Ren*, Theory of Orlicz Spaces (1991)
147. *J. S. Golan and T. Head*, Modules and the Structures of Rings (1991)
148. *C. Small*, Arithmetic of Finite Fields (1991)
149. *K. Yang*, Complex Algebraic Geometry (1991)
150. *D. G. Hoffman et al.*, Coding Theory (1991)
151. *M. O. González*, Classical Complex Analysis (1992)
152. *M. O. González*, Complex Analysis (1992)
153. *L. W. Baggett*, Functional Analysis (1992)
154. *M. Sniedovich*, Dynamic Programming (1992)
155. *R. P. Agarwal*, Difference Equations and Inequalities (1992)
156. *C. Brezinski*, Biorthogonality and Its Applications to Numerical Analysis (1992)
157. *C. Swartz*, An Introduction to Functional Analysis (1992)
158. *S. B. Nadler, Jr.*, Continuum Theory (1992)
159. *M. A. Al-Gwaiz*, Theory of Distributions (1992)
160. *E. Perry*, Geometry: Axiomatic Developments with Problem Solving (1992)
161. *E. Castillo and M. R. Ruiz-Cobo*, Functional Equations and Modelling in Science and Engineering (1992)
162. *A. J. Jerri*, Integral and Discrete Transforms with Applications and Error Analysis (1992)
163. *A. Charlier et al.*, Tensors and the Clifford Algebra (1992)
164. *P. Biler and T. Nadzieja*, Problems and Examples in Differential Equations (1992)
165. *E. Hansen*, Global Optimization Using Interval Analysis (1992)
166. *S. Guerre-Delabrière*, Classical Sequences in Banach Spaces (1992)
167. *Y. C. Wong*, Introductory Theory of Topological Vector Spaces (1992)
168. *S. H. Kulkarni and B. V. Limaye*, Real Function Algebras (1992)
169. *W. C. Brown*, Matrices Over Commutative Rings (1993)
170. *J. Loustau and M. Dillon*, Linear Geometry with Computer Graphics (1993)
171. *W. V. Petryshyn*, Approximation-Solvability of Nonlinear Functional and Differential Equations (1993)

172. *E. C. Young*, Vector and Tensor Analysis: Second Edition (1993)
173. *T. A. Bick*, Elementary Boundary Value Problems (1993)
174. *M. Pavel*, Fundamentals of Pattern Recognition: Second Edition (1993)
175. *S. A. Albeverio et al.*, Noncommutative Distributions (1993)
176. *W. Fulks*, Complex Variables (1993)
177. *M. M. Rao*, Conditional Measures and Applications (1993)
178. *A. Janicki and A. Weron*, Simulation and Chaotic Behavior of α-Stable Stochastic Processes (1994)
179. *P. Neittaanmäki and D. Tiba*, Optimal Control of Nonlinear Parabolic Systems (1994)
180. *J. Cronin*, Differential Equations: Introduction and Qualitative Theory, Second Edition (1994)
181. *S. Heikkilä and V. Lakshmikantham*, Monotone Iterative Techniques for Discontinuous Nonlinear Differential Equations (1994)
182. *X. Mao*, Exponential Stability of Stochastic Differential Equations (1994)
183. *B. S. Thomson*, Symmetric Properties of Real Functions (1994)
184. *J. E. Rubio*, Optimization and Nonstandard Analysis (1994)
185. *J. L. Bueso et al.*, Compatibility, Stability, and Sheaves (1995)
186. *A. N. Michel and K. Wang*, Qualitative Theory of Dynamical Systems (1995)
187. *M. R. Darnel*, Theory of Lattice-Ordered Groups (1995)
188. *Z. Naniewicz and P. D. Panagiotopoulos*, Mathematical Theory of Hemivariational Inequalities and Applications (1995)
189. *L. J. Corwin and R. H. Szczarba*, Calculus in Vector Spaces: Second Edition (1995)
190. *L. H. Erbe et al.*, Oscillation Theory for Functional Differential Equations (1995)
191. *S. Agaian et al.*, Binary Polynomial Transforms and Nonlinear Digital Filters (1995)
192. *M. I. Gil'*, Norm Estimations for Operation-Valued Functions and Applications (1995)
193. *P. A. Grillet*, Semigroups: An Introduction to the Structure Theory (1995)
194. *S. Kichenassamy*, Nonlinear Wave Equations (1996)
195. *V. F. Krotov*, Global Methods in Optimal Control Theory (1996)
196. *K. I. Beidar et al.*, Rings with Generalized Identities (1996)
197. *V. I. Arnautov et al.*, Introduction to the Theory of Topological Rings and Modules (1996)
198. *G. Sierksma*, Linear and Integer Programming (1996)
199. *R. Lasser*, Introduction to Fourier Series (1996)
200. *V. Sima*, Algorithms for Linear-Quadratic Optimization (1996)
201. *D. Redmond*, Number Theory (1996)
202. *J. K. Beem et al.*, Global Lorentzian Geometry: Second Edition (1996)
203. *M. Fontana et al.*, Prüfer Domains (1997)
204. *H. Tanabe*, Functional Analytic Methods for Partial Differential Equations (1997)
205. *C. Q. Zhang*, Integer Flows and Cycle Covers of Graphs (1997)
206. *E. Spiegel and C. J. O'Donnell*, Incidence Algebras (1997)
207. *B. Jakubczyk and W. Respondek*, Geometry of Feedback and Optimal Control (1998)
208. *T. W. Haynes et al.*, Fundamentals of Domination in Graphs (1998)
209. *T. W. Haynes et al.*, Domination in Graphs: Advanced Topics (1998)
210. *L. A. D'Alotto et al.*, A Unified Signal Algebra Approach to Two-Dimensional Parallel Digital Signal Processing (1998)
211. *F. Halter-Koch*, Ideal Systems (1998)
212. *N. K. Govil et al.*, Approximation Theory (1998)
213. *R. Cross*, Multivalued Linear Operators (1998)
214. *A. A. Martynyuk*, Stability by Liapunov's Matrix Function Method with Applications (1998)
215. *A. Favini and A. Yagi*, Degenerate Differential Equations in Banach Spaces (1999)
216. *A. Illanes and S. Nadler, Jr.*, Hyperspaces: Fundamentals and Recent Advances (1999)
217. *G. Kato and D. Struppa*, Fundamentals of Algebraic Microlocal Analysis (1999)
218. *G. X.-Z. Yuan*, KKM Theory and Applications in Nonlinear Analysis (1999)
219. *D. Motreanu and N. H. Pavel*, Tangency, Flow Invariance for Differential Equations, and Optimization Problems (1999)
220. *K. Hrbacek and T. Jech*, Introduction to Set Theory, Third Edition (1999)
221. *G. E. Kolosov*, Optimal Design of Control Systems (1999)
222. *N. L. Johnson*, Subplane Covered Nets (2000)
223. *B. Fine and G. Rosenberger*, Algebraic Generalizations of Discrete Groups (1999)
224. *M. Väth*, Volterra and Integral Equations of Vector Functions (2000)
225. *S. S. Miller and P. T. Mocanu*, Differential Subordinations (2000)

Additional Volumes in Preparation

QUALITATIVE METHODS IN NONLINEAR DYNAMICS

Novel Approaches to Liapunov's Matrix Functions

A. A. Martynyuk

Institute of Mechanics
National Academy of Sciences of Ukraine
Kiev, Ukraine

CRC Press
Taylor & Francis Group
Boca Raton London New York

CRC Press is an imprint of the
Taylor & Francis Group, an **informa** business

CRC Press
Taylor & Francis Group
6000 Broken Sound Parkway NW, Suite 300
Boca Raton, FL 33487-2742

First issued in paperback 2019

© 2002 by Taylor & Francis Group, LLC
CRC Press is an imprint of Taylor & Francis Group, an Informa business

No claim to original U.S. Government works

ISBN-13: 978-0-8247-0735-4 (hbk)
ISBN-13: 978-0-367-39677-0 (pbk)

**Visit the Taylor & Francis Web site at
http://www.taylorandfrancis.com**

**and the CRC Press Web site at
http://www.crcpress.com**

PREFACE

An important place among modern qualitative methods in nonlinear dynamics of systems is occupied by those associated with the development of Poincaré's and Liapunov's ideas for investigating nonlinear systems of differential equations.

Liapunov divides into two categories all methods for the solution of the problem of stability of motion. He includes in the first category those methods that reduce the consideration of the disturbed motion to the determination of the general or particular solution of the equation of perturbed motion. It is usually necessary to search for these solutions in a variety of forms, of which the simplest are those that reduce to the usual method of successive approximations. Liapunov calls the totality of all methods of this first category the "first method".

It is possible, however, to indicate other methods of solution of the problem of stability which do not necessitate the calculation of a particular or the general solution of the equations of perturbed motion, but which reduce to the search for certain functions possessing special properties. Liapunov calls the totality of all methods of this second category the "second method".

During the post-Liapunov period both the first and second Liapunov's methods have been developed considerably. The second method, or the direct Liapunov method, based first on scalar auxiliary function, was replenished with new ideas and new classes of auxiliary functions. This allowed one to apply this fruitful technique in the solution of many applied problems. The ideas of the direct Liapunov method are the source of new modern techniques of qualitative analysis in nonlinear systems dynamics. A considerable number of publications appearing annually in this direction provide a modern tool for qualitative analysis of processes and phenomena in the real world.

The aim of this monograph is to introduce the reader to a new direction in nonlinear dynamics of systems. This direction is closely connected with a new class of matrix-valued function of particular importance in construction of an appropriate Liapunov function for the system under consideration.

It is known that the problem of stability is important not only for the continuous systems modeled by ordinary differential equations. Therefore, in this monograph the methods of qualitative analysis are presented for discrete-time and impulsive systems. Further, in view of the importance of the problem of estimating the domains of asymptotic stability, a new method for its solution is set out in a separate chapter.

The monograph contains five chapter and is arranged as follows.

The first chapter contains all necessary results associated with the method of matrix-valued Liapunov functions. It also provides general information on scalar and vector functions including the cone-valued ones. General theorems on various types of stability of the equilibrium state of the systems cited in this chapter are basic for establishing the sufficient stability tests in subsequent chapters.

The second chapter deals with the construction of matrix-valued functions and corresponding scalar auxiliary Liapunov functions. Here new methods of the initial system decomposition are discussed, including those of hierarchical decomposition. The corresponding sufficient tests for various types of stability and illustrative examples are presented for every case under consideration. Along with the classical notion of stability major attention is paid to new types of motion stability, in particular, to the exponential polystability of separable motions as well as the integral and Lipschitz stability.

The third chapter addresses the methods of stability analysis of discrete-time systems. Our attention is focussed mostly on the development of the method of matrix-valued functions in stability theory of discrete-time systems.

In the fourth chapter the problems of dynamics of nonlinear systems in the presence of impulsive perturbations are discussed. The method of matrix-valued Liapunov functions is adapted here for the class of impulsive systems that were studied before via the scalar Liapunov function. The proposed development of the direct Liapunov method for the given class of systems enables us to make an algorithm constructing the appropriate Liapunov functions and to increase efficiency of this method.

In the final chapter the problem of estimating the domains of asymptotic stability is discussed in terms of the method of matrix-valued Liapunov functions. By means of numerous examples considered earlier by Abdullin, Anapolskii, *et al.* [1], Michel, Sarabudla, *et al.* [1], and Šiljak [1] it is shown that the application of matrix-valued functions involves an essential extension of the domains of asymptotic stability constructed previously.

I wish to acknowledge the essential technical assistance provided by my colleagues in the Stability of Processes Department of S.P.Timoshenko Institute of Mechanics, National Academy of Sciences of Ukraine.

The bibliographical information used in the monograph was checked by CD-ROM Compact MATH, which was kindly provided by Professor, Dr. Bernd Wegner and Mrs. Barbara Strazzabosco from the Zentralblatt MATH.

I express my sincere gratitude to all persons mentioned above. I am also grateful to the staff of Marcel Dekker, Inc., for their initiative and kind assistance.

A. A. Martynyuk

CONTENTS

1

PRELIMINARIES

1.1 Introduction

This chapter contains an extensive overview of the qualitative methods in nonlinear dynamics and is arranged as follows.

Section 1.2 is short and gives information about continuous nonlinear systems that is important for applications in investigation of the mechanical, electrical and electromechanical systems. Also discussed are the equations of perturbed motion of nonlinear systems which are the object of investigation in this monograph.

For the reader's convenience, in Section 1.3 the definitions we use of motion stability of various types are formulated. These formulations result from an adequate description of stability properties of nonlinear and nonautonomous systems.

Section 1.4 deals with three classes of Liapunov functions: scalar, vector and matrix-valued ones, as well as the possibilities of their application in motion stability theory. Along with the well-known results, some new notions are introduced, for example, the notion of the "Liapunov metafunction".

Basic theorems of the comparison principle for SL-class and VL-class of the Liapunov functions are set out in Section 1.5. Also, some important corollaries of the comparison principle related to the results of Zubov are presented here.

Section 1.6 deals with generalization of the main Liapunov and Barbashin-Krasovskii theorems established by the author in terms of matrix-valued functions. Some corollaries of general theorems contain new sufficient stability (instability) tests for the equilibrium state of the system under consideration.

In Section 1.7 the vector and cone-valued functions are applied in the problem of stability with respect to two measures and in stability theory of large scale systems. Detailed discussion of possibilities of these approaches may prove to be useful for many beginners in the field.

In the final Section 1.8, the formulations of theorems of the direct Liapunov method are set out based on matrix-valued functions and intended for application in stability investigation of large scale systems.

Generally, the results of this chapter are necessary to get a clear idea of the results presented in Chapters 2–5. Throughout Chapters 2–5 references to one or an other section of Chapter 1 are made.

1.2 Nonlinear Continuous Systems

1.2.1 General equations of nonlinear dynamics

The systems without nonintegrability differential constraints represent a wide class of mechanical systems with a finite number of degrees of freedom. Let the state of such system in the phase space R^n, $n = 2k$, be determined by the vectors

$$q = (q_1, \ldots, q_k)^{\mathrm{T}} \quad \text{and} \quad \dot{q} = (\dot{q}_1, \ldots, \dot{q}_k)^{\mathrm{T}}.$$

It is known that the general motion equations of such a mechanical system are

(1.2.1) $$\frac{d}{dt}\left(\frac{\partial T}{\partial \dot{q}_s}\right) - \frac{\partial T}{\partial q_s} = U_s, \quad s = 1, 2, \ldots, k.$$

Here T is the kinetic energy of the mechanical system and U_s are the generalized forces.

The system of equations (1.2.1) is simplified, if for the forces affecting the system a force function $U = U(t, q_1, \ldots, q_k)$ exists such that

$$U_s = \frac{\partial U}{\partial q_s}, \quad s = 1, 2, \ldots, k.$$

The simplified system obtained so far,

$$\frac{d}{dt}\left(\frac{\partial (T + U)}{\partial \dot{q}_s}\right) - \frac{\partial (T + U)}{\partial q_s} = 0, \quad s = 1, 2, \ldots, k,$$

can be presented in the canonical form

$$\frac{dq_s}{dt} = \frac{\partial R}{\partial p_s}, \quad \frac{dp_s}{dt} = -\frac{\partial R}{\partial q_s}, \quad s = 1, 2, \ldots, k,$$

where $p_s = \frac{\partial T}{\partial \dot{q}_s}$ and $R = T_2 - T_0 - U$. Here T_0 is the totality of the velocity-independent terms in the expression of the kinetic energy, and T_2 is the totality of the second order terms with respect to velocities.

The qualitative analysis of equations (1.2.1) and its particular cases is the principle point of the investigations in nonlinear dynamics of continuous systems.

1.2.2 Perturbed motion equations

Under certain assumptions the equations (1.2.1) can be represented in the scalar form

$$\frac{dy_i}{dt} = Y_i(t, y_1, \ldots, y_{2k}), \quad i = 1, 2, \ldots, 2k,$$

or in the equivalent vector form

(1.2.2) $$\frac{dy}{dt} = Y(t, y),$$

where* $y = (y_1, y_2, \ldots, y_{2k})^{\mathrm{T}} \in R^{2k}$ and $Y = (Y_1, Y_2, \ldots, Y_{2k})^{\mathrm{T}}$, $Y: \mathcal{T} \times R^{2k} \to R^{2k}$. A motion of (1.2.2) is denoted by $\eta(t; t_0, y_0)$, $\eta(t_0; t_0, y_0) \equiv y_0$, and the reference motion $\eta_r(t; t_0, y_{r0})$. From the physical point of view the reference motion should be realizable by the system. From the mathematical point of view this means that the reference motion is a solution of (1.2.2),

(1.2.3) $$\frac{d\eta_r(t; t_0, y_{r0})}{dt} \equiv Y[t, \eta_r(t; t_0, y_{r0})].$$

Let the Liapunov transformation of coordinates be used,

(1.2.4) $$x = y - y_r,$$

where $y_r(t) \equiv \eta_r(t; t_0, y_{r0})$. Let $f: \mathcal{T} \times R^{2k} \to R^{2k}$ be defined by

(1.2.5) $$f(t, x) = Y[t, y_r(t) + x] - Y[t, y_r].$$

*In Liapunov's notation $y = (q_1, q_2, \ldots, q_k, q_1', q_2', \ldots, q_k')^{\mathrm{T}}$.

It is evident that

$$(1.2.6) \qquad\qquad f(t,0) \equiv 0.$$

Now $(1.2.2) - (1.2.5)$ yield

$$(1.2.7) \qquad\qquad \frac{dx}{dt} = f(t,x).$$

In this way, the behavior of perturbed motions related to the reference motion (in total coordinates) is represented by the behavior of the state deviation x with respect to the zero state deviation. The reference motion in the total coordinates y_i is represented by the zero deviation $x = 0$ in state deviation coordinates x_i. With this in mind, the following result emphasizes complete generality of both Liapunov's second method and results represented by Liapunov [1] for the system $(1.2.7)$. Let $Q\colon R^{2k} \to R^n$, $n = 2k$ is admissible but not required.

In the monograph Grujić, *et al.* [1] the following assertion is proved.

Proposition 1.2.1. *Stability of $x = 0$ of system $(1.2.7)$ with respect to $Q = x$ is necessary and sufficient for stability of the reference motion η_r of system $(1.2.2)$ with respect to every vector function Q that is continuous in y.*

This theorem reduced the problem of the stability of the reference motion of $(1.2.2)$ with respect to Q to the stability problem of $x = 0$ of $(1.2.7)$ with respect to x.

For the sake of clarity we state

Definition 1.2.1. *State x^* of the system $(1.2.7)$ is its equilibrium state over $\mathcal{T}_i$ iff*

$$(1.2.8) \qquad \chi(t; t_0, x^*) = x^*, \quad \text{for all} \quad t \in \mathcal{T}_0, \quad \text{and} \quad t_0 \in \mathcal{T}_i.$$

The expression *"over $\mathcal{T}_i$"* is omitted iff $\mathcal{T}_i = R$.

Proposition 1.2.2. *For $x^* \in R^n$ to be an equilibrium state of the system $(1.2.7)$ over $\mathcal{T}_i$ it is necessary and sufficient that both*

(i) *for every $t_0 \in \mathcal{T}_i$ there is the unique solution $\chi(t; t_0, x^*)$ of $(1.2.7)$, which is defined for all $t_0 \in \mathcal{T}_0$*

and

(ii) *$f(t, x^*) = 0$, for all $t \in \mathcal{T}_0$, and $t_0 \in \mathcal{T}_i$.*

The conditions for existence and uniqueness of the solutions of system (1.2.7) can be found in many well-known books by Dieudonne [1], Hale [1], Hirsch and Smale [1], Simmons [1], Yoshizawa [1], etc.

The next result provides a set of sufficient conditions for the uniqueness of solutions for initial value problem

(1.2.9)
$$\frac{dx}{dt} = f(t, x), \quad x(t_0) = x_0.$$

Proposition 1.2.3. *Let* $\mathcal{D} \subset R^{n+1}$ *be an open and connected set. Assume* $f \in C(\mathcal{D}, R^n)$ *and for every compact* $K \subset \mathcal{D}$, f *satisfies the Lipschitz condition*

$$\|f(t, x) - f(t, y)\| \le L\|x - y\|$$

for all $(t, x), (t, y) \in K$, *where* L *is a constant depending only on* K.

Then (1.2.9) has at most one solution on any interval $[t_0, t_0 + c)$, $c > 0$.

Definition 1.2.2. A *solution* $x(t; t_0, x_0)$ of (1.2.7) defined on the interval (a, b) is said to be *bounded* if there exists $\beta > 0$ such that $\|x(t; t_0, x_0)\| < \beta$ for all $t \in (a, b)$, where β may depend on each solution.

For the system (1.2.7) the following result can be easily demonstrated.

Proposition 1.2.4. *Assume* $f \in C(J \times R^n, R^n)$, *where* $J = (a, b)$ *is a finite or infinite interval. Let every solution of (1.2.7) is bounded. Then every solution of (1.2.7) can be continued on the entire interval* (a, b).

1.3 Definitions of Stability

Consider the differential system (1.2.7), where $f \in C(\mathcal{T}_r \times R^n, R^n)$. Suppose that the function f is smooth enough to guarantee existence, uniqueness and continuous dependence of solutions $\chi(t; t_0, x_0)$ of (1.2.7). We now present various definitions of stability (see Grujić [1] and Grujić, *et al.* [1]).

Definition 1.3.1. The *state* $x = 0$ of the system (1.2.7) is:

(i) *stable with respect to* $\mathcal{T}_i$ iff for every $t_0 \in \mathcal{T}_i$ and every $\varepsilon > 0$ there exists $\delta(t_0, \varepsilon) > 0$, such that $\|x_0\| < \delta(t_0, \varepsilon)$ implies

$$\|\chi(t; t_0, x_0)\| < \varepsilon, \quad \text{for all} \quad t \in \mathcal{T}_0;$$

(ii) *uniformly stable with respect to $\mathcal{T}_0$* iff both (i) holds and for every $\varepsilon > 0$ the corresponding maximal δ_M obeying (i) satisfies

$$\inf[\delta_M(t, \varepsilon) : t \in \mathcal{T}_i] > 0;$$

(iii) *stable in the whole with respect to $\mathcal{T}_i$* iff both (i) holds and

$$\delta_M(t, \varepsilon) \to +\infty \quad \text{as} \quad \varepsilon \to +\infty, \quad \text{for all} \quad t \in \mathcal{T}_i;$$

(iv) *uniformly stable in the whole with respect to $\mathcal{T}_i$* iff both (ii) and (iii) hold;

(v) *unstable with respect to $\mathcal{T}_i$* iff there are $t_0 \in \mathcal{T}_i$, $\varepsilon \in (0, +\infty)$ and $\tau \in \mathcal{T}_0$, $\tau > t_0$, such that for every $\delta \in (0, +\infty)$ there is x_0, $\|x_0\| < \delta$, for which

$$\|\chi(\tau; t_0, x_0)\| \geq \varepsilon.$$

The expression "*with respect to $\mathcal{T}_i$*" is omitted from (i) – (v) iff $\mathcal{T}_i = R$. These stability properties hold as $t \to +\infty$ but not for $t = +\infty$.

Further the definitions on solution attraction are cited. The examples by Hahn [2], Krasovskii [1], and Vinograd [1] showed that the attraction property does not ensure stability.

Definition 1.3.2. The *state* $x = 0$ of the system (1.2.7) is:

(i) *attractive with respect to $\mathcal{T}_i$* iff for every $t_0 \in \mathcal{T}_i$ there exists $\Delta(t_0) > 0$ and for every $\zeta > 0$ there exists $\tau(t_0; x_0, \zeta) \in [0, +\infty)$ such that $\|x_0\| < \Delta(t_0)$ implies $\|\chi(t; t_0, x_0)\| < \zeta$, for all $t \in (t_0 + \tau(t_0; x_0, \zeta), +\infty)$;

(ii) x_0-*uniformly attractive with respect to $\mathcal{T}_i$* iff both (i) is true and for every $t_0 \in \mathcal{T}_i$ there exists $\Delta(t_0) > 0$ and for every $\zeta \in (0, +\infty)$ there exists $\tau_u[t_0, \Delta(t_0), \zeta] \in [0, +\infty)$ such that

$$\sup[\tau_m(t_0; x_0, \zeta) : x_0 \in \mathcal{T}_i] = \tau_u(\mathcal{T}_i, x_0, \zeta);$$

(iii) t_0-*uniformly attractive with respect to $\mathcal{T}_i$* iff (i) is true, there is $\Delta > 0$ and for every $(x_0, \zeta) \in B_\Delta \times (0, +\infty)$ there exists $\tau_u(\mathcal{T}_i, x_0, \zeta) \in [0, +\infty)$ such that

$$\sup[\tau_m(t_0); x_0, \zeta) : t_0 \in \mathcal{T}_i] = \tau_u(\mathcal{T}_i, x_0, \zeta);$$

(iv) *uniformly attractive with respect to T_i* iff both (ii) and (iii) hold, that is, that (i) is true, there exists $\Delta > 0$ and for every $\zeta \in (0, +\infty)$ there is $\tau_u(T_i, \Delta, \zeta) \in [0, +\infty)$ such that

$$\sup [\tau_m(t_0; x_0, \zeta) : (t_0, x_0) \in T_i \times B_\Delta] = \tau(T_i, \Delta, \zeta).$$

(v) The properties (i) – (iv) hold *"in the whole"* iff (i) true for every $\Delta(t_0) \in (0, +\infty)$ and every $t_0 \in T_i$.

The expression *"with respect to T_i"* is omitted iff $T_i = R$.

Definitions 1.3.1 and 1.3.2 enable us to define various types of asymptotic stability as follows.

Definition 1.3.3. The *state* $x = 0$ of the system (1.2.7) is:

(i) *asymptotically stable with respect to T_i* iff it is both stable with respect to T_i and attractive with respect to T_i;

(ii) *equi-asymptotically stable with respect to T_i* iff it is both stable with respect to T_i and x_0-uniformly attractive with respect to T_i;

(iii) *quasi-uniformly asymptotically stable with respect to T_i* iff it is both uniformly stable with respect to T_i and t_0-uniformly attractive with respect to T_i;

(iv) *uniformly asymptotically stable with respect to T_i* iff it is both uniformly stable with respect to T_i and uniformly attractive with respect to T_i;

(v) the properties (i) – (iv) hold *"in the whole"* iff both the corresponding stability of $x = 0$ and the corresponding attraction of $x = 0$ hold in the whole;

(vi) *exponentially stable with respect to T_i* iff there are $\Delta > 0$ and real numbers $\alpha \geq 1$ and $\beta > 0$ such that $\|x_0\| < \Delta$ implies

$$\|\chi(t; t_0, x_0)\| \leq \alpha \|x_0\| \exp[-\beta(t - t_0)],$$

for all $\ t \in T_0, \ $ and for all $\ t_0 \in T_i$.

This holds *"in the whole"* iff it is true for $\Delta = +\infty$.

The expression *"with respect to T_i"* is omitted iff $T_i = R$.

1.4 Scalar, Vector and Matrix-Valued Liapunov Functions

In order that to avoid the problem of nonlinear equations nonintegrability in
their qualitative study, Liapunov [1] suggested to apply auxiliary functions
with the norm properties. The auxiliary function, being a function of one
variable (time) on the system trajectories, allows estimating the distance
from every point of the system integral curve to the origin (to the system
equilibrium state) when time is changing from the fixed value $t_0 \in \mathcal{T}_\tau$.

1.4.1 Auxiliary scalar functions

The simplest type of auxiliary function for system (1.2.7) is the function

$$(1.4.1) \qquad v(t,x) \in C(\mathcal{T}_0 \times R^n, R_+), \qquad v(t,0) = 0.$$

Further all functions (1.4.1) allowing the solution of the problem on stability
(instability) of the equilibrium state $x = 0$ of system (1.2.7) are called the
Liapunov functions.

The construction of the Liapunov functions still remains one of the central problems of stability theory. These functions should satisfy special
requirements such as the property of having a fixed sign, decreasing, radial
unboundedness, etc. The Liapunov functions are often constructed as a
quadratic form of the phase variables whose coefficients are constants or
time functions.

The following definitions are presented according to Gantmacher [1].

Definition 1.4.1. A *matrix* $H = (h_{ij}) \in R^{n \times n}$ is:

(i) *positive (negative) semi-definite* iff its quadratic form $V(x) = x^\mathrm{T} H x$
is positive (negative) semi-definite, respectively;

(ii) *positive (negative) definite* iff its quadratic form $V(x) = x^\mathrm{T} H x$ is
positive (negative) definite, respectively.

Notice that a square matrix A with all real valued elements is (semi-)
definite iff its symmetric part $A_s = \frac{1}{2}(A + A^\mathrm{T})$ is (semi-) definite, and a
square matrix A with complex valued elements is (semi-) definite iff its Her-
mitian part $A_H = \frac{1}{2}(A + A^*)$ is (semi-) definite, where A^* is the transpose
conjugate matrix of the matrix A.

Now, the fundamental theorem of the stability theory – the *Liapunov
matrix theorem* – can be stated in the form.

Theorem 1.4.1. *In order that real parts of all eigenvalues of a matrix A, $A \in R^{n \times n}$, be negative it is necessary and sufficient that for any positive definite symmetric matrix G, $G \in R^{n \times n}$, there exists the unique solution H, $H \in R^{n \times n}$, of the (Liapunov) matrix equation*

$$(1.4.2) \qquad A^{\mathrm{T}} H + H A = -G,$$

which is also positive definite symmetric matrix.

If all the characteristic roots of A have negative real parts we can solve the matrix equations (1.4.2) in closed form (see Zubov [3], and Hahn [2])

$$H = \int_0^\infty e^{s A^{\mathrm{T}}} G e^{s A} \, ds.$$

For solving the Liapunov matrix equation (1.4.2), see also Aliev and Larin [1], Barbashin [2], Barnett and Storey [1], etc.

1.4.2 Comparison functions

Comparison functions are used as upper or lower estimates of the function V and its total time derivative. They are usually denoted by φ, $\varphi \colon R_+ \to R_+$. The main contributor to the investigation of properties of and use of the comparison functions is Hahn [2]. What follows is mainly based on his definitions and results.

Definition 1.4.2. *A function φ, $\varphi \colon R_+ \to R_+$, belongs to*

(i) *the class $K_{[0,\alpha)}$, $0 < \alpha \le +\infty$, iff both it is defined, continuous and strictly increasing on $[0,\alpha)$ and $\varphi(0) = 0$;*

(ii) *the class K iff (i) holds for $\alpha = +\infty$, $K = K_{[0,+\infty)}$;*

(iii) *the class KR iff both it belongs to the class K and $\varphi(\zeta) \to +\infty$ as $\zeta \to +\infty$;*

(iv) *the class $L_{[0,\alpha)}$ iff both it is defined, continuous and strictly decreasing on $[0,\alpha)$ and $\lim [\varphi(\zeta) \colon \zeta \to +\infty] = 0$;*

(v) *the class L iff (iv) holds for $\alpha = +\infty$, $L = L_{[0,+\infty)}$.*

Let φ^{-1} denote the *inverse function* of φ, $\varphi^{-1}[\varphi(\zeta)] \equiv \zeta$. The next result was established by Hahn [2].

Proposition 1.4.1.

(1) If $\varphi \in K$ and $\psi \in K$ then $\varphi(\psi) \in K$;
(2) If $\varphi \in K$ and $\sigma \in L$ then $\varphi(\sigma) \in L$;
(3) If $\varphi \in K_{[0,\alpha)}$ and $\varphi(\alpha) = \xi$ then $\varphi^{-1} \in K_{[0,\xi)}$;
(4) If $\varphi \in K$ and $\lim[\varphi(\zeta) : \zeta \to +\infty] = \xi$ then φ^{-1} is not defined on $(\xi, +\infty]$;
(5) If $\varphi \in K_{[0,\alpha)}$, $\psi \in K_{[0,\alpha)}$ and $\varphi(\zeta) > \psi(\zeta)$ on $[0,\alpha)$ then $\varphi^{-1}(\zeta) < \psi^{-1}(\zeta)$ on $[0,\beta]$, where $\beta = \psi(\alpha)$.

Definition 1.4.3. A *function* φ, $\varphi: R_+ \times R_+ \to R_+$, belongs to:

(i) the *class* $KK_{[0;\alpha,\beta)}$ iff both $\varphi(0,\zeta) \in K_{[0,\alpha)}$ for every $\zeta \in [0,\beta)$ and $\varphi(\zeta,0) \in K_{[0,\beta)}$ for every $\zeta \in [0,\alpha)$;
(ii) the *class* KK iff (i) holds for $\alpha = \beta = +\infty$;
(iii) the *class* $KL_{[0;\alpha,\beta)}$ iff both $\varphi(0,\zeta) \in K_{[0,\alpha)}$ for every $\zeta \in [0,\beta)$ and $\varphi(\zeta,0) \in L_{[0,\beta)}$ for every $\zeta \in [0,\alpha)$;
(iv) the *class* KL iff (iii) holds for $\alpha = \beta = +\infty$;
(v) the *class* CK iff $\varphi(t,0) = 0$, $\varphi(t,u) \in K$ for every $t \in R_+$;
(vi) the *class* $\mathcal{M}$ iff $\varphi \in C(R_+ \times R^n, R_+)$, inf $\varphi(t,x) = 0$, $(t,x) \in R_+ \times R^n$;
(vii) the *class* $\mathcal{M}_0$ iff $\varphi \in C(R_+ \times R^n, R_+)$, $\inf_x \varphi(t,x) = 0$ for each $t \in R_+$;
(viii) the *class* Φ iff $\varphi \in C(K, R_+)$: $\varphi(0) = 0$, and $\varphi(w)$ is increasing with respect to cone K.

Definition 1.4.4. Two *functions* $\varphi_1, \varphi_2 \in K$ (or $\varphi_1, \varphi_2 \in KR$) are said to be *of the same order of magnitude* if there exist positive constants α, β, such that

$$\alpha\varphi_1(\zeta) \le \varphi_2(\zeta) \le \beta\varphi_1(\zeta) \quad \text{for all} \quad \zeta \in [0,\zeta_1] \quad (\text{or for all} \quad \zeta \in [0,\infty)).$$

In terms of the comparison function's existence, the special properties of functions (1.4.1) or the function

$$(1.4.3) \qquad\qquad v(x) \in C(R^n, R_+), \quad v(0) = 0,$$

applied in the analysis of the autonomous system

$$(1.4.4) \qquad\qquad \frac{dx}{dt} = g(x), \quad g(0) = 0,$$

where $x \in R^n$, $g \in C(R^n, R^n)$, are specified in the following way.

Definition 1.4.5. A *function* $v: R^n \to R$ is

(i) *positive semi-definite* iff there is a time-invariant neighborhood $\mathcal{N}$ of $x = 0$, $\mathcal{N} \subseteq R^n$, such that
    (a) v is continuous on $\mathcal{N}$: $v \in C(\mathcal{N}, R)$;
    (b) v is non-negative on $\mathcal{N}$: $v(x) \geq 0$ for all $x \in \mathcal{N}$;
    (c) v vanishes at the origin: $v(0) = 0$;
(ii) *positive semi-definite on a neighborhood* S *of* $x = 0$ iff (i) holds for $\mathcal{N} = S$;
(iii) *positive semi-definite in the whole* iff (i) holds for $\mathcal{N} = R^n$;
(iv) *negative semi-definite (on a neighborhood* S *of* $x = 0$ *or in the whole)* iff $(-v)$ is positive semi-definite (on the neighborhood S or in the whole), respectively.

Remark 1.4.1. It is to be noted that a function v defined by $v(x) = 0$ for all $x \in R^n$ is both positive and negative semi-definite. This ambiguity can be avoided by introducing the notion of strictly positiveness (negativeness).

Definition 1.4.6. A *function* $v: R^n \to R$ is said to be *strictly positive (negative) semi-definite* iff both it is positive (negative) semi-definite and there is $y \in \mathcal{N}$ such that $v(y) > 0$ $(v(y) < 0)$, respectively.

The H is strictly positive (negative) semi-definite iff $v(x) = x^{\mathrm{T}} H x$ is strictly positive (negative) semi-definite, respectively.

Definition 1.4.7. A *function* $v: R^n \to R$ is:

(i) *positive definite* if there is a time-invariant neighborhood $\mathcal{N}$, $\mathcal{N} \subseteq R^n$, or $x = 0$ such that both it is positive semi-definite on $\mathcal{N}$ and $v(x) > 0$ for all $(x \neq 0) \in \mathcal{N}$;
(ii) *positive definite on a neighborhood* S *of* $x = 0$ iff (i) holds for $\mathcal{N} = S$;
(iii) *positive definite in the whole* iff (i) holds for $\mathcal{N} = R^n$;
(iv) *negative definite (on a neighborhood* S *of* $x = 0$ *or in the whole)* iff $(-v)$ is positive definite (on the neighborhood S or in the whole, respectively).

Hahn [2] proved.

Proposition 1.4.2. *Necessary and sufficient for positive definiteness of* v *on a neighborhood* $\mathcal{N}$ *of* $x = 0$ *is existence of comparison functions* $\varphi_i \in K_{[0,\alpha)}$, $i = 1, 2$, *where* $\alpha = \sup\{\|x\|: x \in \mathcal{N}\}$, *such that both* $v(x) \in C(\mathcal{N})$ *and* $\varphi_1(\|x\|) \leq v(x) \leq \varphi_2(\|x\|)$ *for all* $x \in \mathcal{N}$.

Definition 1.4.8. A *function* $v: R \times R^n \to R$ is:

(i) *positive semi-definite on* $\mathcal{T}_\tau = [\tau, +\infty)$, $\tau \in R$, iff there is a time-invariant connected neighborhood $\mathcal{N}$ of $x = 0$, $\mathcal{N} \subseteq R^n$, such that
  (a) v is continuous in $(t, x) \in \mathcal{T}_\tau \times \mathcal{N}$: $v(t, x) \in C(\mathcal{T}_\tau \times \mathcal{N}, R)$;
  (b) v is non-negative on $\mathcal{N}$: $v(t, x) \geq 0$ for all $(t, x) \in \mathcal{T}_\tau \times \mathcal{N}$;
  (c) v vanishes at the origin: $v(t, 0) = 0$ for all $t \in \mathcal{T}_\tau$;
  (d) iff the conditions (a)–(c) holds and for every $t \in \mathcal{T}_\tau$ there is $y \in \mathcal{N}$ such that $v(t, y) > 0$, then v is strictly positive semi-definite on $\mathcal{T}_\tau$;

(ii) *positive semi-definite on* $\mathcal{T}_\tau \times \mathcal{S}$ iff (i) holds for $\mathcal{N} = \mathcal{S}$;

(iii) *positive semi-definite in the whole on* $\mathcal{T}_\tau$ iff (i) holds for $\mathcal{N} = R^n$;

(iv) *negative semi-definite (in the whole) on* $\mathcal{T}_\tau$ (on $\mathcal{T}_\tau \times \mathcal{N}$) iff $(-v)$ is positive semi-definite (in the whole) on $\mathcal{T}_\tau$ (on $\mathcal{T}_\tau \times \mathcal{N}$), respectively.

The expression "*on* $\mathcal{T}_\tau$" is omitted iff all corresponding requirements hold for every $\tau \in R$.

Definition 1.4.9. A *function* $v: R \times R^n \to R$ is:

(i) *positive definite on* $\mathcal{T}_\tau$, $\tau \in R$, iff there is a time-invariant connected neighborhood $\mathcal{N}$ of $x = 0$, $\mathcal{N} \subseteq R^n$, such that both it is positive semi-definite on $\mathcal{T}_\tau \times \mathcal{N}$ and there exists a positive definite function w on $\mathcal{N}$, $w: R^n \to R$, obeying $w(x) \leq v(t, x)$ for all $(t, x) \in \mathcal{T}_\tau \times \mathcal{N}$;

(ii) *positive definite on* $\mathcal{T}_\tau \times \mathcal{S}$ iff (i) holds for $\mathcal{N} = \mathcal{S}$;

(iii) *positive definite in the whole on* $\mathcal{T}_\tau$ iff (i) holds for $\mathcal{N} = R^n$;

(iv) *negative definite (in the whole) on* $\mathcal{T}_\tau$ (on $\mathcal{T}_\tau \times \mathcal{N}$) iff $(-v)$ is positive definite (in the whole) on $\mathcal{T}_\tau$ (on $\mathcal{T}_\tau \times \mathcal{N}$), respectively.

The expression "*on* $\mathcal{T}_\tau$" is omitted iff all corresponding requirements hold for every $\tau \in R$.

The following result is obtained directly from Proposition 1.4.2 and Definition 1.4.8.

Proposition 1.4.3. *Necessary and sufficient for a function* $v: R \times R^n \to R$ *to be positive definite on* $\mathcal{T}_\tau \times \mathcal{N}$ *when* $\mathcal{N}$ *is a time-invariant neighborhood of* $x = 0$ *is that (a) and (c) of Definition 1.4.8 are fulfilled and there is* $\varphi \in K_{[0, \alpha]}$, *where* $\alpha = \sup\{\|x\|: x \in \mathcal{N}\}$, *such that*

$$v(t, x) = v_+(t, x) + \varphi(\|x\|) \quad \text{for all} \quad \mathcal{T}_\tau \times \mathcal{N},$$

where $v_+(t, x)$ *is positive semi-definite on* $\mathcal{T}_\tau$.

Definition 1.4.10. *Set* $v_\zeta(t)$ *is the* largest connected neighborhood of $x = 0$ *at* $t \in R$ *which can be associated with a function* v, $v \colon R \times R^n \to R$, *so that* $x \in v_\zeta(t)$ *implies* $v(t, x) < \zeta$.

Definition 1.4.11. *A* function $v \colon R \times R^n \to R$ *is:*

(i) *decreasing on* $\mathcal{T}_\tau$, $\tau \in R$, *iff there is a time-invariant neighborhood* $\mathcal{N}$ *of* $x = 0$ *and a positive definite function* w *on* $\mathcal{N}$, $w \colon R^n \to R$, *such that* $v(t, x) \leq w(x)$ *for all* $(t, x) \in \mathcal{T}_\tau \times \mathcal{N}$;

(ii) *decreasing on* $\mathcal{T}_\tau \times \mathcal{S}$ *iff* (i) *holds for* $\mathcal{N} = \mathcal{S}$;

(iii) *decreasing in the whole on* $\mathcal{T}_\tau$ *iff* (i) *holds for* $\mathcal{N} = R^n$.

The expression "*on* $\mathcal{T}_\tau$" is omitted iff all corresponding conditions hold for every $\tau \in R$.

Definition 1.4.11 implies.

Proposition 1.4.4. *Necessary and sufficient for* v *to be decreasing on* $\mathcal{T}_\tau \times \mathcal{N}$ *when* $\mathcal{N}$ *is a time-invariant neighborhood of* $x = 0$ *is existence of a comparison function* $\varphi \in K_{[0,\alpha)}$, *where* $\alpha = \sup\{\|x\| \colon x \in \mathcal{N}\}$, *such that*

$$v(t, x) = v_-(t, x) + \varphi(\|x\|) \quad \text{for all} \quad \mathcal{T}_\tau \times \mathcal{N},$$

where $v_-(t, x)$ *is negative semi-definite on* $\mathcal{T}_\tau$.

Barbashin and Krasovskii [1,2] discovered the concept of radially unbounded functions. They showed the necessity of it for asymptotic stability in the whole.

Definition 1.4.12. *A* function $v \colon R \times R^n \to R$ *is:*

(i) *radially unbounded on* $\mathcal{T}_\tau$, $\tau \in R$, *iff* $\|x\| \to +\infty$ *implies* $v(t, x) \to +\infty$ *for all* $t \in \mathcal{T}_\tau$;

(ii) *radially unbounded iff* $\|x\| \to +\infty$ *implies* $v(t, x) \to +\infty$ *for all* $t \in \mathcal{T}_\tau$ *for all* $\tau \in R$.

The next can be easily verified (see Hahn [2]).

Proposition 1.4.5. *Necessary and sufficient for a positive definite in the whole* (*on* $\mathcal{T}_\tau$) *function* v *to be radially unbounded is that there exists* $\varphi \in KR$ *obeying, respectively,* $v(t, x) \geq \varphi(\|x\|)$ *for all* $x \in R^n$ *and for all* $t \in R$ (*for all* $t \in \mathcal{T}_\tau$).

For the details see Barbashin and Krasovskii [1,2], Grujić, *et al.* [1], Hahn [2], Martynyuk [9], etc.

1.4.3 Vector Liapunov functions

We return back to system (1.2.7) and assume that for it the vector function

$$(1.4.5) \qquad V(t,x) = (v_1(t,x), v_2(t,x), \ldots, v_m(t,x))^{\mathrm{T}}$$

is constructed in some way, whose components $v_s \in C(\mathcal{T}_\tau \times R^n, R_+)$, $s = 1, 2, \ldots, m$. For the function (1.4.5) to be suitable for stability analysis of the equilibrium state $x = 0$ of system (1.2.7) it is necessary for it to possess the norm type properties (see Definitions 1.4.7 – 1.4.12). The presence of such properties of function (1.4.5) is established in terms of one of the following functions (see Lakshmikantham, Matrosov, *et al.* [1])

$$(1.4.6) \qquad v(t,x) = \max_{s \in [1,m]} v_s(t,x);$$

$$(1.4.7) \qquad \begin{aligned} & v(t,x,\alpha) = \alpha^{\mathrm{T}} V(t,x), \quad \alpha \in R^m, \\ & \alpha_i = \text{const}, \quad i = 1, 2, \ldots, m; \end{aligned}$$

$$(1.4.8) \qquad v(t,x) = \sum_{i=1}^{m} v_i(t,x);$$

$$(1.4.9) \qquad v(t,x) = Q(V(t,x)), \quad Q(0) = 0,$$

$Q \in C(R_+^m, R_+)$, the function $Q(u)$ is nondecreasing in u. Since the functions (1.4.6) – (1.4.9) are scalar and are constructed in terms of the vector function (1.4.5), the special properties of the vector function (1.4.5) are established according to Definitions 1.4.7 – 1.4.12.

Remark 1.4.2. Properties of positive definiteness, decrease and radial unboundedness of the function (1.4.5) follow from the algebraic inequalities, provided that the components $v_s(t,x)$ of the vector function (1.4.5) satisfy the conditions

$$(1.4.10) \qquad a_{i1} \psi_{i1}^{\frac{1}{2}}(\|x_i\|) \le v_i(t,x) \le a_{i2} \psi_{i2}^{\frac{1}{2}}(\|x_i\|), \quad i = 1, 2, \ldots, m,$$

where $a_{i1}, a_{i2} > 0$ and ψ_{i1} and ψ_{i2} are of class $K(KR)$, $x_i \in R^{n_i}$, $n_1 + \cdots + n_m = n$, $x = (x_1^{\mathrm{T}}, \ldots, x_m^{\mathrm{T}})^{\mathrm{T}}$, $\|x_i\| = (x_i^{\mathrm{T}} x_i)^{\frac{1}{2}}$.

The conditions (1.4.10) are the broadest ones under which the algebraic conditions of the property of having a fixed sign can be established for the vector function (1.4.5).

The assumptions on the components $v_i(t, x)$ of the vector function are known being other than (1.4.10):

(a) $\varphi_{1i}(\|x_i\|) \leq v_i(t, x_i) \leq \varphi_{2i}(\|x_i\|)$, for all $(t, x_i) \in \mathcal{T}_\tau \times R^{n_i}$, where $\varphi_{1i}, \varphi_{2i}$ are of class K (KR), $i = 1, 2, \ldots, m$;

(b) $\eta_i\|x_i\| \leq v_i(t, x_i) \leq \kappa_i\|x_i\|$, for all $(t, x_i) \in \mathcal{T}_\tau \times R^{n_i}$, where η_i and κ_i are positive constants, $i = 1, 2, \ldots, m$,

(see Michel and Miller [1], Šiljak [1], etc.).

1.4.4 Matrix-valued metafunction

Assume that for system (1.2.7) the two-indexes system of functions

$$(1.4.11) \qquad \Pi(t, x) = \begin{pmatrix} v_{11}(t, x) & \cdots & v_{1k}(t, x) \\ \cdots & \ddots & \cdots \\ v_{l1}(t, x) & \cdots & v_{lk}(t, x) \end{pmatrix}$$

is constructed, where $v_{ij} \in C(\mathcal{T} \times R^n, R)$, $i = 1, 2, \ldots, k$; $j = 1, 2, \ldots, l$.

Definition 1.4.13. A *function* $\Pi \colon \mathcal{T}_\tau \times R^n \to R^{k \times l}$ is called the *matrix-valued metafunction*, if one of the Liapunov functions can be constructed based on it, namely, a scalar, vector or simple matrix-valued one, which solves the problem on stability of the equilibrium state $x = 0$ of system (1.2.7).

The properties of having a fixed sign of metafunction (1.4.11) are established by a general rule in terms of one of the functions

$$(1.4.12) \qquad v_\Pi(t, x) = \max_{\substack{1 \leq i \leq k \\ 1 \leq j \leq l}} v_{ij}(t, x),$$

$$(1.4.13) \qquad v_\Pi(t, x, \alpha, \beta) = \alpha^{\mathrm{T}} \Pi(t, x) \beta,$$

where $\alpha \in R^l$, $\alpha = \text{const} \neq 0$, $\beta \in R^k$, $\beta = \text{const} \neq 0$;

$$(1.4.14) \qquad v_\Pi(t, x) = \sum_{i=1}^{k} \sum_{j=1}^{l} v_{ij}(t, x),$$

$$(1.4.15) \qquad v_\Pi(t, x) = \Phi(\Pi(t, x)),$$

where $\Phi \in C(R^{k \times l}, R_+)$, $\Phi(0) = 0$, $\Phi(s) > 0$ for $s > 0$, and $\lim_{s \to \infty} \Phi(s) = +\infty$.

Since the functions $v_\Pi(t, \cdot)$ determined by (1.4.12) – (1.4.15) are scalar, the ordinary technique of the Liapunov functions method is used to check their property of having a fixed sign, decreasing and radially unboundedness.

Remark 1.4.3. If $k = l = m$ in (1.4.11), then $\Pi(t, x)$ becomes an ordinary matrix-valued function $U(t, x)$

$$(1.4.16) \qquad U(t, x) = \begin{pmatrix} v_{11}(t, x) & \dots & v_{1m}(t, x) \\ \dots & \ddots & \dots \\ v_{m1}(t, x) & \dots & v_{mm}(t, x) \end{pmatrix},$$

where $U \colon \mathcal{T}_\tau \times R^n \to R^{m \times m}$.

The property of having a fixed sign, decreasing and radial unboundedness of the matrix-valued function (1.4.16) is established, provided that the elements $v_{sk}(t, x)$, $s, k = 1, 2, \dots, m$, satisfy the estimates

$$\underline{a}_{ss} \psi_{s1}^2(\|x_s\|) \leq v_{ss}(t, x) \leq \bar{a}_{ss} \psi_{s2}^2(\|x_s\|)$$
$$\text{for all} \quad (t, x) \in \mathcal{T}_0 \times \mathcal{N} \quad (\text{for all} \quad (t, x) \in \mathcal{T}_\tau \times R^n),$$

for all $s = 1, 2, \dots, m$, and (cf. Djordjević [2])

$$\underline{a}_{sr} \psi_{s1}(\|x_s\|) \psi_{r1}(\|x_r\|) \leq v_{sr}(t, x) \leq \bar{a}_{sr} \psi_{s2}(\|x_s\|) \psi_{r2}(\|x_r\|)$$
$$\text{for all} \quad (t, x) \in \mathcal{T}_0 \times \mathcal{N} \quad (\text{for all} \quad (t, x) \in \mathcal{T}_\tau \times R^n),$$

when all $s \neq r$.

We shall formulate the assertions on the property of having a fixed sign of the matrix metafunction similar to how it has been done for the ordinary matrix-valued function (see Martynyuk [5 – 7, 20]).

Proposition 1.4.6. *A metafunction $\Pi \colon \mathcal{T}_\tau \times R^n \to R^{k \times l}$ is positive definite on $\mathcal{T}_\tau$, $\tau \in R$ iff there exists $\alpha \in R^l$, $\beta \in R^k$, and $a \in K$, and it can be written as*

$$v_\Pi(t, x, \alpha, \beta) = \alpha^T \Pi_+(t, x) \beta + a(\|x\|),$$

where $\Pi_+(t, x)$ is positive semi-definite on $\mathcal{T}_\tau$.

Proposition 1.4.7. *A metafunction* $\Pi: \mathcal{T}_\tau \times R^n \to R^{k \times l}$ *is decreasing on* $\mathcal{T}_\tau$, $\tau \in R$ *iff there exists* $\alpha \in R^l$, $\beta \in R^k$, *and* $b \in K$, *and it can be written as*

$$v_\Pi(t, x, \alpha, \beta) = \alpha^T \Pi_-(t, x)\beta + b(\|x\|),$$

where $\Pi_-(t, x)$ *is negative semi-definite on* $\mathcal{T}_\tau$.

Proposition 1.4.8. *A metafunction* $\Pi: \mathcal{T}_0 \times R^n \to R^{k \times l}$ *is radially unbounded in the whole (on* $\mathcal{T}_\tau$) *iff it can be written as*

$$v_\Pi(t, x, \alpha, \beta) = \alpha^T \Pi_+(t, x)\beta + c(\|x\|),$$

where $\Pi_+(t, x)$ *is positive semi-definite in the whole (on* $\mathcal{T}_\tau$), $\alpha \in R^l$, $\beta \in R^k$, *and* $c \in KR$.

Remark 1.4.4. If $k = l = m$, the vectors α and β are replaced by one vector $y \in R^m$ and Propositions 1.4.6 – 1.4.8 become the known ones (see Martynyuk [20]).

1.5 Comparison Principle

In this section we formulate the basic comparison results in terms of Liapunov-like functions and the theory of differential inequalities that are necessary for our later discussion (see also Yoshizawa [1], Szarski [1], etc.).

For system (1.2.7) we shall consider a continuous function $v(t, x)$ defined on an open set in $\mathcal{T}_\tau \times \mathcal{N}$. We assume that $v(t, x)$ satisfies *locally Lipschitz condition* with respect to x that is, for each point in $\mathcal{T}_\tau \times \mathcal{N}$ there are a neighborhood $\mathcal{T}_\tau \times \mathcal{S}$ and a positive number $L > 0$ such that

$$|v(t, x) - v(t, y)| \leq L\|x - y\|$$

for any $(t, x) \in \mathcal{T}_\tau \times \mathcal{S}$, $(t, y) \in \mathcal{T}_\tau \times \mathcal{S}$.

Definition 1.5.1. Let v be a continuous (either scalar, vector or matrix-valued) function, $v: \mathcal{T}_\tau \times R^n \to R^{s \times s}$, $v(t, x) \in C(\mathcal{T}_\tau \times \mathcal{N})$, and let solutions χ of the system (1.2.7) exist and be defined on $\mathcal{T}_\tau \times \mathcal{N}$. Then, for all $(t, x) \in \mathcal{T}_\tau \times \mathcal{N}$,

  (i) $D^+ v(t, x) = \limsup \left\{ \frac{v[t+\theta, \chi(t+\theta; t, x)] - v(t, x)}{\theta} : \theta \to 0^+ \right\}$ is the *upper right Dini derivative* of v along the motion χ at (t, x);

(ii) $D_+v(t,x) = \liminf\left\{\frac{v[t+\theta,\chi(t+\theta;t,x)]-v(t,x)}{\theta} : \theta \to 0^+\right\}$ is the *lower right Dini derivative* of v along the motion χ at (t,x);

(iii) $D^-v(t,x) = \limsup\left\{\frac{v[t+\theta,\chi(t+\theta;t,x)]-v(t,x)}{\theta} : \theta \to 0^-\right\}$ is the *upper left Dini derivative* of v along the motion χ at (t,x);

(iv) $D_-v(t,x) = \liminf\left\{\frac{v[t+\theta,\chi(t+\theta;t,x)]-v(t,x)}{\theta} : \theta \to 0^-\right\}$ is the *lower left Dini derivative* of v along the motion χ at (t,x).

(v) The function v has Eulerian derivative $\dot{v}$, $\dot{v}(t,x) = \frac{d}{dt}v(t,x)$, at (t,x) along the motion χ iff

$$D^+v(t,x) = D_+v(t,x) = D^-v(t,x) = D_-v(t,x) = Dv(t,x)$$

and then $\dot{v}(t,x) = Dv(t,x)$.

If v is a scalar function and differentiable at (t,x) then (see Liapunov [1])

$$\dot{v}(t,x) = \frac{\partial v}{\partial t} + (\operatorname{grad} v)^{\mathrm{T}} f(t,x),$$

where

$$\operatorname{grad} v = \left(\frac{\partial v}{\partial x_1}, \frac{\partial v}{\partial x_2}, \ldots, \frac{\partial v}{\partial x_n}\right)^{\mathrm{T}}.$$

Effective application of D^+v in the framework of the second Liapunov method is based on the next result by Yoshizawa [1], which enables calculation of D^+v without utilizing system motion themselves.

Theorem 1.5.1. *Let v be continuous and locally Lipshitzian in x over $\mathcal{T}_\tau \times S$ and S be an open set. Then,*

$$D^+v(t,x)\big|_{(1.2.7)} = \limsup\left\{\frac{v[t+\theta, x+\theta f(t,x)] - v(t,x)}{\theta} : \theta \to 0^+\right\}$$

holds along solutions χ of the system (1.2.7) at $(t,x) \in \mathcal{T}_\tau \times S$.

D^*v will mean that both D^+v and D_+v can be used.

The system of equations (1.2.7) is considered with the matrix-valued function $U(t,x)$.

Definition 1.5.2. All *scalar function* of the type

(1.5.1) $v(t,x,\alpha) = \alpha^{\mathrm{T}}U(t,x)\alpha, \quad \alpha \in R^m,$

where $U \in C(\mathcal{T}_\tau \times \mathcal{N}, R^{m\times m})$, are attributed to the *class SL*.

The vector α can be determined in several ways (see Martynyuk [12]) and its choice can effect the property of having a fixed sign of function (1.5.1).

By Definition 1.5.1 for function (1.5.1) when all $(t, x) \in T_\tau \times \mathcal{N}$ the total derivative is calculated by virtue of system (1.2.7)

$$(1.5.2) \qquad D^+v(t, x, \alpha)\big|_{(1.2.7)} = \alpha^T D^+ U(t, x)\alpha,$$

where $D^+U(t, x)$ is calculated element-wise.

Let us consider the following scalar differential equation

$$(1.5.3) \qquad \frac{du}{dt} = g(t, u), \quad u(t_0) = u_0 \geq 0, \quad t_0 \in R \ (t_0 \in T_\tau),$$

where $g \in C(R \times R, R)$ (or $g \in C(T_\tau \times R, R)$) and $g(t, 0) = 0$ for all $t \in T_0$.

Definition 1.5.3. Let $\gamma(t)$ be a solution of (1.5.3) existing on some interval $J = [t_0, t_0 + \alpha)$, $0 < \alpha \leq +\infty$, $t_0 \in R$ $(t_0 \in T_\tau)$. Then $\gamma(t)$ is said to be *the maximal solution of* (1.5.3) if for every solution $u(t) = u(t; t_0, x_0)$ of (1.5.3) existing on J, the following inequalities hold

$$(1.5.4) \qquad u(t) \leq \gamma(t), \quad t \in \mathcal{G}, \quad t_0 \in R \ (t_0 \in T_\tau).$$

A minimal solution is defined similarly by reversing the inequality (1.5.4).

Proposition 1.5.1. *Let* $U \colon T_\tau \times \mathcal{N} \to R^{m \times m}$, $U(t, x)$ *be locally Lipschitzian in* x. *Assume that*

(1) *function* $g \in C(T_\tau \times R^n \times R_+, R)$, $g(t, 0, 0) = 0$ *exists for all* $t \in R_+$ *such that*

$$D^+v(t, x, \alpha)\big|_{(1.2.7)} \leq g(t, x, v(t, x, \alpha)) \quad \text{for all} \quad (t, x, \alpha) \in T_\tau \times \mathcal{N} \times R^m;$$

(2) *solution* $x(t) = x(t; t_0, x_0)$ *of system* (1.2.7) *is definite and continuous for all* $(t; t_0, x_0) \in T_0 \times T_\tau \times \mathcal{N}$;

(3) *maximal solution of the comparison equation*

$$\frac{du}{dt} = g(t, x, u), \quad u(t_0) = u_0, \quad x(t_0) = x_0$$

exist for all $t \in T_\tau$.

Then the estimate

$$v(t, x(t), \alpha) \leq r(t; t_0, x_0, u_0) \quad \text{for all} \quad t \in T_\tau$$

holds whenever $v(t_0, x_0, \alpha) \leq u_0$.

For the proof see monographs by Lakshmikantham, Leela, *et al.* [1].

Proposition 1.5.2. Let $U: \mathcal{T}_\tau \times \mathcal{N} \to R^{m \times m}$, $U(t, x)$ be locally Lipschitzian in x. Assume that

(i) function $g \in C(\mathcal{T}_\tau \times R^n \times R_+, R)$ exists such that

$$D^+ v(t, x, \alpha)\big|_{(1.2.7)} \geq g(t, x, v(t, x, \alpha)) \quad \text{for all} \quad (t, x, \alpha) \in \mathcal{T}_\tau \times \mathcal{N} \times R^m;$$

(ii) solution of system (1.2.7) is definite and continuous for $(t; t_0, x_0) \in \mathcal{T}_0 \times \mathcal{T}_\tau \times \mathcal{N}$;

(iii) minimal solution $r^-(t; t_0, x_0, \omega_0)$ of the comparison equation

$$\frac{d\omega}{dt} = g(t, x, \omega), \quad \omega(t_0) = \omega_0 \geq 0$$

exists for all $t \in \mathcal{T}_\tau$.

Then inequality $v(t, x_0, \alpha) \geq \omega_0$ yields the estimate

$$v(t, x(t), \alpha) \geq r^-(t; t_0, x_0, \omega_0)$$

for all $t \in \mathcal{T}_\tau$.

Propositions 1.5.1 and 1.5.2 are a scalar version of the principle of comparison with the matrix-valued function.

In the monograph by Zubov [4] the following assertions are proved.

Corollary 1.5.1. Let

(i) function (1.5.1) obey the bilateral inequality

$$\varphi_1(t)\rho^2(t) \leq v(t, x, \alpha) \leq \varphi_2(t)\rho^2(t),$$

where $\varphi_i(t) > 0$ for all $t \in \mathcal{T}_0$ and $\rho(t) = (x^T(t; t_0, x_0)x(t; t_0, x_0))^{\frac{1}{2}}$,

(ii) function $g(t, x, v)$ satisfy the estimates

$$-\psi_1(t)\rho^2(t) \leq g(t, x, v) \leq -\psi_2(t)\rho^2(t),$$

where $\psi_i(t) \geq 0$ for all $t \in \mathcal{T}_0$ and functions $\psi_i(t)/\varphi_i(t)$, $i = 1, 2$ are integrable.

Then for the solutions of system (1.2.7) the estimates

$$\rho_0 \varphi_1^{\frac{1}{2}}(t_0)\varphi_2^{-\frac{1}{2}}(t) \exp\left(-\frac{1}{2}\int_{t_0}^t \frac{\psi_1(\tau)}{\varphi_1(\tau)}\, d\tau\right) \leq \rho(t)$$

$$\leq \rho_0 \varphi_2^{\frac{1}{2}}(t_0)\varphi_1^{-\frac{1}{2}}(t) \exp\left(-\frac{1}{2}\int_{t_0}^t \frac{\psi_2(\tau)}{\varphi_2(\tau)}\, d\tau\right)$$

are valid for all $t \in \mathcal{T}_0$ and $t_0 \in \mathcal{T}_i$.

Corollary 1.5.2. *Let*

(i) *function (1.5.1) obey the bilateral inequality*

$$\varphi_1(t)\rho^{l_1}(t) \le v(t, x, \alpha) \le \varphi_2(t)\rho^{l_2}(t),$$

where $\varphi_i(t)$ are piece-wise continuous positive functions given for all $t \in \Delta = [t_0, t_0 + T]$, $l_1 \ge l_2$ are positive numbers;

(ii) *function $g(t, x, v)$ satisfy the estimates*

$$-\psi_1(t)\rho^{k_1}(t) \le g(t, x, v) \le -\psi_2(t)\rho^{k_2}(t),$$

where $\psi_i(t)$, $i = 1, 2$ are positive piece-wise continuous functions for all $t \in \Delta$, $k_1 \le k_2$.

Then for the solutions of system (1.2.7) the estimates

$$\left\{ \varphi_2^{-1}(t)v_0 \left[\left(1 + (\lambda_1 - 1)v_0^{\lambda_1 - 1} \int_{t_0}^{t} \psi_1(\tau)\varphi_1^{-\lambda_1}(\tau)d\tau \right) \right]^{\frac{1}{1-\lambda_1}} \right\}^{\frac{1}{\lambda_2}} \le \rho(t)$$

$$\le \left\{ \varphi_1^{-1}(t)v_0 \left[\left(1 + (\lambda_2 - 1)v_0^{\lambda_2 - 1} \int_{t_0}^{t} \psi_2(\tau)\varphi_2^{-\lambda_2}(\tau)d\tau \right) \right]^{\frac{1}{1-\lambda_2}} \right\}^{\frac{1}{\lambda_1}}$$

are valid for all $t \in \Delta$, $v_0 = v(t_0, x_0, \alpha)$, $\lambda_i = \frac{k_i}{l_i}$, $\lambda_i > 1$ for $i = 1, 2$.

Corollary 1.5.3. *Let both conditions (i) and (ii) of Corollary 1.5.2 be satisfied and $\lambda_i > 1$, $i = 1, 2$. Then for $l_1 = l_2 = l$ the solutions of system (1.2.7) satisfy the estimate*

$$\left[\varphi_2^{-1}(t)v_0 \left(1 + (\lambda_1 - 1)v_0^{\lambda_1 - 1} \int_{t_0}^{t} f_1(\tau)d\tau \right)^{\frac{1}{1-\lambda_1}} \right]^{\frac{1}{l}} \le \rho(t)$$

$$\le \left[\varphi_1^{-1}(t)v_0 \left(1 + (\lambda_2 - 1)v_0^{\lambda_2 - 1} \int_{t_0}^{t} f_2(\tau)d\tau \right)^{\frac{1}{1-\lambda_2}} \right]^{\frac{1}{l}}$$

for all $t \in \Delta$, where

$$f_1 = \begin{cases} \psi_1(t)\varphi_1^{-\lambda_1}(t) & \text{if } \psi_1(t) \ge 0, \\ \psi_1(t)\varphi_2^{-\lambda_1}(t) & \text{if } \psi_1(t) < 0; \end{cases}$$

$$f_2 = \begin{cases} \psi_2(t)\varphi_2^{-\lambda_2}(t) & \text{if } \psi_2(t) \ge 0, \\ \psi_2(t)\varphi_1^{-\lambda_2}(t) & \text{if } \psi_2(t) < 0. \end{cases}$$

Corollary 1.5.4. *Let both conditions (i) – (ii) of Corollary 1.5.2 be satisfied and* $\lambda_1 = \lambda_2 = 1$. *Then for* $\psi_i \geq 0$, $i = 1, 2$ *every solution of system (1.2.7) satisfies the estimates*

$$\left[\varphi_2^{-1}(t)v_0 \exp\left(-\int_{t_0}^{t} \psi_1(s)\varphi_1^{-1}(s)\,ds \right) \right]^{\frac{1}{l_2}} \leq \rho(t)$$

$$\leq \left[\varphi_1^{-1}(t)v_0 \exp\left(-\int_{t_0}^{t} \psi_2(s)\varphi_2^{-1}(s)\,ds \right) \right]^{\frac{1}{l_1}}$$

for all $t \in \Delta$.

Definition 1.5.4. All *vector functions* of the type

(1.5.5) $$L(t, x, b) = AU(t, x)b,$$

are attributed to the *class VL*.

Here $U \in C(\mathcal{T}_\tau \times \mathcal{N}, R^{m \times m})$, A is constant $m \times m$-matrix, and b is m-vector.

For the vector function (1.5.5) we calculate

(1.5.6) $$D^+L(t, x, b)\big|_{(1.2.7)} = AD^+U(t, x)b$$

for all $(t, x, b) \in \mathcal{T}_\tau \times \mathcal{N} \times R^m_+$.

Proposition 1.5.3. *Let* $U \colon \mathcal{T}_\tau \times \mathcal{N} \to R^{m \times m}$, $U(t, x)$ *be locally Lipschitzian in* x. *Assume that*

(1) *a constant* $m \times m$-*matrix* A, *a vector* $b \in R^m_+$, *a vector* $y \in R^m$ *and a function* $a \in K$ *exist such that*

$$y^T L(t, x, b) \geq a(\|x\|)$$

for all $(t, x, b) \in \mathcal{T}_\tau \times \mathcal{N} \times R^m_+$;

(2) *vector function* $G \in C(\mathcal{T}_\tau \times R^n \times R^m, R^m)$ *is such that* $G(t, x, u)$ *is quasimonotone nondecreasing in* u *for any* $t \in \mathcal{T}_\tau$ *and*

$$D^+L(t, x, b)\big|_{(1.2.7)} \leq G(t, x, L(t, x, b));$$

(3) *solution* $x(t) = x(t; t_0, x_0)$ *of system (1.2.7) is definite and continuous for* $(t; t_0, x_0) \in \mathcal{T}_\tau \times \mathcal{T}_i \times \mathcal{N}$ *and the maximal solution* $\omega^+(t; t_0, \omega_0)$ *of the comparison system*

$$\frac{d\omega}{dt} = G(t, x, \omega), \quad \omega(t_0) = \omega_0$$

exists for all $t \in \mathcal{T}_\tau$.

Then the inequality $L(t_0, x_0, b) \leq \omega_0$ implies the estimate

(1.5.7) $$L(t, x(t), b) \leq \omega^+(t; t_0, x_0, \omega_0)$$

for all $t \in \mathcal{T}_\tau$.

Besides, estimate (1.5.7) holds component-wise.

The proof of Proposition 1.5.3 is similar to that of Theorem 3.1.2. from Lakshmikantham, Leela, *et al.* [1].

1.6 Liapunov-Like Theorems

There are several directions in stability theory to search for new conditions which weaken one of other suppositions of the original Liapunov's theorems. We recall the classification of these directions:

(1) search of minimal weak assumptions on the properties of auxiliary functions (semi-definite functions, integral positive functions, etc.);

(2) modification of assumptions on the properties of total derivative of scalar function along solutions of perturbed motion equations;

(3) construction and application of multicomponent auxiliary functions (vector, matrix-valued, metafunctions).

It is natural to expect the development of both the first and the second directions within the framework of the third one.

Further on this section basic theorems of the direct Liapunov method are set out in terms of the matrix-valued functions. Also, main definitions of the class of matrix-valued functions are presented, here.

1.6.1 Matrix-valued function and its properties

Together with the system (1.2.7) we shall consider a two-indices system of functions

(1.6.1) $$U(t, x) = [v_{ij}(t, x)], \quad i, j = 1, 2, \ldots, m,$$

where $v_{ii} \in C(\mathcal{T}_\tau \times R^n, R_+)$, and $v_{ij} \in C(\mathcal{T}_\tau \times R^n, R)$ for all $i \neq j$. Moreover the next conditions are making

(i) $v_{ij}(t, x)$ are locally Lipschitzian in x;

(ii) $v_{ij}(t, 0)$ for all $t \in R$ $(t \in \mathcal{T}_\tau)$, $i, j = 1, 2, \ldots, m$;

(iii) $v_{ij}(t, x) = v_{ji}(t, x)$ in any open connected neighborhood $\mathcal{N}$ of point $x = 0$ for all $t \in R_+$ $(t \in \mathcal{T}_\tau)$.

Let $y \in R^m$, $y \neq 0$, be given. By means of the vector y and matrix-valued function (1.6.1) we introduce the following function

$$v(t, x, y) = y^T U(t, x) y.$$

The following definitions will be used throughout the book, which are based on the corresponding results by Djordjević [3], Grujić [2], Hahn [2], Krasovskii [1], Liapunov [1], and Martynyuk [3–7].

Definition 1.6.1. The *matrix-valued function* $U: \mathcal{T}_\tau \times R^n \to R^{m \times m}$ is:

- (i) *positive semi-definite on* $\mathcal{T}_\tau = [\tau, +\infty)$, $\tau \in R$ iff there are time-invariant connected neighborhood $\mathcal{N}$ of $x = 0$, $\mathcal{N} \subseteq R^n$, and vector $y \in R^m$, $y \neq 0$ such that
    - (a) $v(t, x, y)$ is continuous in $(t, x) \in \mathcal{T}_\tau \times \mathcal{N} \times R^m$;
    - (b) $v(t, x, y)$ is non-negative on $\mathcal{N}$, $v(t, x, y) \geq 0$ for all $(t, x, y \neq 0) \in \mathcal{T}_\tau \times \mathcal{N} \times R^m$ and
    - (c) vanishes at the origin: $v(t, 0, y) = 0$ for all $(t, y) \in \mathcal{T}_\tau \times R^m$;
    - (d) iff the conditions (a) – (c) hold and for every $t \in \mathcal{T}_\tau$, there is $w \in \mathcal{N}$ such that $v(t, w, y) > 0$, then v is strictly positive semi-definite on $\mathcal{T}_\tau$;
- (ii) *positive semi-definite on* $\mathcal{T}_\tau \times \mathcal{S}$ iff (i) holds for $\mathcal{N} = \mathcal{S}$;
- (iii) *positive semi-definite in the whole on* $\mathcal{T}_\tau$ iff (i) holds for $\mathcal{N} = R^n$;
- (iv) *negative semi-definite (in the whole) on* $\mathcal{T}_\tau$ *(on* $\mathcal{T}_\tau \times \mathcal{N}$) iff $(-v)$ is positive semi-definite (in the whole) on $\mathcal{T}_\tau$ (on $\mathcal{T}_\tau \times \mathcal{N}$), respectively.

The expression "*on* $\mathcal{T}_\tau$" is omitted iff all corresponding requirements hold for every $\tau \in R$.

Definition 1.6.2. The *matrix-valued function* $U: \mathcal{T}_\tau \times R^n \to R^{m \times m}$ is:

- (i) *positive definite on* $\mathcal{T}_\tau$, $\tau \in R$, iff there are a time-invariant connected neighborhood $\mathcal{N}$ of $x = 0$, $\mathcal{N} \subseteq R^n$ and a vector $y \in R^m$, $y \neq 0$ such that it is both positive semi-definite on $\mathcal{T}_\tau \times \mathcal{N}$ and there exists a positive definite function w on $\mathcal{N}$, $w: R^n \to R_+$, obeying $w(x) \leq v(t, x, y)$ for all $(t, x, y) \in \mathcal{T}_\tau \times \mathcal{N} \times R^m$;
- (ii) *positive definite on* $\mathcal{T}_\tau \times \mathcal{S}$ iff (i) holds for $\mathcal{N} = \mathcal{S}$;
- (iii) *positive definite in the whole on* $\mathcal{T}_\tau$ iff (i) holds for $\mathcal{N} = R^n$;
- (iv) *negative definite (in the whole) on* $\mathcal{T}_\tau$ *(on* $\mathcal{T}_\tau \times \mathcal{N} \times R^m$) iff $(-v)$ is positive definite (in the whole) on $\mathcal{T}_\tau$ (on $\mathcal{T}_\tau \times \mathcal{N} \times R^m$), respectively;

(v) *weakly decreasing on* T_τ if there exists a $\Delta_1 > 0$ and a function $a \in CK$ such that $v(t,x,y) \le a(t,\|x\|)$ as soon as $\|x\| < \Delta_1$ and $(t,y) \in T_\tau \times R^m$;

(vi) *asymptotically decreasing on* T_τ if there exists a $\Delta_2 > 0$ and a function $b \in KL$ such that $v(t,x,y) \le b(t,\|x\|)$ as soon as $\|x\| < \Delta_2$ and $(t,y) \in T_\tau \times R^m$.

The expression "*on* T_τ" is omitted iff all corresponding requirements hold for every $\tau \in R$.

Proposition 1.6.1. *The matrix-valued function* $U: R \times R^n \to R^{m \times m}$ *is positive definite on* T_τ, $\tau \in R$ *iff it can be written as*

$$y^T U(t,x)y = y^T U_+(t,x)y + a(\|x\|),$$

where $U_+(t,x)$ *is a positive semi-definite matrix-valued function and* $a \in K$.

Definition 1.6.3. (cf. Grujić, et al. [1]). *Set* $v_\zeta(t)$ *is the largest connected neighborhood of* $x = 0$ *at* $t \in R$ *which can be associated with a function* $U: R \times R^n \to R^{m \times m}$ *so that* $x \in v_\zeta(t)$ *implies* $v(t,x,y) < \zeta$, $y \in R^m$.

Definition 1.6.4. *The matrix-valued function* $U: R \times R^n \to R^{s \times s}$ *is:*

(i) *decreasing on* T_τ, $\tau \in R$, iff there is a time-invariant neighborhood $\mathcal{N}$ of $x = 0$ and a positive definite function w on $\mathcal{N}$, $\overline{w}: R^n \to R$, such that $y^T U(t,x)y \le \overline{w}(x)$, for all $(t,x) \in T_\tau \times \mathcal{N}$;

(ii) *decreasing on* $T_\tau \times S$ iff (i) holds for $\mathcal{N} = S$;

(iii) *decreasing in the whole on* T_τ iff (i) holds for $\mathcal{N} = R^n$.

The expression "*on* T_τ" is omitted iff all corresponding conditions still hold for every $\tau \in R$.

Proposition 1.6.2. *The matrix-valued function* $U: R \times R^n \to R^{m \times m}$ *is decreasing on* T_τ, $\tau \in R$, *iff it can be written as*

$$y^T U(t,x)y = y^T U_-(t,x)y + b(\|x\|), \quad (y \ne 0) \in R^m,$$

where $U_-(t,x)$ *is a negative semi-definite matrix-valued function and* $b \in K$.

Definition 1.6.5. *The matrix-valued function* $U: R \times R^n \to R^{m \times m}$ *is:*

(i) *radially unbounded on* T_τ, $\tau \in R$, iff $\|x\| \to \infty$ implies $y^T U(t,x)y \to +\infty$, for all $t \in T_\tau$, $y \in R^m$;

(ii) *radially unbounded*, iff $\|x\| \to \infty$ implies $y^T U(t,x)y \to +\infty$, for all $t \in T_\tau$, and for all $\tau \in R$, $y \in R^m$, $y \ne 0$.

Proposition 1.6.3. *The matrix-valued function* $U: \mathcal{T}_\tau \times R^n \to R^{m \times m}$ *is radially unbounded in the whole (on $\mathcal{T}_\tau$) iff it can be written as*

$$y^{\mathrm{T}}U(t,x)y = y^{\mathrm{T}}U_+(t,x)y + a(\|x\|) \quad \text{for all} \quad x \in R^n,$$

where $U_+(t,x)$ *is a positive semi-definite matrix-valued function in the whole (on $\mathcal{T}_\tau$) and* $a \in KR$.

For the proof of Proposition 1.6.1 – 1.6.3 see Martynyuk [9, 20].

1.6.2 A version of the original theorems of Liapunov

The following results are useful in the subsequent sections.

Proposition 1.6.4. *Suppose $m(t)$ is continuous on (a,b). Then $m(t)$ is nondecreasing (nonincreasing) on (a,b) iff*

$$D^+m(t) \geq 0 \quad (\leq 0) \quad \text{for every} \quad t \in (a,b),$$

where

$$D^+m(t) = \limsup\{[m(t+\theta) - m(t)]\theta^{-1}: \theta \to 0^+\}.$$

Following Liapunov [1], Persidskii [1], and Yoshizawa [1] the next result follows (see Martynyuk [20]).

Theorem 1.6.1. *Let the vector function f in system (1.2.7) be continuous on $R \times \mathcal{N}$ (on $\mathcal{T}_\tau \times \mathcal{N}$). If there exist*

(1) *an open connected time-invariant neighborhood $S \subseteq \mathcal{N}$ of point $x = 0$;*

(2) *a positive definite on $\mathcal{N}$ (on $\mathcal{T}_\tau \times \mathcal{N}$) matrix-valued function $U(t,x)$ and vector $y \in R^m$ such that function $v(t,x,y) = y^{\mathrm{T}}U(t,x)y$ is locally Lipschitzian in x and $D^+v(t,x,y) \leq 0$.*

Then

(a) *the state $x = 0$ of system (1.2.7) is stable (on $\mathcal{T}_\tau$), provided $U(t,x)$ is weakly decreasing on $\mathcal{N}$ (on $\mathcal{T}_\tau \times \mathcal{N}$);*

(b) *the state $x = 0$ of system (1.2.7) is uniformly stable (on $\mathcal{T}_\tau$), provided $U(t,x)$ is decreasing on $\mathcal{N}$ (on $\mathcal{T}_\tau \times \mathcal{N}$).*

Corollary 1.6.1. *Assume that the functions $v_{ij}(t,x)$ in (1.6.1) are continuously differentiable for all $(t,x) \in \mathcal{T}_\tau \times \mathcal{N}$, and*

$$Dv(t,x,y)\big|_{(1.2.7)} = \sum_{i=1}^{m}\sum_{j=1}^{m} y_i y_j Dv_{ij}(t,x)\big|_{(1.2.7)}. \qquad (1.6.2)$$

If there exist functions $\Phi_{ij}(t,x,y)$, $\Phi_{ij}(t,0,y) = 0$ for $y \neq 0$ for all $i,j = 1,2,\ldots,m$, such that

$$y_i y_j Dv_{ij}(t,x)\big|_{(1.2.7)} \leq \Phi_{ij}(t,x,y) \qquad (1.6.3)$$

for all $(t,x) \in \mathcal{T}_\tau \times \mathcal{N}$, then (cf. Djordjević [3])

$$Dv_M(t,x,y)\big|_{(1.2.7)} \leq \sum_{i=1}^{m}\sum_{j=1}^{m} \Phi_{ij}(t,x,y) = e^{\mathrm{T}}\Phi(t,x,y)e, \qquad (1.6.4)$$

where $e = (1,1,\ldots,1)^{\mathrm{T}} \in R_+^m$.

Further we denote

$$B(t,x,y) = \frac{1}{2}\left[\Phi(t,x,y) + \Phi^{\mathrm{T}}(t,x,y)\right] \qquad (1.6.5)$$

and assume that there exist comparison functions $w_1(\|x\|),\ldots,w_m(\|x\|)$ of class K and matrix $\hat{B}(t,x,y)$ such that

$$e^{\mathrm{T}}B(t,x,y)e \leq w^{\mathrm{T}}(\|x\|)\hat{B}(t,x,y)w(\|x\|) \qquad (1.6.6)$$

for all $(t,x,y) \in \mathcal{T}_\tau \times \mathcal{N} \times R^m$.
Compile the equation

$$\det\left[\hat{B}(t,x,y) - \lambda E\right] = 0, \qquad (1.6.7)$$

where E is an $m \times m$ identity matrix. Designate the roots of this equation by $\lambda_i = \lambda_i(t,x,y)$, $i = 1,2,\ldots,m$. It is easy to verify that $Dv_M(t,x,y) \leq 0$ in domain $\mathcal{T}_\tau \times \mathcal{N} \times R^m$ if

$$\lambda_i(t,x,y) \leq 0 \quad \text{for all} \quad i = 1,2,\ldots,m, \qquad (1.6.8)$$

and all $(t,x,y) \in \mathcal{T}_\tau \times \mathcal{N} \times R^m$.
Conditions (1.6.8) together with the other conditions of Theorem 1.6.1 is a sufficient test for stability and uniform stability of the state $x = 0$ of system (1.2.7).

Theorem 1.6.2. *Let the vector function f in system (1.2.7) be continuous on $R \times R^n$ (on $\mathcal{T}_\tau \times R^n$). If there exist radially unbounded positive definite in the whole matrix-valued function $U \in C(R \times R^n, R^{m \times m})$ (or $U \in C(\mathcal{T}_\tau \times R^n, R^{m \times m})$ (on $\mathcal{T}_\tau$) and vector $y \in R^m$ such that the function $v(t, x, y) = y^T U(t, x) y$ is locally Lipschitzian in x and*

$$D^+ v(t, x, y)\big|_{(1.2.7)} \leq 0 \quad \text{for all} \quad (t, x) \in R \times R^n$$
$$\text{(for all} \quad (t, x) \in \mathcal{T}_\tau \times R^n) \, .$$

Then

(a) *the state $x = 0$ of system (1.2.7) is stable in the whole (on $\mathcal{T}_\tau$), provided $U(t, x)$ is weakly decreasing in the whole (on $\mathcal{T}_\tau$);*

(b) *the state $x = 0$ of system (1.2.7) is uniformly stable in the whole (on $\mathcal{T}_\tau$), provided $U(t, x)$ is decreasing in the whole (on $\mathcal{T}_\tau$).*

Remark 1.6.1. If f is locally Lipschitzian on $R \times \mathcal{N}$ (on $\mathcal{T}_\tau$) then U in the preceding theorems is also locally Lipschitzian on $R \times \mathcal{N}$ (on $\mathcal{T}_\tau$) which enables effective calculation of $D^+ U$ via Theorem 1.6.1.

Following Grujić, *et al.* [1], Hahn [2], Liapunov [1], Massera [1, 2], Yoshizawa [1] the next result follows (see Martynyuk [20]).

Theorem 1.6.3. *Let the vector function f in system (1.2.7) be continuous on $R \times \mathcal{N}$ (on $\mathcal{T}_\tau \times \mathcal{N}$). If there exist*

(1) *open connected time-invariant neighborhood $\mathcal{G} \subseteq \mathcal{N}$ of the point $x = 0$;*

(2) *positive definite on $\mathcal{G}$ (on $\mathcal{T}_\tau \times \mathcal{G}$) matrix-valued function $U(t, x)$, a vector $y \in R^m$ and positive definite on $\mathcal{G}$ function ψ such that the function $v(t, x, y) = y^T U(t, x) y$ is locally Lipschitzian in x and*

$$D^+ v(t, x, y)\big|_{(1.2.7)} \leq -\psi(x) \quad \text{for all} \quad (t, x, y) \in R \times \mathcal{G} \times R^m$$
$$\text{(for all} \quad (t, x, y) \in \mathcal{T}_\tau \times \mathcal{G} \times R^m) \, .$$

Then

(a) *iff $U(t, x)$ is weakly decreasing on $\mathcal{G}$ (on $\mathcal{T}_\tau \times \mathcal{G}$), the state $x = 0$ of system (1.2.7) is asymptotically stable (on $\mathcal{T}_\tau$);*

(b) *iff $U(t, x)$ is decreasing on $\mathcal{G}$ (on $\mathcal{T}_\tau \times \mathcal{G}$), the state $x = 0$ of system (1.2.7) is uniformly asymptotically stable (on $\mathcal{T}_\tau$).*

According to Barbashin and Krasovskii [1, 2] and Grujić, *et al.* [1], and the preceding proof in which we choose $\theta \in KR$ it is easy to prove (see Martynyuk [20]).

Theorem 1.6.4. *Let the vector function f in system (1.2.7) be continuous on $R \times R^n$ (on $\mathcal{T}_\tau \times R^n$). If there exist radially unbounded positive definite in the whole matrix-valued function $U(t, x) \in C(R \times R^n, R^{m \times m})$ (or $U(t, x) \in C(\mathcal{T}_\tau \times R^n, R^{m \times m})$ (on $\mathcal{T}_\tau$), a vector $y \in R^m$ and a positive definite in the whole function θ, such that the function*

$$v(t, x, y) = y^T U(t, x) y$$

is locally Lipschitzian in x and

$$D^+ v(t, x, y)\big|_{(1.2.7)} \le -\theta(x) \quad \text{for all} \quad (t, x, y) \in R \times R^n \times R^m$$
$$(\text{for all} \quad (t, x, y) \in \mathcal{T}_\tau \times R^n \times R^m).$$

Then

(a) *iff $U(t, x)$ is weakly decreasing in the whole (on $\mathcal{T}_\tau$), the state $x = 0$ of system (1.2.7) is asymptotically stable in the whole (on $\mathcal{T}_\tau$);*

(b) *iff $U(t, x)$ is decreasing in the whole (on $\mathcal{T}_\tau$), the state $x = 0$ of system (1.2.7) is uniformly asymptotically stable in the whole (on $\mathcal{T}_\tau$).*

Corollary 1.6.2. *Assume in domain $\mathcal{T}_\tau \times \mathcal{N}$ the roots $\lambda_i(t, x, y)$ of (1.6.7) obey one of the conditions*

$$(1.6.9) \qquad \lambda_i(t, x, y) \le -\delta, \quad \delta = \text{const} > 0, \quad i = 1, 2, \ldots, m$$

or

$$\lambda_i(t, x, y) \le -\delta(t), \qquad \int\limits_{t_0}^{+\infty} \delta(s)\, ds = +\infty.$$

Then, $Dv_M(t, x, y)\big|_{(1.2.7)} < 0$ and the solution $x = 0$ of system (1.2.7) is asymptotically stable.

If conditions (1.6.9) holds for all $(t, x) \in \mathcal{T}_\tau \times R^n$, the equilibrium state $x = 0$ of system (1.2.7) is asymptotically stable in the whole (on $\mathcal{T}_\tau$).

Corollary 1.6.3. *(cf. Zubov [1]). Let $\hat{B}(\infty, x, y) = C$, where C is an $m \times m$ constant matrix and the equation*

$$\det [C - \nu E] = 0$$

have the numbers $\nu_1, \ldots, \nu_m$ as its solutions. If the matrix $\hat{B}(t, x, y)$ is continuous for $x = 0$ and $t = \infty$ and

(1.6.10) $\mathrm{Re}\, \nu_i < -\delta, \quad \delta = \mathrm{const} > 0, \quad i = 1, 2, \ldots, m,$

then $Dv_M(t, x, y) \leq -\beta\theta(\|x\|)$, where $\theta(\|x\|)$ is some function of class K.

Proof. In view of (1.6.5) – (1.6.7) the expression (1.6.9) is reduced as

$$Dv_M(t, x, y)\big|_{(1.2.7)} \leq w^{\mathrm{T}}(\|x\|)Cw(\|x\|) + w^{\mathrm{T}}(\|x\|)[\hat{B}(t, x, y) - C]w(\|x\|).$$

The continuity of matrix $\hat{B}(t, x, y)$ implies that $[\hat{B}(t, x, y) - C] \to 0$ as $\|x\| \to 0$. In this case, $T > 0$ and $\delta > 0$ are found for any $\varepsilon > 0$ such that $\|\hat{B}(t, x, y) - C\| < \varepsilon$ whenever $t > T$ and $\|x\| < \delta$. Therefore

$$\begin{aligned}
Dv_M(t, x, y)\big|_{(1.2.7)} &\leq \lambda_M(C)w^{\mathrm{T}}(\|x\|)w^{\mathrm{T}}(\|x\|) + \varepsilon w^{\mathrm{T}}(\|x\|)w(\|x\|) \\
&= (-\lambda_M(C) + \varepsilon)w^{\mathrm{T}}(\|x\|)w(\|x\|).
\end{aligned}$$

Hence, it follows that $\varepsilon > 0$ should be chosen so that $-\lambda_M(C) + \varepsilon < -\beta < 0$. Since $w(\|x\|)$ is of class K, $\theta(\|x\|)$ of class K is found such that $\theta(\|x\|) \geq w^{\mathrm{T}}(\|x\|)w(\|x\|)$. Finally we obtain $Dv_M(t, x, y) < -\beta\theta(\|x\|)$. Condition (1.6.10) together with the other conditions of Theorem 1.6.4 ensure asymptotic (uniform asymptotic) stability of the state $x = 0$ of the system (1.2.7).

Following He and Wang [1], and Krasovskii [1] it is easy to prove the following result (see Martynyuk [14, 20]).

Theorem 1.6.5. *Let the vector function f in system (1.2.7) be continuous on $R \times \mathcal{N}$ (on $\mathcal{T}_\tau \times \mathcal{N}$). If there exist*

(1) *an open connected time-invariant neighborhood $\mathcal{G} \subseteq \mathcal{N}$ of point $x = 0$;*

(2) *a matrix-valued function $U(t, x)$ and a vector $y \in R^m$ such that the function $v(t, x, y) = y^{\mathrm{T}}U(t, x)y$ is locally Lipschitzian in x;*

(3) *functions $\varphi_1, \varphi_2 \in K$ and a positive real number η_1 and positive integer p such that*

$$\eta_1\|x\|^p \leq v(t, x, y) \leq \varphi_1(\|x\|) \quad \text{for all} \quad (t, x, \eta \neq 0) \in R \times \mathcal{G} \times R^m$$

and

$$D^+v(t, x, y)\big|_{(1.2.7)} \leq -\varphi_2(\|x\|) \quad \text{for all} \quad (t, x, \eta \neq 0) \in R \times \mathcal{G} \times R^m$$

$$(\text{for all} \quad (t, x, \eta \neq 0) \in \mathcal{T}_\tau \times \mathcal{G} \times R^m).$$

Then, iff the comparison functions φ_1 and φ_2 are of the same magnitude, the state $x = 0$ of system (1.2.7) is exponentially stable (on $\mathcal{T}_\tau$).

Remark 1.6.2. The statement of Theorem 1.6.5 remains valid, if $\varphi_1(\|x\|) = \eta_2\|x\|^p$ and $\varphi_2(\|x\|) = \eta_3\|x\|^p$, $\eta_2, \eta_3 = \mathrm{const} > 0$.

Theorem 1.6.6. *Let the vector function f in system (1.2.7) be continuous on $R \times R^n$ (on $\mathcal{T}_\tau \times R^n$). If there exist*

  (1) *radially unbounded positive definite in the whole matrix-valued function $U(t,x) \in C(R \times R^n, R^{m \times m})$ (or $U(t,x) \in C(\mathcal{T}_\tau \times R^n, R^{m \times m})$) (on $\mathcal{T}_\tau$) and vector $y \in R^m$ such that the function*

$$v(t,x,\eta) = y^T U(t,x)y$$

  is locally Lipschitzian in x;
  (2) *functions ψ_1, $\psi_2 \in KR$, a positive real number η_1 and positive integer q such that*

$$\eta_2\|x\|^q \le v(t,x,y) \le \psi_1(\|x\|) \quad \text{for all} \quad (t,x,y \ne 0) \in R \times R^n \times R^m$$
$$\text{and for all} \quad (t,x,y \ne 0) \in \mathcal{T}_\tau \times R^n \times R^m$$

  and

$$D^+ v(t,x,y)\big|_{(1.2.7)} \le -\psi_2(\|x\|) \quad \text{for all} \quad (t,x,y \ne 0) \in R \times R^n \times R^m$$
$$\text{and for all} \quad (t,x,y \ne 0) \in \mathcal{T}_\tau \times R^n \times R^m.$$

Then, if the comparison functions ψ_1, ψ_2 are of the same magnitude, the state $x = 0$ of system (1.2.7) is exponentially stable in the whole (on $\mathcal{T}_\tau$).

Remark 1.6.3. The assertion of Theorem 1.6.6 remains valid, if $\varphi_1(\|x\|) = \eta_2\|x\|^q$ and $\varphi_2(\|x\|) = \eta_3\|x\|^q$.

Proposition 1.6.5. *In order that the state $x = 0$ of system (1.2.7) be exponentially stable (on $\mathcal{T}_\tau$) in the whole, it is necessary and sufficient for it to be exponentially stable (on $\mathcal{T}_\tau$) and uniformly asymptotically stable in the whole (on $\mathcal{T}_\tau$).*

Following Zubov [4] we shall formulate and prove a result on instability (see Martynyuk [20]).

Theorem 1.6.7. *Let the vector function f in system (1.2.7) be continuous on $R \times \mathcal{N}$ (on $\mathcal{T}_\tau \times \mathcal{N}$). If there exist*

(1) *an open connected time-invariant neighborhood $\mathcal{G} \subseteq \mathcal{N}$ of the point $x = 0$;*

(2) *a matrix-valued function $U(t,x) \in C^{1,1}(R \times \mathcal{G}, R^{m \times m})$ or $U(t,x) \in C^{1,1}(\mathcal{T}_\tau \times \mathcal{G}, R^{m \times m})$ and a vector $y \in R^m$ such that the function $v(t,x,y) = y^T U(t,x)y$ is strictly positive semi-definite (on $\mathcal{T}_\tau$) and satisfies the relation*

$$\frac{dv}{dt} = \lambda v + \widetilde{\theta}(x), \quad \lambda = \lambda(t,x),$$

where $\widetilde{\theta}(x)$ is a positive semi-definite function on $\mathcal{G}$;

(3) *a number $\varepsilon > 0$ such that when $\delta > 0$ ($\delta < \varepsilon$) for continuous on $\mathcal{T}_0 \times R \times \mathcal{G}$ (on $\mathcal{T}_0 \times \mathcal{T}_\tau \times \mathcal{G}$) solution $\chi(t;t_0,x_0)$ of system (1.2.7) which satisfies the condition $\|x_0\| < \delta$, $v(t_0,x_0) > 0$ implies $\|\chi(t;t_0,x_0)\| < \varepsilon$ for all $t \in R$ (for all $t_0 \in \mathcal{T}_\tau$) the inequality*

$$|v(t, \chi(t;t_0,x_0), y)| \geq v(t_0,x_0,y) \exp\left(\int\limits_{t_0}^{t} \lambda(s)\,ds\right)$$

does not hold for all $t \geq t_0$, $t_0 \in R$ ($t_0 \in \mathcal{T}_\tau$), $t \in \mathcal{T}_0$.

Then and only then the state $x = 0$ of system (1.2.7) is unstable (on $\mathcal{T}_\tau$).

We return back to system (1.4.4) and set out one instability test for the equilibrium state $x = 0$.

Let

$$v(x,\eta) = \eta^T U(x)\eta, \quad \eta \in R_+^m, \quad \eta \neq 0,$$

where $U(x) \in C^1(R^n, R^{m \times m})$, $U(x)$ is a matrix-valued function with the elements $v_{ij}(x)$, $i,j = 1,2,\ldots,m$.

We assume that the functions $\theta_{ij}(x,\eta) \colon \theta_{ij}(0,\eta) = 0$, $\theta_{ij}(x,\eta) \neq 0$ for $x \neq 0$, $\eta \neq 0$, $\theta_{ij} \colon R^n \times R^m \to R$ for all $i,j = 1,2,\ldots,m$ such that

$$\eta_i \eta_j Dv_{ij}(x)|_{(1.4.4)} \geq \theta_{ij}(x,\eta), \quad i,j = 1,2,\ldots,m.$$

In view of (1.6.2) and (1.6.10) one gets (cf. Djordjević [3])

$$Dv_m(x,\eta)\big|_{(1.4.4)} \geq \sum_{i=1}^{m}\sum_{j=1}^{m} \theta_{ij}(x,\eta) = e^T\theta(x,\eta)e.$$

Designate

$$H(x, \eta) = \frac{1}{2} [\theta(x, \eta) + \theta^{\mathrm{T}}(x, \eta)]$$

and assume that there exist functions $w_1(\|x\|), \ldots, w_m(\|x\|)$ of class K and a matrix $\hat{H}(x, \eta)$ such that

$$e^{\mathrm{T}} H(x, \eta) e \geq w^{\mathrm{T}}(\|x\|) \hat{H}(x, \eta) w(\|x\|).$$

Compile the equation

$$\det [\hat{H}(x, \eta) - \lambda E] = 0$$

and assume that

(1.6.11) $\operatorname{Re} \lambda_i(x, \eta) > 0$ for all $(x, \eta) \in \mathcal{N} \times R^m$, $i = 1, 2, \ldots, m$.

Moreover, it is easy to verify that

$$Dv_m(x, \eta)\big|_{(1.4.4)} > 0 \quad \text{for all} \quad (x, \eta) \in \mathcal{N} \times R^m.$$

This means that condition (1.6.11) together with the other conditions of Theorem 1.6.7 are a sufficient instability test for the state $x = 0$ of system (1.4.4).

1.7 Advantages of Cone-Valued Liapunov Functions

Let B denote a compact metric space, and $(R^n, \|\cdot\|)$ be an n-dimensional Euclidean space with any convenient norm $(\|\cdot\|)$ and scalar product. Cartesian product $B \times R^n = E$ with projection $p: E \to B$ is a phase space for a given comparison system.

Definition 1.7.1. A *proper subset* $K \subset R^n$ is said to be a *cone* if:

(i) $\alpha K \subset K$, for all $\alpha \geq 0$;

(ii) $K + K \subset K$;

(iii) $K = \bar{K}$;

(iv) $K \cap (-K) = \{0\}$ and

(v) $\inf K = K^0$ is nonempty.

Here $\bar{K}$ denotes the closure of K and K^0 denotes the interior of K. The order relation on R^n induced by the cone K is defined as follows: let $u_1, u_2 \in K$, then

$$u_1 \leq u_2 \quad \text{iff} \quad (u_1 - u_2) \in K \quad \text{and}$$

$$u_1 < u_2 \quad \text{iff} \quad (u_1 - u_2) \in K^0.$$

Definition 1.7.2. The *set K^** is called the *adjoint cone* if K^* defined by $K^* = \{\varphi \in R^n : \varphi(x) \geq 0 \text{ for all } x \in K\}$, where $\varphi(x)$ denotes the scalar product (φ, x) is called the adjoint cone and satisfies the properties (i) – (v) of Definition 1.7.1.

Definition 1.7.3. A *function* $g : R_+ \times R^n \to R^n$ is said to be *quasi-monotone* in u relative to the cone K, for each $t \in R_+$ if for all (t, u), (t, v) $\in R_+ \times R^n$ and $u - v \in K$ imply that there exists $z \in K_0^*$ such that $(z, u - v) = 0$ and $(z, g(t, u) - g(t, v)) \geq 0$.

1.7.1 Stability with respect to two measures

We demonstrate the application of cone-valued function in investigating the stability of system (1.2.7) (not necessarily the state $x = 0$) with respect to two different measures. Following Lakshmikantham, Leela, *et al.* [1], we will use classes of comparison functions (v) – (viii) from Definition 1.4.3.

For system (1.2.7) the notion of stability with respect to two measures is formulated as follows.

Definition 1.7.4. The *solution* $y(t; t_0, y_0)$ of system (1.2.2) is (ρ_0, ρ)-*equi-stable*, if given $\varepsilon > 0$ and $t_0 \in R_+$, there exists a function $\delta = \delta(t_0, \varepsilon) > 0$ continuous in t_0 for every value of ε and such that

$$\rho(t, y(t; t_0, y_0)) < \varepsilon \quad \text{for all} \quad t \geq t_0$$

whenever $\rho_0(t_0, y_0) < \delta$.

Based on this definition, it is easy to formulate many definitions of stability, boundedness, and practical stability, provided an appropriate choice of the measures ρ_0 and ρ from the classes $\mathcal{M}_0$ and $\mathcal{M}$, respectively.

Further the dynamical properties of system (1.2.2) are associated with the dynamical properties of the comparison system

$$(1.7.1) \qquad \frac{dw}{dt} = g(t, w), \quad w(t_0) = w_0 \geq 0,$$

where $g \in C(R_+ \times K, R^m)$.

Definition 1.7.5. Let q_0 and q be of class Φ. We claim that the *solution* $w(t; t_0, w_0)$ of system (1.7.1) is (q_0, q)-*equi-stable*, if given $\beta > 0$ and $t_0 \in$

R_+, there exists a positive function $\alpha = \alpha(t_0, \beta) > 0$ continuous in t_0 for every β and such that

$$q(w(t; t_0, w_0)) < \beta \quad \text{for all} \quad t \geq 0$$

whenever $q_0(w_0) < \alpha$.

Within the framework of the direct Liapunov method and principle of comparison with the cone-valued function $L \in C(R_+ \times R^n \times R^m, K)$, $L(t, x, y)$ is locally Lipschitzian in x with respect to cone K, its total derivative $D^+ L(t, x, y)$ is considered as well as the majorizing vector function $g(t, w)$, $g \in C(R_+ \times K, R^m)$, $g(t, w)$ is quasimonotone nondecreasing in w with respect to cone K for every $t \in R_+$.

The next statement is a general result of the principle of comparison with the cone-valued function.

Theorem 1.7.1. *For system (1.2.2) it is assumed that*

(1) *for given measures ρ_0 and ρ of class $\mathcal{M}$ there exist a function φ of the class K and a constant $\Delta > 0$ such that $\rho(t, x) \leq \varphi(\rho_0(t, x))$ whenever $\rho_0(t, x) < \Delta$;*

(2) *for given measures q_0 and q of class Φ there exist a function ψ of the class K and a constant $\Delta_1 > 0$ such that $q(w) \leq \psi(q_0(w))$ whenever $q_0(w) < \Delta_1$;*

(3) *there exist functions $L(t, x, y) = AU(t, x)y$, $L \in C(R_+ \times R^n \times R^m, K)$, $L(t, x, y)$ is locally Lipschitzian in x with respect to cone K and functions $g \in C(R_+ \times K, R^m)$, $g(t, w)$ is quasimonotone nondecreasing in w with respect to cone K for every $t \in R_+$ such that*

$$D^+ L(t, x, y)\big|_{(1.2.2)} \leq g(t, L(t, x, y))$$

for all $(t, x) \in S(\rho, h) = \{\rho \in \mathcal{M} : \rho(t, x) \leq h\}$;

(4) *there exist constants $\Delta_3 > 0$ and $\Delta_4 > 0$ and functions a and b of class K such that*

   (a) *$b(\rho(t, x)) \leq q(c^T L(t, x, y))$ for all $\rho(t, x) < \Delta_3$, $c \in R_+^m$, $c > 0$;*

   (b) *$q_0(c^T L(t, x, y)) \leq a(\rho_0(t, x))$ for all $\rho_0(t, x) < \Delta_4$.*

Then the dynamical (ρ_0, ρ)-properties of solutions of system (1.2.2) follow from the dynamical (q_0, q)-properties of solutions of comparison system (1.7.1).

This theorem as well as the theorems of the type characteristic for the comparison principle is proved for a given specific dynamical property of solutions of the comparison system (1.7.1), for example, (q_0, q)-equistability, (q_0, q)-uniform stability, etc. (see Lakshmikantham and Papageorgiou [1]).

Example 1.7.1. Consider the system

(1.7.2)
$$\frac{dx}{dt} = -xy^2 e^{-t} H(t, x, y),$$
$$\frac{dy}{dt} = \beta e^{-t} x^2 y H(t, x, y) + \frac{1}{2} y e^{-t},$$

where $H(t, x, y) \geq 0$ is a continuous function in the domain $S(\rho, h)$. We take $L_1 = x^2$ and $L_2 = e^{-t} y^2$ so that

(1.7.3) $D^+ L_1(x)\big|_{(1.7.2)} \leq 0,$ $D^+ L_2(y)\big|_{(1.7.2)} \leq -\beta D^+ L_1(x),$

where $\beta > 0$.

Designate $q(w) = w_1$, $q_0(w) = w_2 + (1 + \beta) w_1$, $\rho_0 = x^2 + y^2$ and $\rho = x^2$.

The application of the method of vector functions does not allow studying (ρ_0, ρ)-equi-stability of system (1.7.2).

Consider the cone $K \triangleq \{V = d_1 w_1 + d_2 w_2, \ w_i \geq 0, \ i = 1, 2\}$, where $d_1 = (1, -\beta)^T$, and $d_2 = (0, 1)^T$.

It is easy to find that $D^+ V(t, x)\big|_{(1.7.2)} \leq 0$ in the domain $S(\rho, h)$, and

$$q(V(t, x)) = L_1(t, x) \geq b(\rho(t, x)) \quad \text{if} \quad \rho(t, x) < \Delta_3;$$
$$q_0(V(t, x)) = L_2(t, x) + (1 + \beta) L_1(t, x) \leq a(t, \rho_0(t, x))(1 + \beta)$$
$$\text{if} \quad \rho_0(t, x) < \Delta_4.$$

By Theorem 1.7.1 the solution of system (1.7.2) is (ρ_0, ρ)-equi-stable.

1.7.2 Stability analysis of large scale system

Assume that the mathematical decomposition of system (1.2.7) is carried out into m interconnected subsystems

(1.7.4) $\dfrac{dx_i}{dt} = f_i(t, x_i) + g_i(t, x_1, ..., x_m), \quad i = 1, 2, ..., m,$

where $x_i \in R^{n_i}$, $f_i \in C(\mathcal{T}_\tau \times R^{n_i}, R^{n_i})$, $g_i \in C(\mathcal{T}_\tau \times R^{n_1} \times ... \times R^{n_m}, R^{n_i})$, $n_1 + n_2 + \cdots + n_m = n$, $f_i(t, 0) = 0$, $g_i(t, 0, ..., 0) = 0$, $i = 1, 2, ..., m$. The functions $f_i(t, x_i)$ in (1.7.4) represent isolated decoupled subsystems

(1.7.5) $\dfrac{dx_i}{dt} = f_i(t, x_i), \quad x_i(t_0) = x_{i0}, \quad i = 1, 2, ..., m.$

Let for the subsystems (1.7.5) the functions $v_i \in C(\mathcal{T}_\tau \times \mathcal{N}_i, R_+)$ exist, $\mathcal{N}_i \subseteq R^{n_i}$, which satisfy the Lipschitz condition

$$|v_i(t, x_i) - v_i(t, y_i)| \leq L_i \|x_i - y_i\|, \quad i = 1, 2, \ldots, m$$

for all (t, x_i), $(t, y_i) \in \mathcal{T}_\tau \times \mathcal{N}_i$, $L_i > 0$, $i = 1, 2, \ldots, m$.

The interconnection functions $g_i(t, x_1, \ldots, x_m)$ between the subsystems (1.7.5) may be described in various ways, for example (see Šiljak [1])

$$g_i(t, x) = g_i(t, e_{i1} x_1, e_{i2} x_2, \ldots, e_{im} x_m),$$

where $e_{ij} \in C(\mathcal{T}_\tau \times R^n, [0, 1])$ are the elements of $m \times m$ interconnection matrix $E = (e_{ij})$ of fundamental interactions.

The $m \times m$ fundamental interconnection matrix $\bar{E} = (\bar{e}_{ij})$ corresponding to the system (1.2.7) has one row and one column for each subsystem, and the elements $\bar{e}_{ij}$ are defined as

$$\bar{e}_{ij} = \begin{cases} 1, & x_j \text{ occurs in } g_i(t, x), \\ 0, & x_j \text{ does not occur in } g_i(t, x). \end{cases}$$

Therefore, $\bar{E}$ is the standard interconnection matrix, which has binary elements $\bar{e}_{ij}$: 1 if j-subsystem of (1.7.4) can act on i-subsystem of (1.7.4), and 0 if j-subsystem of (1.7.4) cannot act on i-subsystem of (1.7.4).

Following Šiljak [1] an $m \times m$ interconnection matrix $E = (e_{ij})$ is said to be generated by an $m \times m$ fundamental interconnection matrix $\bar{E} = (\bar{e}_{ij})$ if $\bar{e}_{ij} = 0$ implies $e_{ij} = 0$ for all $i, j = 1, 2, \ldots, m$

We recall some notions used in the subsequent presentation.

Definition 1.7.6. (Newman [1]). An $m \times m$-*matrix* $A = (a_{ij})$ is the *Metzler matrix*, iff

$$a_{ij} = \begin{cases} < 0, & i = j \\ \geq 0, & i \neq j, \end{cases}$$

for all $(i, j) \in [1, m]$.

Note that according to Kotelyanskii [1] and Sevastyanov [1] the Metzler $m \times m$-matrix A is stable, i.e. $\mathrm{Re}\,\lambda_i(A) < 0$, if

$$(-1)^k \begin{vmatrix} a_{11} & a_{12} & \cdots & a_{1k} \\ \cdots & \cdots & \ddots & \cdots \\ a_{k1} & a_{k2} & \cdots & a_{kk} \end{vmatrix} > 0 \quad \text{for all} \quad k = 1, 2, \ldots, m.$$

Definition 1.7.7. (Šiljak [1]). The *equilibrium state* $x = 0$ of system (1.7.4) is *uniformly connective stable*, if for any $\varepsilon > 0$ a $\delta(\varepsilon) > 0$ exists such that

$$\|x(t; t_0, x_0)\| < \varepsilon \quad \text{for all} \quad t \in \mathcal{T}_\tau \quad \text{whenever} \quad \|x_0\| < \delta \quad \text{for all} \quad E \in \bar{E}.$$

Definition 1.7.8. (Šiljak [1]). The *equilibrium state* $x = 0$ of system (1.7.4) is *uniformly asymptotically connective stable*, if it is connective stable and, besides a $\mu > 0$ exists such that whenever $\|x_0\| < \mu$, $\lim_{t \to \infty} \|x(t; t_0, x_0)\| = 0$ for all $E \in \bar{E}$.

In order to analyse the connective stability of system (1.7.4) Ladde [1] used the comparison system

$$(1.7.6) \qquad \frac{du}{dt} = A(u)W(u), \quad u(t_0) = u_0 \geq 0,$$

where $u \in R_+^m$, $W(u) = (w_1(u_1), w_2(u_2), \ldots, w_m(u_m))^\mathrm{T}$ and $w_i \in SK = \{z \in C(R_+, R_+) \colon z(0) = 0, z(r)$ is strictly increasing in $r\}$, $i = 1, 2, \ldots, m$.

Ladde further assumed that $A(u)$ is an $m \times m$ matrix function defined on R_+^m into $R^{m \times m}$ with coefficients defined by

$$a_{ij}(u) = \begin{cases} -q_j(u_j) + \bar{e}_{ij} q_{jj}(u_j) & \text{for} \quad i = j, \\ \bar{e}_{ij} q_{ij}(u) & \text{for} \quad i \neq j, \end{cases}$$

where $q_i \colon R_+ \to R_+$, and $q_{ij} \in C(R_+^m, R_+)$.

The following results is due to Ladde [1] (see also Akpan [1]).

Theorem 1.7.2. *Assume that:*

(1) (a) $q_j(u_j) > q_{jj}(u_j)$, $u_j \in R_+$, $j = 1, 2, \ldots, m$,

   (b) $|q_{jj}(u)| = d_j^{-1} \sum\limits_{i=1, i \neq j} d_j |q_{ij}(u)| \geq q$, $u \in R_+^m$,

   $i, j = 1, 2, \ldots, m$, $q > 0$, $d_i > 0$;

(2) (a) $w_{i1}(\|x_i\|) \leq v_i(t, x_i) \leq w_{i2}(\|x_i\|)$, $i = 1, 2, \ldots, m$,

   (b) $D^+ v_i(t, x_i)\big|_{(1.7.5)} \leq q_i(v_i(t, x_i)) w_{i3}(v_i(t, x_i))$,

   $i = 1, 2, \ldots, m$;

(3) $\|g_i(t, x_1, \ldots, x_m)\| \leq \sum\limits_{j=1}^{m} \bar{e}_{ij} q_{ij}(V(t, x)) w_{j3}(v_i(t, x_j))$;

(4) $f_i(t, 0) = g_i(t, 0, \ldots, 0) = 0$ for all $t \in R_+$, $i = 1, 2, \ldots, m$.

Then the equilibrium state $x = 0$ of system (1.7.4) is asymptotically stable.

Theorem 1.7.2 is a typical result in the stability analysis of large scale system via the method of vector Liapunov functions.

Comment 1. Condition (2)(a) of Theorem 1.7.2 means that the functions $v_i(t, x_i)$, $i = 1, 2, \ldots, m$, are positive definite, decreasing and radially unbounded (in investigation of stability in the whole of the state $x = 0$ of system (1.7.5)). The assumption on $A(u)$ contained in (1.7.7) implies that the comparison matrix function $A(u)$ must be Metzler with quasi-dominant main diagonal property. This means that if the comparison matrix is neither Metzler nor possesses the restrictive quasi-dominant diagonal property, then the method fails to yield the required stability results.

Comment 2. Condition (3) of Theorem 1.7.2 established the limits of changing of the interaction functions $g_i(t, x)$ between the subsystems (1.7.5) without distinguishing their stabilizing or destabilizing effect on the dynamics of the whole system (1.7.4).

Comment 3. Condition (4) of Theorem 1.7.2 means that the state $x = 0$ is the only equilibrium state of system (1.7.4) and in this state the subsystems (1.7.5) do not interact one with the other, because $g_i(t, 0, ..., 0) = 0$ for all $i = 1, 2, \ldots, m$. Therefore, in this state the interconnections do not effect the dynamical behavior of system (1.7.4), whereas the subsystems (1.7.5) possess the property of asymptotic stability (on $\mathcal{T}_\tau$). However, because of the physical continuity principle the stability analysis of system (1.7.4) is made actually in the presence of small interactions in the neighborhood of the equilibrium state $x = 0$.

We shall present now a typical result obtained via application by cone-valued function.

Assumption 1.7.1. *There exists a vector cone-valued function*

$$L \in C(R_+ \times \mathcal{D}, K), \quad \mathcal{D} \subseteq \mathcal{N} \cap K,$$

where K is an arbitrary cone in R_+^m, functions $q_{ij}(u)$, $q_{ij} \in C(P, R)$, $H_i(u)$, $H \in C(P, P)$ and matrix $A(u)$ with the elements $a_{ij} = \bar{e}_{ij} q_{ij}(u)$, where $\bar{e}_{ij}$ are the entries of the fundamental interconnection matrix $\bar{E}$, such that

(1) $D^+ L(t, x)\big|_{(1.7.5)} \leq -\alpha k \|x\|_m$ *for all $(t, x) \in \mathcal{T}_\tau \times \mathcal{D}$, where $\alpha \in R$, $k \in C(K, K)$ and $\| \cdot \|_m = (\|x_1\|, \ldots, \|x_m\|)^T$;*

(2) $D^+L(t,x)\big|_{(1.7.4)} - D^+L(t,x)\big|_{(1.7.5)} \leq \sum_{i,j=1}^{m} e_{ij}q_{ij}(L(t,x))H(L(t,x))$

$\quad + \alpha k\|x\|_m$ for all $(t,x) \in \mathcal{T}_\tau \times \mathcal{D};$

(3) $f_i(t,x_i^*) = 0$, and $g(t,x_1^*,...,x_m^*) \neq 0$ for $x_i^* \neq 0$,

$\quad$ but $f(t,0) + g(t,0,...,0) = 0$ for all $t \in \mathcal{T}_\tau.$

Theorem 1.7.3. *Assume that*

(1) *all conditions of Assumption 1.7.1 are satisfied;*

(2) *the vector $c \in K$ exists such that the system of inequalities*

$$A(c)W(c) < 0$$

has the solution.

Then the equilibrium state $x = 0$ of system (1.7.4) is uniformly asymptotically connective stable (on $\mathcal{T}_\tau$).

The proof of this theorem is based on the results the papers of Akpan [1], and Martynyuk and Obolenskii [1,2].

We focus our attention on some peculiarities of Theorem 1.7.3.

Comment 4. Since the value α in condition (1) of Assumption 1.7.1 can be $\alpha > 0$, $\alpha = 0$ or $\alpha < 0$, the subsystems (1.7.5) can be asymptotically stable (on $\mathcal{T}_\tau$), stable (on $\mathcal{T}_\tau$) or unstable (on $\mathcal{T}_\tau$), respectively.

Comment 5. Condition (2) of Assumption 1.7.1 does not require smallness of the interaction functions $g_i(t,x)$, $i = 1,2,\ldots,m$, between the subsystems in the neighborhood of the equilibrium state $x = 0$.

Comment 6. Condition (3) of Assumption 1.7.1 means that the interaction functions $g_i(t,x)$, $i = 1,2,\ldots,m$ may not vanish in the equilibrium state of system (1.7.4) and this means that they can stabilize or destabilize system (1.7.4) in the neighborhood of the equilibrium state.

Comment 7. Condition (3) of Theorem 1.7.3 is a necessary and sufficient (see Martynyuk and Obolenskii [1]) condition for uniform asymptotic stability of state $u = 0$ of the comparison system

$$\frac{du}{dt} = A(u)W(u),$$

where $u \in K$ and $W(u) = (H_1(u),\ldots,H_m(u))^{\mathrm{T}}.$

It should also be noted that the requirement to the comparison matrix to be Metzler with quasi-dominant main diagonal property is completely dropped in Theorem 1.7.2.

1.8 Liapunov's Theorems for Large Scale Systems in General

1.8.1 Why are matrix-valued Liapunov functions needed?

In the context of qualitative analysis of large scale systems both ordinary vector function and cone-valued function are associated with the fundamental property of the comparison system, the property of quasimonotonicity. As known, this property of the comparison system is not necessary in stability analysis of its solutions; however it is needed for the Chaplygin-Ważewski theorem (or its generalization) to be applicable in the estimation of the components changing any vector function along solutions of the system under consideration. The direct Liapunov's method based on the matrix-valued function is the most suitable for stability investigation of large scale systems. This method has the following advantages:

(a) it does not require the application of quasimonotone comparison systems;

(b) it extends the class of auxiliary functions suitable for construction of the appropriate Liapunov function;

(c) it allows to the greatest extent to take into account the effect of the connections between the subsystems on the dynamics of the whole system;

(d) it allows to take into account the dynamical properties of the compositions of (i, j)-couples of subsystems of the first level decomposition of the initial system.

The sections deals with the theorems on stability based on the matrix-valued function.

Matrix Liapunov functions make it possible to establish easily verified stability conditions for the state $x = 0$ of system (1.2.7) in terms of the property of having a fixed sign of special matrices.

The application of the matrix-valued function $U: \mathcal{T}_\tau \times R^n \to R^{m \times m}$, $U(t, x) = [v_{ij}(t, x)]$ with the elements

$$v_{ii}(t, x) \in C(\mathcal{T}_\tau \times R^n, R_+), \quad i = 1, 2, \ldots, m$$

and

$$v_{ij}(t, x) \in C(\mathcal{T}_\tau \times R^n, R), \quad i \neq j$$

is based on the potential possibility to construct functions $v_{ii}(t, x)$, $i = 1, 2, \ldots, m$, for subsystems (1.7.5) and functions $v_{ij}(t, x)$, $i \neq j$, $i, j =$

$1, 2, \ldots, m$ taking into account either the interaction functions $g_i(t, x)$, $i = 1, 2, \ldots, m$ or the dynamics of compositions of (i, j)-couples of subsystems (1.7.5).

In the presence of the matrix-valued function constructed in such a way, its further application in the framework of the direct Liapunov method is carried out in two ways, either by construction of a scalar function or by construction of a vector function (including the cone-valued one). Below, the main theorems of the method of matrix Liapunov functions are presented in the framework of scalar approach.

1.8.2 Stability and instability of large scale systems

Some particularization of conditions in Theorems 1.6.1 – 1.6.7 provides a version of theorems of the method of matrix-valued Liapunov functions available for application in stability investigation of large scale systems.

Theorem 1.8.1. *Let the vector function f in system (1.2.7) be continuous on $R \times \mathcal{N}$ (on $\mathcal{T}_\tau \times \mathcal{N}$). If there exist*

(1) *an open connected time-invariant neighborhood $\mathcal{G} \subset \mathcal{N}$ of the point $x = 0$;*

(2) *a matrix-valued function $U \in C(R \times \mathcal{N}, R^{m \times m})$ and a vector $y \in R^m$ such that the function $v(t, x, y) = y^{\mathrm{T}} U(t, x) y$ is locally Lipschitzian in x for all $t \in R$ ($t \in \mathcal{T}_\tau$);*

(3) *functions $\psi_{i1}, \psi_{i2}, \psi_{i3} \in K$, $\widetilde{\psi}_{i2} \in CK$, $i = 1, 2, \ldots, m$;*

(4) *$m \times m$ matrices $A_j(y)$, $j = 1, 2, 3$, $\widetilde{A}_2(y)$ such that*

(a)
$$\psi_1^{\mathrm{T}}(\|x\|) A_1(y) \psi_1(\|x\|) \leq v(t, x, y) \leq \widetilde{\psi}_2^{\mathrm{T}}(t, \|x\|) \widetilde{A}_2(y) \widetilde{\psi}_2(t, \|x\|)$$
$$\text{for all} \quad (t, x, y) \in R \times \mathcal{G} \times R^m$$
$$(\text{for all} \quad (t, x, y) \in \mathcal{T}_\tau \times \mathcal{G} \times R^m);$$

(b)
$$\psi_1^{\mathrm{T}}(\|x\|) A_1(y) \psi_1(\|x\|) \leq v(t, x, y) \leq \psi_2^{\mathrm{T}}(\|x\|) A_2(y) \psi_2(\|x\|)$$
$$\text{for all} \quad (t, x, y) \in R \times \mathcal{G} \times R^m$$
$$(\text{for all} \quad (t, x, y) \in \mathcal{T}_\tau \times \mathcal{G} \times R^m);$$

(c)
$$D^+ v(t, x, y)\big|_{(1.2.7)} \leq \psi_3^{\mathrm{T}}(\|x\|) A_3(y) \psi_3(\|x\|)$$
$$\text{for all} \quad (t, x, y) \in R \times \mathcal{G} \times R^m$$
$$(\text{for all} \quad (t, x, y) \in \mathcal{T}_\tau \times \mathcal{G} \times R^m).$$

Then, if the matrices $A_1(y)$, $A_2(y)$, $\tilde{A}_2(y)$, $(y \neq 0) \in R^m$ are positive definite and $A_3(y)$ is negative semi-definite, then

(a) the state $x = 0$ of system (1.2.7) is stable (on $\mathcal{T}_\tau$), provided condition (4)(a) is satisfied;

(b) the state $x = 0$ of system (1.2.7) is uniformly stable (on $\mathcal{T}_\tau$), provided condition (4)(b) is satisfied.

Remark 1.8.1. If the elements $v_{ij}(t, \cdot)$, $i, j = 1, 2, \ldots, m$ of the matrix-function $U(t, x)$ satisfy the estimates

$$\underline{\gamma}_{ij}\psi_{ij}(\|x_i\|)\psi_{ji}(\|x_j\|) \leq v_{ij}(t, x) \leq \overline{\gamma}_{ij}\psi_{ij}(\|x_i\|)\psi_{ji}(\|x_j\|),$$

where $\gamma_{ii}, \overline{\gamma}_{ij} > 0$, $\underline{\gamma}_{ij}$ and $\overline{\gamma}_{ij}$ are constants for $i \neq j$, $(\psi_{i,j}\psi_{j,i}) \in K(KR)$-class, then for the function $v(t, x, y)$ in estimate 4(a)

$$\psi_1^T(\|x\|)Y^T\underline{G}Y\psi_1(\|x\|) \leq v(t, x, y) \leq \psi_2^T(\|x\|)Y^T\overline{G}Y\psi_2(\|x\|),$$

where

$$\psi_1(\|x\|) = (\psi_{11}(\|x_1\|), \ldots, \psi_{1m}(\|x_m\|))^T,$$

$$\psi_2(\|x\|) = (\psi_{22}(\|x_1\|), \ldots, \psi_{m2}(\|x_m\|))^T,$$

$$Y = \text{diag}[y_1, \ldots, y_m],$$

$$\underline{G} = [\underline{\gamma}_{ij}], \quad \overline{G} = [\overline{\gamma}_{ij}], \quad i, j = 1, 2, \ldots, m.$$

Remark 1.8.2. The construction of the matrix $A_3(y)$ in estimate 4(c) is quite a difficult problem and is associated with the form of decomposition and aggregation of the system under investigation.

In Chapter 2 some methods of constructing the estimates of 4(c) type will be presented.

Theorem 1.8.2. Let the vector function f in system (1.2.7) be continuous on $R \times R^n$ (on $\mathcal{T}_\tau \times R^n$). If there exist

(1) a matrix-valued function $U \in C(R \times R^n, R^{m \times m})$ $(U \in C(\mathcal{T}_\tau \times R^n, R^{m \times m}))$ and a vector $y \in R^m$ such that the function $v(t, x, y) = y^T U(t, x)y$ is locally Lipschitzian in x for all $t \in R$ $(t \in \mathcal{T}_\tau)$;

(2) functions $\varphi_{1i}, \varphi_{2i}, \varphi_{3i} \in KR$, $\tilde{\varphi}_{2i} \in CKR$, $i = 1, 2, \ldots, m$;

(3) $m \times m$ matrices $B_j(y)$, $j = 1, 2, 3$, $\tilde{B}_2(y)$ such that

(a)
$$\varphi_1^{\mathrm{T}}(\|x\|)B_1(y)\varphi_1(\|x\|) \leq v(t, x, y) \leq \tilde{\varphi}_2^{\mathrm{T}}(t, \|x\|)\tilde{B}_2(y)\tilde{\varphi}_2(t, \|x\|)$$
$$\text{for all} \quad (t, x, y) \in R \times R^n \times R^m$$
$$(\text{for all} \quad (t, x, y) \in \mathcal{T}_\tau \times R^n \times R^m);$$

(b)
$$\varphi_1^{\mathrm{T}}(\|x\|)B_1(y)\varphi_1(\|x\|) \leq v(t, x, y) \leq \varphi_2^{\mathrm{T}}(\|x\|)B_2(y)\varphi_2(\|x\|)$$
$$\text{for all} \quad (t, x, y) \in R \times R^n \times R^m$$
$$(\text{for all} \quad (t, x, y) \in \mathcal{T}_\tau \times R^n \times R^m);$$

(c)
$$D^+v(t, x, y)\big|_{1.2.7)} \leq \varphi_3^{\mathrm{T}}(\|x\|)B_3(y)\varphi_3(\|x\|)$$
$$\text{for all} \quad (t, x, y) \in R \times R^n \times R^m$$
$$(\text{for all} \quad (t, x, y) \in \mathcal{T}_\tau \times R^n \times R^m).$$

Then, provided that matrices $B_1(y)$, $B_2(y)$ and $\tilde{B}_2(y)$, for all $(y \neq 0) \in R^m$ are positive definite and matrix $B_3(y)$ is negative definite,

(a) under condition (3)(a) the state $x = 0$ of system (1.2.7) is stable in the whole (on $\mathcal{T}_\tau$);
(b) under condition (3)(b) the state $x = 0$ of system (1.2.7) is uniformly stable in the whole (on $\mathcal{T}_\tau$).

The proof of this theorem is similar to that of Theorem 1.8.1 (see also Martynyuk [21]).

Theorem 1.8.3. Let the vector function f in system (1.2.7) be continuous on $R \times \mathcal{N}$ (on $\mathcal{T}_\tau \times \mathcal{N}$). If there exist

(1) an open connected time-invariant neighborhood $\mathcal{G} \subset \mathcal{N}$ of the point $x = 0$;
(2) a matrix-valued function $U \in C(R \times \mathcal{N}, R^{m \times m})$ ($U \in C(\mathcal{T}_\tau \times \mathcal{N}, R^{m \times m})$) and a vector $y \in R^m$ such that the function $v(t, x, y) = y^{\mathrm{T}}U(t, x)y$ is locally Lipschitzian in x for all $t \in R$ ($t \in \mathcal{T}_\tau$);
(3) functions $\eta_{1i}, \eta_{2i}, \eta_{3i} \in K$, $\tilde{\eta}_{2i} \in CK$, $i = 1, 2, \ldots, m$;
(4) $m \times m$ matrices $C_j(y)$, $j = 1, 2, 3$, $\tilde{C}_2(y)$ such that

(a)
$$\eta_1^{\mathrm{T}}(\|x\|)C_1(y)\eta_1(\|x\|) \leq v(t, x, y) \leq \tilde{\eta}_2^{\mathrm{T}}(t, \|x\|)\tilde{C}_2(y)\tilde{\eta}_2(t, \|x\|)$$
$$\text{for all} \quad (t, x, y) \in R \times \mathcal{G} \times R^m$$
$$(\text{for all} \quad (t, x, y) \in \mathcal{T}_\tau \times \mathcal{G} \times R^m);$$

$$\eta_1^T(\|x\|)C_1(y)\eta_1(\|x\|) \leq v(t,x,y) \leq \eta_2^T(\|x\|)C_2(y)\eta_2(\|x\|)$$

(b) $\qquad$ for all $\quad (t,x,y) \in R \times \mathcal{G} \times R^m$

$\qquad\qquad$ (for all $\quad (t,x,y) \in \mathcal{T}_\tau \times \mathcal{G} \times R^m$);

$$D^*v(t,x,y)\big|_{(1.2.7)} \leq \eta_3^T(\|x\|)C_3(y)\eta_3(\|x\|) + m\,(t,\,\eta_3(\|x\|))$$

(c) $\qquad$ for all $\quad (t,x,y) \in R \times \mathcal{G} \times R^m$

$\qquad\qquad$ (for all $\quad (t,x,y) \in \mathcal{T}_\tau \times \mathcal{G} \times R^m$),

where function $m(t,\cdot)$ satisfies the condition

$$\lim \frac{|m\,(t,\eta_3(\|x\|))\,|}{\|\eta_3\|} = 0 \quad \text{as} \quad \|\eta_3\| \to 0$$

uniformly in $t \in R$ $(t \in \mathcal{T}_\tau)$.

Then, provided the matrices $C_1(y)$, $C_2(y)$, $\tilde{C}_2(y)$ are positive definite and matrix $C_3(y)$ $(y \neq 0) \in R^m$ is negative definite, then

(a) under condition (4)(a) the state $x = 0$ of the system (1.2.7) is asymptotically stable (on $\mathcal{T}_\tau$);

(b) under condition (4)(b) the state $x = 0$ of the system (1.2.7) is uniformly asymptotically stable (on $\mathcal{T}_\tau$).

Theorem 1.8.4. Let the vector function f in system (1.2.7) be continuous on $R \times R^n$ (on $\mathcal{T}_\tau \times R^n$) and conditions (1) – (3) of Theorem 1.8.2 are satisfied.

Then, provided that matrices $B_1(y)$, $B_2(y)$ and $\tilde{B}_2(y)$ are positive definite and matrix $B_3(y)$ for all $(y \neq 0) \in R^m$ is negative definite,

(a) under condition (3)(a) of Theorem 1.8.2 the state $x = 0$ of system (1.2.7) is asymptotically stable in the whole (on $\mathcal{T}_\tau$);

(b) under condition (3)(b) of Theorem 1.8.2 the state $x = 0$ of system (1.2.7) is uniformly asymptotically stable in the whole (on $\mathcal{T}_\tau$).

For the proof of this theorem as well as the preceding one see Martynyuk [6].

Theorem 1.8.5. Let the vector function f in system (1.2.7) be continuous on $R \times R^n$ (on $\mathcal{T}_\tau \times R^n$). If there exist

(1) a matrix-valued function $U \in C\,(R \times R^n, R^{m \times m})$ $(U \in C(\mathcal{T}_\tau \times R^n, R^{m \times m}))$ and a vector $y \in R^m$ such that the function $v(t,x,y) = y^T U(t,x)y$ is locally Lipschitzian in x for all $t \in R$ (for all $t \in \mathcal{T}_\tau$);

(2) functions ν_{2i}, $\nu_{3i} \in KR$, $i = 1, 2, \ldots, m$, a positive real number $\Delta_2 > 0$ and a positive integer q;

(3) $m \times m$ matrices H_2, H_3 such that

$$\Delta_2 \|x\|^q \leq v(t,x,y) \leq \nu_2^T(\|x\|)H_2(y)\nu_2(\|x\|)$$
(a)   for all $(t,x,y \neq 0) \in R \times R^n \times R^m$
(for all $(t,x,y) \in \mathcal{T}_\tau \times R^n \times R^m$);

$$D^*v(t,x,y)\big|_{(1.2.7)} \leq \nu_3^T(\|x\|)H_3(y)\nu_3(\|x\|)$$
(b)   for all $(t,x,y \neq 0) \in R \times R^n \times R^m$
(for all $(t,x,y \neq 0) \in \mathcal{T}_\tau \times R^n \times R^m$).

Then, if the matrix $H_2(y)$ for all $(y \neq 0) \in R^m$ is positive definite, the matrix $H_3(y)$ for all $(y \neq 0) \in R^m$ is negative definite and functions ν_{2i}, ν_{3i} are of the same magnitude, the state $x = 0$ of system (1.2.7) is exponentially stable in the whole (on $\mathcal{T}_\tau$).

For the proof see Martynyuk [20, 21].

Theorem 1.8.6. Let the vector function f in system (1.2.7) be continuous on $R \times \mathcal{N}$ (on $\mathcal{T}_\tau \times \mathcal{N}$). If there exist

(1) an open connected time-invariant neighborhood $\mathcal{G} \subset \mathcal{N}$ of the point $x = 0$;

(2) a matrix-valued function $U \in C^{1,1}(R \times \mathcal{N}, R^{m \times m})$ ($U \in C^1(\mathcal{T}_\tau \times \mathcal{N}, R^{m \times m})$) and a vector $y \in R^m$;

(3) functions ψ_{1i}, ψ_{2i}, $\psi_{3i} \in K$, $i = 1, 2, \ldots, m$, $m \times m$ matrices $A_1(y)$, $A_2(y)$, $G(y)$ and a constant $\Delta > 0$ such that

$$\psi_1^T(\|x\|)A_1(y)\psi_1(\|x\|) \leq v(t,x,y) \leq \psi_2^T(\|x\|)A_2(y)\psi_2(\|x\|)$$
(a)   for all $(t,x,y) \in R \times \mathcal{G} \times R^m$
(for all $(t,x,y) \in \mathcal{T}_\tau \times \mathcal{G} \times R^m$);

$$Dv(t,x,y)\big|_{(1.2.7)} \geq \psi_3^T(\|x\|)G(y)\psi_3(\|x\|)$$
(b)   for all $(t,x,y) \in R \times \mathcal{G} \times R^m$
(for all $(t,x,y) \in \mathcal{T}_\tau \times \mathcal{G} \times R^m$);

(4) point $x = 0$ belong to $\partial\mathcal{G}$;

(5) $v(t,x,y) = 0$ on $\mathcal{T}_0 \times (\partial\mathcal{G} \cap B_\Delta)$, where $B_\Delta = \{x \colon \|x\| < \Delta\}$.

Then, if matrices $A_1(y)$, $A_2(y)$ and $G(y)$ for all $(y \neq 0) \in R^m$ are positive definite, the state $x = 0$ of system (1.2.7) is unstable (on T_τ).

For the proof see Martynyuk [21].

1.9 Notes

1.2. The problems of modern nonlinear dynamics are both complex and interdisciplinary (see, for example, Leitmann, *et al.* [1], Sivasundaram and Martynyuk [1], and Sivasundaram [1], etc.). The Liapunov's second method has gained increasing significance and has given a decisive impetus for modern development of qualitative methods in nonlinear dynamics of continuous, discrete-time, and other systems (see Michel, Wang, *et al.* [1]).

1.3. Our presentation of the accepted Definitions 1.2.1, 1.3.1–1.3.3, 1.4.5–1.4.12 is based on the results by Grujić [1], and Grujić, *et al.* [1]. For the details see also Barbashin and Krasovskii [1, 2], Coppel [1], Demidovich [1], Hahn [2], Hirsch and Smale [1], Krasovskii [1], Liapunov [1], Massera [1], Nemytskii and Stepanov [1], Yoshizawa [1], Zubov [4], etc.

1.4. Classical results of qualitative analysis of systems in terms of the direct Liapunov method based on the scalar auxiliary function can be found in the well-known monographs and manuals by Barbashin [2], Bhatia and Szegö [1], Chetaev [1], Hahn [2], Kalman and Bertman [1], Krasovskii [1], Lakshmikantham, Leela, *et al.* [1], Liapunov [1], Zubov [3], etc.

The results obtained via the vector Liapunov functions (see Azbelev [1], Bellman [2], Matrosov [1], and Mel'nikov [1]) and the ideas of the Kamke-Chaplygin-Ważewski comparison principle (see Hahn [2], Lakshmikantham and Leela [1], Rouche, Habets, *et al.* [1], and Szarski [1], etc.) in its modern interpretation are summarized by Abdullin, Anapolskii, *et al.* [1], Lakshmikantham, Matrosov, *et al.* [1], Martynyuk [1], Michel and Miller [1], Michel, Wang, *et al.* [1], Šiljak [1], etc.

For recent results obtained while developing the method of matrix Liapunov functions see Martynyuk and Slyn'ko [1, 2], Peng [1], Shaw [1], etc. Monograph by Martynyuk [20] exposes the main principles of the Liapunov matrix functions method and some of its applications to singularly perturbed and stochastic systems.

In the presentation of the material of this section we also incorporated the results by Grujić, *et al.* [1], Martynyuk [6, 7], and Zubov [4].

1.5. All necessary data on the comparison principle are given according to Lakshmikantham, Leela, *et al.* [1], and Yoshizawa [1]. Corollaries 1.5.1 – 1.5.4 are obtained based on the results by Zubov [4].

1.6. The proposed generalization of the original theorems of the classical stability theory is based on the application of the class of matrix-valued functions (see Martynyuk [3, 15, 20, 21]). In Corollaries 1.6.1 – 1.6.4 some results by Djordjević [3], and Zubov [1] are used.

1.7. General information on cone-valued functions is given according to Lakshmikantham and Leela [2]. For several results and references on cone-valued functions see Akinyele [1], Lakshmikantham, Leela, *et al.* [1], Martynyuk and Obolenskii [1, 2]. The notion of stability with respect to two measures is presented following Lakshmikantham, Leela, *et al.* [1], Lakshmikantham and Salvadori [1], Movchan [1]. Theorem 1.7.1 is a somewhat generalization of Theorem 3.1 by Lakshmikantham and Papageorgiou [1]. The notion of connected stability was introduced into stability theory by Šiljak [1, 2]. Theorem 1.7.2 is due to Ladde [1]. Theorem 1.7.3 is based on the results by Akpan [1] and Martynyuk and Obolenskii [1, 2].

1.8. Theorems 1.8.1 – 1.8.6 are due to Martynyuk [20, 21]. Estimations of the elements of matrix-valued function in Remark 1.8.1 incorporate some results by Djordjević [3].

2

QUALITATIVE ANALYSIS OF CONTINUOUS SYSTEMS

2.1 Introduction

General results of qualitative analysis of nonlinear systems motions presented in Section 1.6 and 1.8 are set out in terms of existence of an appropriate Liapunov matrix-valued function. Similarly to the classical version of the direct Liapunov method the efficiency of the results application in the investigation of concrete systems dynamics depends on the successful solution of the problem of constructing an appropriate matrix-valued function.

In this chapter we reveal some methods of solution of this problem. It is natural that some ideas developed earlier for constructing the scalar Liapunov functions proved to be useful. This chapter consists of seven sections in which we successively discuss various assumptions on dynamical properties of subsystems and work out new approaches to solve the problem of qualitative analysis of nonlinear systems.

In Section 2.2 the system of differential equations under consideration is assumed to admit mixed hierarchical decomposition. In terms of this assumption proposed is a solution to the problem of forming a two-index system of functions that is a suitable basis for constructing the Liapunov function.

In Section 2.3 a homogeneous hierarchical decomposition of system is employed which was proposed by Ikeda and Šiljak [2]. A general method of the Liapunov function construction developed in this section is used to analyze the motions of linear and nonlinear nonautonomous systems.

In Section 2.4 the method of motion stability analysis is discussed in terms of one level decomposition of the initial nonautonomous or autonomous system. Based on the matrix-valued Liapunov functions constructed in this section new stability criteria are established.

In Section 2.5 we develop the method of constructing the matrix-valued function by employing the idea of the initial system extension in terms of the overlapping decomposition of Ikeda and Šiljak [1].

Section 2.6 gives an account of results obtained in one of new directions of the development of nonlinear dynamics of system associated with the investigation of motion polystability. Namely, we set out various existence criteria for the exponential polystability of motion of nonautonomous and time-invariant systems with separating motions. Also, the possibility of studying the polystability of motion with respect to the first approximation is stated.

Section 2.7 deals with the problem of integral and Lipschitz stability of motion. The motion stability conditions established here are more flexible as compared with those obtained in terms of the scalar Liapunov function.

2.2 Nonlinear Systems with Mixed Hierarchy of Subsystems

2.2.1 Mixed hierarchical structures

We consider the dynamical system described by the equations

$$(2.2.1) \qquad \frac{dx}{dt} = f(t, x), \quad x(t_0) = x_0,$$

where $x \in R^n$, $f \in C(R_+ \times R^n, R^n)$ and the solution $x(t; t_0, x_0)$ of which exists for all initial values of $(t_0, x_0) \in R \times R^n$.

We recall that if $f(t, 0) = 0$ for all $t \in R$ (for all $t \in \mathcal{T}_\tau$) then the state $x = 0$ is the unique equilibrium state of the system (2.2.1). The system (2.2.1) can be interpreted as a physical composition of some systems or as a large scale system admitting mathematical decomposition into several free subsystems

$$(2.2.2) \qquad \sigma_i: \quad \frac{dx_i}{dt} = g_i(t, x_i), \quad i = 1, 2, ..., m,$$

where $x_i \in R^{n_i}$, $g_i \in C(R \times R^{n_i}, R^{n_i})$ are united into system (2.2.1) by link functions

$$h_i: \quad h_i = (t, x_1, \ldots, x_s), \quad h_i \in C(R \times R^{n_1} \times \cdots \times R^{n_s}, R^{n_i}).$$

Pair (σ_k, h_k) describes completely the fixed kth system

$$(2.2.3) \qquad \frac{dx_k}{dt} = g_k(t, x_k) + h_k(t, x_1, \ldots, x_s)$$

in the totality of systems

$$(2.2.4) \qquad \frac{dx_i}{dt} = g_i(t, x_i) + h_i(t, x_1, \ldots, x_s), \quad i = 1, \ldots, m.$$

The transformation of system (2.2.1) to the form of (2.2.4) is called the *mathematical first level decomposition* of system (2.2.1).

Remark 2.2.1. The stability theory of large scale systems developed in terms of vector Liapunov functions, incorporates equations (2.2.4) i.e. only first level decomposition of system (2.2.1) is used (see Abdullin, Anapolskii, *et al.* [1], Grujić, *et al.* [1], Martynyuk [1], Michel and Miller [1], Šiljak [1], etc.).

Together with the first level decomposition of system (2.2.1) we single out couples (i, j) for all $(i \neq j) \in [1, m]$ of free subsystems

$$(2.2.5) \qquad \sigma_{ij}: \quad \begin{cases} \frac{dx_i}{dt} = q_i(t, x_i, x_j), \\ \frac{dx_j}{dt} = q_j(t, x_i, x_j), \end{cases}$$

where $x_i \in R^{n_i}$, $x_j \in R^{n_j}$, $q_i \in C(R \times R^{n_i} \times R^{n_j}, R^{n_i})$, $q_j \in C(R \times R^{n_i} \times R^{n_j}, R^{n_j})$.

We designate $(x_i, x_j) = [0, \ldots, 0, x_i^{\mathrm{T}}, \ldots, x_j^{\mathrm{T}}, 0, \ldots, 0]^{\mathrm{T}}$ for all $(i \neq j) \in [1, m]$. Then

$$q_i = f_i(t, 0, \ldots, 0, x_i, \ldots, x_j, 0, \ldots, 0),$$
$$q_j = f_j(t, 0, \ldots, 0, x_i, \ldots, x_j, 0, \ldots, 0).$$

Further we designate

$$(2.2.6) \qquad \begin{aligned} h_i^*(t, x_1, \ldots, x_s) &= f_i(t, x_1, \ldots, x_s) \\ &\quad - f_i(t, 0, \ldots, 0, x_i, \ldots, x_j, 0, \ldots, 0), \\ h_j^*(t, x_1, \ldots, x_s) &= f_j(t, x_1, \ldots, x_s) \\ &\quad - f_j(t, 0, \ldots, 0, x_i, \ldots, x_j, 0, \ldots, 0), \end{aligned}$$

where

$$h_i^* \in C(R \times R^{n_1} \times R^{n_2} \times \cdots \times R^{n_s}, R^{n_i}),$$
$$h_j^* \in C(R \times R^{n_1} \times R^{n_2} \times \cdots \times R^{n_s}, R^{n_j}).$$

In view of (2.2.5) and (2.2.6) system (2.2.1) in the form of interacting couples (i, j) of subsystems is presented as

(2.2.7)
$$\frac{dx_i}{dt} = q_i(t, x_i, x_j) + h_i^*(t, x_1, \ldots, x_s),$$
$$\frac{dx_j}{dt} = q_j(t, x_i, x_j) + h_j^*(t, x_1, \ldots, x_s), \quad \forall (i \neq j) \in [1, s].$$

Transformation of system (2.2.1) to the form of (2.2.7) is called the *mathematical second level decomposition* of system (2.2.1).

If

$$h_i^* = h_i^*(t, x_1, \ldots, x_{i-1}, x_{i+1}, \ldots, x_{j-1}, x_{j+1}, \ldots, x_m),$$
$$h_j^* = h_j^*(t, x_1, \ldots, x_{i-1}, x_{i+1}, \ldots, x_{j-1}, x_{j+1}, \ldots, x_m),$$

then system (2.2.1) allows *complete second level decomposition* in the form of equations (2.2.7). System (2.2.1) transformed to the form (2.2.7) is a mixed hierarchical structure of subsystems (2.2.4).

Dynamical properties of solutions of system (2.2.1) depend on the properties of solutions (2.2.2) and (2.2.5) and link functions h_i, h_i^*, h_j^*, for $(i \neq j) \in [1, m]$. This makes possible the application of hierarchical matrix Liapunov function for the analysis of system (2.2.1).

2.2.2 Hierarchical matrix function structure

We construct for subsystem (2.2.2) the functions

(2.2.8) $v_{ii} \in C(R_+ \times R^{n_i}, R_+), \quad i = 1, 2, \ldots, m,$

where $v_{ii}(t, 0) = 0$ for all $t \in R_+$, $i = 1, 2, \ldots, s$ and functions v_{ii} are locally Lipschitzian in x_i.

For couples (i, j) of (2.2.5) we construct the functions

(2.2.9) $v_{ij} \in C(R_+ \times R^{n_i} \times R^{n_j}, R)$ for all $(i \neq j) = 1, 2, \ldots, m,$

where $v_{ij}(t, 0, 0) = 0$ for all $t \in R_+$ and $v_{ij}(t, x_i, x_j)$ are locally Lipschitzian in x_i and x_j. In terms of the functions (2.2.8) and (2.2.9) we formulate the matrix-valued function

(2.2.10) $U(t, x) = [v_{ij}(t, \cdot)], \quad i, j = 1, 2, \ldots, m,$

where $U \in C(R_+ \times R^{n_i} \times R^{n_j}, R^{m \times m})$. The *matrix-valued function* (2.2.10) is *hierarchical* and takes into account first and second level decompositions of the system (2.2.1).

In order for the matrix-valued function (2.2.10) to be applicable in stability investigation of system (2.2.1), some assumptions on the functions (2.2.8) and (2.2.9) are necessary.

Assumption 2.2.1. *There exists*

(1) *an open time-invariant neighborhood $\mathcal{N}_i$ of the point $x_i = 0$, $\mathcal{N}_i \subseteq R^{n_i}$, $i = 1, 2, \ldots, m$;*

(2) *the functions $\varphi_i \in K(KR)$, $\psi_i \in K(KR)$, $\varphi_{ij} \in K(KR)$, $\psi_{ij} \in K(KR)$ such that*

    (a) *$\varphi_i(\|x_i\|) \le v_{ii}(t, x_i) \le \psi_i(\|x_i\|)$ for all $i = 1, 2, \ldots, m$ for all $(t, x_i) \in R_+ \times \mathcal{N}_i$ or (for all $(t, x) \in R_+ \times R^{n_i}$);*

    (b) *$\varphi_{ij}(\|x_{ij}\|) \le v_{ij}(t, x_i, x_j) \le \psi_{ij}(\|x_{ij}\|)$ for all $(i \ne j) = 1, 2, \ldots, m$, for all $(t, x_i, x_j) \in R_+ \times \mathcal{N}_i \times \mathcal{N}_j$ or (for all $(t, x_i, x_j) \in R_+ \times R^{n_i} \times R^{n_j}$).*

Proposition 2.2.1. *If all conditions of Assumption 2.2.1 are satisfied, then hierarchical matrix-valued function (2.2.10) is positive definite, decreasing and radially unbounded.*

Proof. By means of the vector $e = (1, \ldots, 1)^{\mathrm{T}} \in R_+^m$ we construct the function

$$(2.2.11) \qquad v(t, x, e) = e^{\mathrm{T}} U(t, x) e,$$

the coordinate form of which is

$$v(t, x) = \sum_{i=1}^{m} v_{ii}(t, x_i) + \sum_{\substack{i, j=1 \\ i \ne j}}^{m} v_{ij}(t, x_i, x_j).$$

Under the conditions of Assumption 2.2.1 we have

$$\sum_{i=1}^{m} \varphi_i(\|x_i\|) + \sum_{\substack{i, j=1}}^{m} \varphi_{ij}(\|x_{ij}\|)$$

$$\le v(t, x) \le \sum_{i=1}^{m} \psi_i(\|x_i\|) + \sum_{\substack{i, j=1 \\ i \ne j}}^{m} \psi_{ij}(\|x_{ij}\|)$$

for all $(t, x_i, x_j) \in R_+ \times \mathcal{N}_i \times \mathcal{N}_j$. Hence, it follows that matrix-valued function (2.2.10) is positive definite, decreasing and radially unbounded. In fact, there exist functions $\alpha_1, \alpha_2 \in K(KR)$ such that

$$\alpha_1(\|x\|) \le \sum_{i=1}^{m} \varphi_i(\|x_i\|) \quad \text{and} \quad \alpha_2(\|x\|) \ge \sum_{i=1}^{m} \psi_i(\|x_i\|)$$

and functions β_1, $\beta_2 \in K(KR)$ such that

$$\beta_1(\|x\|) \leq \sum_{\substack{i,j=1 \\ i \neq j}}^{m} \varphi_{ij}(\|x_{ij}\|) \quad \text{and} \quad \beta_2(\|x\|) \geq \sum_{\substack{i,j=1 \\ i \neq j}}^{m} \psi_{ij}(\|x_{ij}\|).$$

Therefore,

$$\alpha_1(\|x\|) + \beta_1(\|x\|) \leq v(t,x) \leq \alpha_2(\|x\|) + \beta_2(\|x\|)$$

for all $(t,x) \in R_+ \times R^n$.

Since α_1, α_2 and β_1, $\beta_2 \in K(KR)$, functions γ_1, $\gamma_2 \in K(KR)$ will be found, such that

$$\gamma_1(\|x\|) \leq \alpha_1(\|x\|) + \beta_1(\|x\|) \quad \text{and} \quad \gamma_2(\|x\|) \geq \alpha_2(\|x\|) + \beta_2(\|x\|)$$

and then

$$\gamma_1(\|x\|) \leq v(t,x) \leq \gamma_2(\|x\|) \quad \text{for all} \quad (t,x) \in R_+ \times R^n.$$

Remark 2.2.2. Conditions of Assumption 2.2.1 are "too sufficient" for hierarchical matrix-valued function (2.2.10) be positive definite decreasing and radially unbounded.

Assumption 2.2.2. *There exists*

(1) *an open time-invariant neighborhood $\mathcal{N}_i$ of the point $x_i = 0$, $\mathcal{N}_i \subseteq R^{n_i}$, $i = 1, 2, \ldots, m$;*

(2) *vector $y \in R^m$, $y \neq 0$ and functions (2.2.8) and (2.2.9) such that*
   (a) $v_1(t, x_1, \ldots, x_s, y) = \sum_{i=1}^{m} y_i^2 v_{ii}(t, x_i)$ *and*
   (b) $v_2(t, x_1, \ldots, x_s, y) = \sum_{\substack{i,j=1 \\ i \neq j}}^{m} y_i y_j v_{ij}(t, x_i, x_j)$

are positive definite, decreasing and radially unbounded for all $(t, x_i) \in R_+ \times \mathcal{N}_i$ (for all $(t, x_i, x_j) \in R_+ \times R^{n_i} \times R^{n_j}$).

Proposition 2.2.2. *If all conditions of Assumption 2.2.2 are satisfied, then hierarchical matrix-valued function (2.2.10) is positive definite, decreasing and radially unbounded.*

Proof. Since

(2.2.12) $$y^T U(t,x)y = v_1(t,x,y) + v_2(t,x,y)$$

for all $(t,x) \in R_+ \times \mathcal{N}$ (for all $(t,x) \in R_+ \times R^n$), one can apply to the function (2.2.12) all the arguments used in the proof of Proposition 2.2.1.

Remark 2.2.3. By conditions (2)(a) and (2)(b) of Assumption 2.2.2 functions $v_{ii}(t,x_i)$ and $v_{ij}(t,x_i,x_j)$ $(i \neq j)$ can be positive semi-definite, while functions $v_1(t,x,y)$ and $v_2(t,x,y)$ will posses all mentioned properties.

Assumption 2.2.3. *There exists*

(1) *an open time invariant neighborhood $\mathcal{N}_i$ of point $x_i = 0$, for all $i = 1, 2, \ldots, m$;*

(2) *a vector $\eta \in R_+^m$, $\eta > 0$ and positive semi-definite functions $u_i(x_i)$, $u_i(0) = 0$, $w_i(x_i)$, $w_i(0) = 0$, $i = 1, 2, \ldots, m$, positive constants $\underline{c}_{ii} > 0$, $\overline{c}_{ii} > 0$ and arbitral constants $\underline{c}_{ij}$, $\overline{c}_{ij}$ $(i \neq j)$ such that*

   (a) *$\underline{c}_{ii} u_i(x_i) \leq v_{ii}(t,x_i) \leq \overline{c}_{ii} w_i(x_i)$ for all $(t,x_i) \in R_+ \times \mathcal{N}_i$ (for all $(t,x_i) \in R_+ \times R^{n_i}$)*

   (b) *$\underline{c}_{ij} u_i(x_i) u_j(x_j) \leq v_{ij}(t,x_i,x_j) \leq \overline{c}_{ij} w_i(x_i) w_j(x_j)$ for all $(t,x_i,x_j) \in R_+ \times \mathcal{N}_i \times \mathcal{N}_j$ (for all $(t,x_i,x_j) \in R_+ \times R^{n_i} \times R^{n_j}$).*

Proposition 2.2.3. *If all conditions of Assumption 2.2.3 are satisfied, and matrices*

$$A = [\underline{c}_{ij}], \quad B = [\overline{c}_{ij}], \quad i, j = 1, 2, \ldots, m$$

are positive definite, then hierarchical matrix-valued function (2.2.10) is positive definite and decreasing.

Proof. In view of the estimate from below in inequalities (2)(a), (2)(b) we get for the function $v(t,x,\eta) = \eta^{\mathrm{T}} U(t,x)\eta$ that

$$(2.2.13) \qquad v(t,x,\eta) \geq u^{\mathrm{T}} H^{\mathrm{T}} A H u,$$

for all $(t,x) \in R_+ \times \mathcal{N}$ or (for all $(t,x) \in R_+ \times R^n$). Here

$$u = (u_1(x_1), \ldots, u_m(x_m))^{\mathrm{T}}, \quad H = \mathrm{diag}\,[\eta_1, \ldots, \eta_m].$$

Similarly

$$(2.2.14) \qquad v(t,x,\eta) \leq w^{\mathrm{T}} H^{\mathrm{T}} B H w,$$

where $w = (w_1(x_1), \ldots, w_m(x_m))^{\mathrm{T}}$.

According to inequalities (2.2.13) and (2.2.14) hierarchical matrix-valued function (2.2.10) is positive definite, provided matrix A is positive definite and decreasing, provided matrix B is positive definite.

Remark 2.2.4. If in Proposition 2.2.3 positive semi-definite functions $u_i(x_i)$ and $w_i(x_i)$ are replaced by the functions

$$u_i = \varphi_i \in K(KR), \quad w_i = \psi_i \in K(KR), \quad i = 1, 2, \ldots, m$$

or $u_i = w_i = \|x_i\|$, $i = 1, 2, \ldots, m$, then under conditions of Assumption 2.2.3 the proposition remains valid.

Thus, Assumptions 2.2.1 – 2.2.3 contain combination of conditions sufficient for hierarchical matrix-valued function (2.2.10) to be positive definite, decreasing and radially unbounded. Conditions of these assumptions together with decomposition forms define the structure of hierarchical matrix-valued function.

2.2.3 Structure of hierarchical matrix function derivative

The application of hierarchical matrix function (2.2.10) in the stability investigation of equilibrium state $x = 0$ of system (2.2.1) involves construction of some estimates for both functions (2.2.8) and (2.2.9) and the derivatives of these functions along solutions of the systems under consideration. We recall that in case of scalar Liapunov function construction for system (2.2.1) in the investigation on uniformly in t_0, x_0 asymptotic stability, Liapunov's theorem is applied with addition and conversion (see Hahn [2], Krasovskii [1], Zubov [3], etc.). This theorem can be formulated as follows.

For unperturbed motion $x = 0$ to be uniform in t_0, x_0 asymptotically stable, it is necessary and sufficient that there exists positive definite and decreasing function v, the derivative of which along solutions of system (2.2.1) is negative definite.

In this case the problem of Liapunov function construction for system (2.2.1) is reduced to searching for partial solution of the equation

$$(2.2.15) \qquad \frac{\partial v}{\partial t} + \left(\frac{\partial v}{\partial x}\right)^{\mathrm{T}} f(t, x) = w(t, x), \quad v(t, 0) = 0,$$

where $w(t, x)$ is negative definite, $w(t, 0) = 0$ for all $t \in R_+$. This solution must be of definite sign in the sense of Liapunov and decreasing for prescribed sign definite function $w(t, x)$.

Assumption 2.2.4. *Independent subsystems (2.2.2) of first level decomposition and link functions* $h_i(t, x) = h_i(t, x_1, \ldots, x_m)$ *are such that*

(1) *there exist functions* $v_{ii} \in C(R_+ \times R^{n_i}, R_+)$, *satisfying conditions of Assumption 2.2.1;*

(2) *there exist functions* $\varphi_i \in K$ *and constants* p_{ii}^0, μ_{ik}, $i, k = 1, \ldots, m$ *for which the following conditions are satisfied*

    (a) $D_t^+ v_{ii} + (D_{x_i}^+ v_{ii})^{\mathrm{T}} f_i(t, x_i) \leq p_{ii}^0 \varphi_i(\|x_i\|);$

    (b) $(D_{x_i}^+ v_{ii})^{\mathrm{T}} h_i(t, x) \leq \varphi_i^{1/2}(\|x_i\|) \sum\limits_{k=1}^m \mu_{ik} \varphi_k^{1/2}(\|x_k\|)$

    for all $(t, x) \in R_+ \times \mathcal{N}$, $\mathcal{N} = \mathcal{N}_1 \times \ldots \times \mathcal{N}_m$.

We note that conditions (1) and (2)(a) imply uniform in t_0, x_{i0} asymptotic stability of ith subsystem (1) if $p_{ii}^0 < 0$, $i \in [1, m]$. If $p_{ii}^0 = 0$, the state $x_i = 0$ of subsystem (2.2.3) is uniformly stable, and it is unstable for $p_{ii}^0 > 0$.

Assumption 2.2.5. *Independent* (i, j) *couples of subsystems (2.2.5) of the second level decomposition and link functions (2.2.6) are such that*

(1) *there exist functions* v_{ii} *satisfying the conditions of Assumption 2.2.1;*

(2) *there exist functions* $\varphi_i \in K$ *and constants* p_{ij}^1, p_{ij}^2, p_{ij}^3, ν_{kp}^{ij}; $i, j, k, p = 1, 2, \ldots, m$, *for which the following conditions are satisfied*

    (a) $D_t^+ v_{ij} + (D_{x_{ij}}^+ v_{ij})^{\mathrm{T}} q_{ij}(t, x_{ij}) \leq p_{ij}^1 \varphi_i(\|x_i\|)$
        $+ 2 p_{ij}^2 \varphi_i^{1/2}(\|x_i\|) \varphi_j^{1/2}(\|x_j\|) + p_{ij}^3 \varphi_j(\|x_j\|);$

    (b) $(D_{x_{ij}}^+ v_{ij})^{\mathrm{T}} H_{ij}(t, x) \leq \sum\limits_{k, p=1}^s \nu_{kp}^{ij} \varphi_k^{1/2}(\|x_k\|) \varphi_p^{1/2}(\|x_p\|)$

    for all $(t, x_{ij}) \in R_+ \times \mathcal{N}_i \times \mathcal{N}_j$, *for all* $i \neq j$, $(i, j) = 1, 2, \ldots, m$.

Here the following notations are adopted

$$x_{ij} = (x_i^{\mathrm{T}}, x_j^{\mathrm{T}})^{\mathrm{T}} \quad \text{for all} \quad (i \neq j) \in [1, m];$$

$$q_{ij} = (q_i^{\mathrm{T}}(t, x_{ij}), \quad q_j^{\mathrm{T}}(t, x_{ij}))^{\mathrm{T}}, \quad H_{ij} = ((h_i^*(t, x))^{\mathrm{T}}, (h_j^*(t, x))^{\mathrm{T}})^{\mathrm{T}},$$

$$\mathcal{N}_i \subseteq R^{n_i}, \quad \mathcal{N}_j \subseteq R^{n_j}.$$

We note that the dynamical properties of couples (i, j) of subsystems (2.2.5) are connected with the properties of matrices

(2.2.16) $$Q_{ij} = \begin{pmatrix} p_{ij}^1 & p_{ij}^2 \\ p_{ij}^2 & p_{ij}^3 \end{pmatrix}.$$

Namely, if matrices (2.2.16) for all $(i \neq j) = 1, 2, \ldots, m$ are negative semi-definite (negative definite), then states $x_i = x_j = 0$ of couples (i, j) of subsystems are stable (uniformly asymptotically stable).

Proposition 2.2.4. *If all conditions of Assumptions 2.2.4 and 2.2.5 are satisfied, then for the function*

$$\eta^{\mathrm{T}} D^+ U(t, x) \eta = D^+ v(t, x, \eta)$$

estimate

$$(2.2.17) \qquad D^+ v(t, x, \eta) \leq \varphi^{\mathrm{T}}(\|x\|) S \varphi(\|x\|)$$

holds.

Here $\varphi(\|x\|) = (\varphi_1^{1/2}(\|x_1\|), \ldots, \varphi_m^{1/2}(\|x_m\|))^{\mathrm{T}}$, $S = \frac{1}{2}(B + B^{\mathrm{T}})$ *and elements* b_{ij} *of matrix* B *are defined by the expressions*

$$b_{qq} = \eta_q^2 (p_{qq}^0 + \mu_{qq}) + \eta_q \left(\sum_{\substack{i,j=1 \\ i \neq j}}^{m} \eta_j p_{qj}^1 + \sum_{\substack{i,j=1 \\ i \neq j}}^{m} \eta_i p_{iq}^3 \right) + \sum_{i,j=1}^{m} \eta_i \eta_j \nu_{qq}^{ij};$$

$$b_{ql} = \eta_q^2 \mu_{ql} + 2\eta_q \eta_l p_{ql}^2 + \sum_{\substack{i,j=1 \\ i \neq j}}^{m} \eta_i \eta_j \nu_{ql}^{ij}.$$

Proof. The function $D^+ v(t, x, \eta)$ in coordinate form reads

$$(2.2.18) \qquad D^+ v(t, x, \eta) = \sum_{i,j=1}^{s} D^+ v_{ij}(t, \cdot) \eta_i \eta_j, \qquad i, j = 1, 2, \ldots, m.$$

Substituting into (2.2.18) estimates from Assumptions 2.2.4 and 2.2.5 we get estimate (2.2.17).

Proposition 2.2.5. *If in conditions of Assumptions 2.2.4 and 2.2.5 inequalities (2)(a) and (2)(b) are satisfied with reversed sign and constants* $\tilde{p}_{ii}^0$, $\tilde{\mu}_{ik}$, $\tilde{p}_{ij}^1$, $\tilde{p}_{ij}^2$, $\tilde{p}_{ij}^3$, ν_{kp}^{ij}, $i, j, k, p = 1, 2, \ldots, m$, *then for the function* $D^+ v(t, x)$ *the estimate from below is*

$$(2.2.19) \qquad D^+ v(t, x, \eta) \geq \varphi^{\mathrm{T}}(\|x\|) \tilde{S} \varphi(\|x\|),$$

where $\tilde{S} = \frac{1}{2}(\tilde{B} + \tilde{B}^{\mathrm{T}})$ and elements $\tilde{b}_{ij}$ are defined similarly to elements b_{ij} of matrix B.

Remark 2.2.5. In conditions (2)(a) and (2)(b) of Assumptions 2.2.4 and 2.2.5 it is possible to use instead of functions $\varphi_i \in KR$, $i \in [1, m]$ the positive semi-definite functions $u_i(x_i)$ or the Euclidean norms of the vectors $x_i \in R^{n_i}$, $i = 1, 2, \ldots, m$.

2.2.4 Stability and instability conditions

There are conditions under which hierarchical matrix-valued function $U(t, x)$ is definite and decreasing and the estimate (2.2.17) allow us to establish existence of dynamical properties of certain type of the state $x = 0$ of system (2.2.1).

Theorem 2.2.1. *Let vector function f in system (2.2.1) be continuous on $R_+ \times \mathcal{N}$ (on $R_+ \times R^n$). If the following conditions are satisfied*

(1) *all conditions of Assumptions 2.2.3, 2.2.4 and 2.2.5 hold;*
(2) *matrices A and B (see Assumption 2.2.3) are positive definite;*
(3) *matrix $S \in R^{m \times m}$ in inequality (2.2.17) is*
    (a) *negative semi-definite;*
    (b) *negative definite;*
(4) *matrix $\tilde{S} \in R^{m \times m}$ in inequality (2.2.19) is*
    (a) *positive definite.*

Then

(a) *conditions (1), (2) and (3)(a) imply uniform stability of the state $x = 0$ of system (2.2.1);*
(b) *conditions (1), (2) and (3)(b) imply uniform asymptotic stability of the state $x = 0$ of system (2.2.1);*
(c) *conditions (1), (2) and (4)(a) imply instability of the state $x = 0$ of system (2.2.1).*

If, in addition, in conditions of Assumptions 2.2.3, 2.2.4 and 2.2.5 $\mathcal{N}_i = R^{n_i}$, $i = 1, 2, \ldots, m$ and the functions

$$u_i(x_i) = \varphi_i(\|x_i\|) \in KR, \qquad w_i(x_i) = \psi_i(\|x_i\|) \in KR$$

then

(d) *conditions (1), (2) and (3)(a) imply uniform stability in the whole of the state $x = 0$ of system (2.2.1);*
(e) *conditions (1), (2) and (3)(b) imply uniform asymptotic stability in the whole of the state $x = 0$ of system (2.2.1).*

60   2. CONTINUOUS SYSTEMS

Proof. Assertion (a) of Theorem 2.2.1 results from Theorem 1.8.2, since under conditions (1),(2) and (3)(a) of Theorem 2.2.1 all conditions of Theorem 1.8.2 are satisfied. Namely, matrix-valued function $v(t,x)$ is positive definite and decreasing and its derivative $D^+v(t,x)$ is negative semidefinite. Assertions (b) and (c) of Theorem 2.2.1 follow from Theorems 1.8.4 and 1.8.6. Assertions (d) and (e) of Theorem 2.2.1 are proved in terms of Theorems 1.6.2 and 1.6.3.

2.2.5 Linear autonomous system

We consider a *linear autonomous system*

$$(2.2.20) \qquad \frac{dx}{dt} = Ax, \quad x(t_0) = x_0,$$

where $A = \{A_{ij}\}$ for all $i, j = 1, 2, \ldots, m$ is a block matrix $n \times n$ with blocks A_{ij} of dimensions $n_i \times n_j$, $\sum_{i=1}^{m} n_i = n$. Vector $x = (x_1^T, \ldots, x_m^T)^T$ has subvectors $x_i \in R^{n_i}$, $i \in [1, m]$ as its components.

We get for the system (2.2.20) in the result of first level decomposition

$$(2.2.21) \qquad \frac{dx_i}{dt} = A_{ii}x_i + \sum_{\substack{k=1 \\ (k \neq i)}}^{m} A_{ik}x_k, \quad i = 1, 2, \ldots, m.$$

Besides, from the system (2.2.21) we obtain obviously the independent subsystems

$$\frac{dx_i}{dt} = A_{ii}x_i, \quad x_i(t_0) = x_{i0}, \quad i = 1, 2, \ldots, m$$

and the interconnections functions

$$h_i(x) = \sum_{\substack{k=1 \\ (k \neq i)}}^{m} A_{ik}x_k, \quad i = 1, 2, \ldots, m.$$

Complete second level decomposition of the system (2.2.20) is always possible and results in couples (i, j) of subsystems

$$\frac{dx_i}{dt} = A_{ii}x_i + A_{jj}x_j + \sum_{\substack{k=1 \\ (k \neq i, k \neq j)}}^{m} A_{ik}x_k,$$

$$(2.2.22)$$

$$\frac{dx_j}{dt} = A_{jj}x_j + A_{ii}x_i + \sum_{\substack{k=1 \\ (k \neq i, k \neq j)}}^{m} A_{jk}x_k.$$

System (2.2.22) is rewritten as

$$(2.2.23) \qquad \frac{dx_{ij}}{dt} = \bar{A}_{ij}x_{ij} + \sum_{\substack{k=1 \\ (k \neq i,\, k \neq j)}}^{m} \bar{A}_{ij}^k x_k,$$

where $x_{ij} = (x_i^T, x_j^T)^T$, $x_{ij} \in R^{n_i} \times R^{n_j}$, and block matrices $\bar{A}_{ij}$ and $\bar{A}_{ij}^k$ with dimensions $(n_i + n_j) \times (n_i + n_j)$ and $(n_i + n_j) \times n_k$ are defined as

$$(2.2.24) \qquad \bar{A}_{ij} = \begin{pmatrix} A_{ii} & A_{ij} \\ A_{ji} & A_{jj} \end{pmatrix},$$

$$(2.2.25) \qquad \bar{A}_{ij}^k = (A_{ik}^T, A_{jk}^T)^T.$$

Elements (2.2.8) of the main diagonal of matrix-valued function (2.2.10) are defined as

$$(2.2.26) \qquad v_{ii}(x_i) = x_i^T B_{ii} x_i \quad \text{for all} \quad x_i \in R^{n_i},$$

where

$$(2.2.27) \qquad A_{ii}^T B_{ii} + B_{ii} A_{ii} = C_{ii}, \quad i = 1, 2, \ldots, m.$$

Here C_{ii} are symmetric matrices $n_i \times n_i$. Matrices $C_{ii} \leq 0$ ($C_{ii} < 0$), if independent subsystems of first level decomposition are stable (asymptotically stable). Nondiagonal elements (2.2.9) of matrix-valued function (2.2.10) are defined as

$$(2.2.28) \qquad \begin{aligned} v_{ij}(x_i, x_j) &= x_{ij}^T B_{ij} x_{ij}, \quad (i < j) \in [1, m] \\ &\text{for all} \quad x_{ij} \in R^{n_i} \times R^{n_j}, \\ v_{ij}(x_{ij}) &= v_{ji}(x_{ji}) \quad \text{for all} \quad (i \neq j) \in [1, m], \end{aligned}$$

where

$$(2.2.29) \qquad \bar{A}_{ij}^T B_{ij} + B_{ij} \bar{A}_{ij} = C_{ij} \quad (i < j) \in [1, m].$$

Here C_{ij} are symmetric matrices $(n_i + n_j) \times (n_i + n_j)$. Matrices $C_{ij} \leq 0$ ($C_{ij} < 0$), if free couple (i, j) of second level decomposition are stable (asymptotically stable).

We write down symmetric matrices C_{ij} and B_{ij} in block form

$$C_{ij} = \begin{pmatrix} C_{ij}^i & \bar{C}_{ij} \\ \bar{C}_{ij}^T & C_{ij}^j \end{pmatrix}, \qquad B_{ij} = \begin{pmatrix} B_{ij}^i & \bar{B}_{ij} \\ \bar{B}_{ij}^T & B_{ij}^j \end{pmatrix}.$$

Here C_{ij}^i and B_{ij}^i are symmetric matrices with dimensions $n_i \times n_i$, C_{ij}^j and B_{ij}^j are matrices $n_j \times n_j$ and $\bar{B}_{ij}$ and $\bar{C}_{ij}$ are matrices $n_i \times n_j$.

We introduce the designations

$$\sigma_{pp} = \eta_p^2 C_{pp} + \eta_p \sum_{\substack{j=1 \\ j \neq p}}^{m} \eta_j (C_{pj}^p + C_{jp}^j), \quad p = 1, 2, \ldots, m,$$

where $\eta_p > 0$,

$$\sigma_{pq} = 2 \left[\eta_p \eta_q \bar{C}_{pq} + \eta_p \sum_{\substack{j=q \\ j \neq p}}^{m} \eta_j (A_{qp}^T B_{qj}^q + A_{jp}^T B_{qj}^T) \right.$$

$$+ \eta_q \sum_{\substack{j=q \\ j \neq p}}^{m} \eta_j (A_{jp}^T \bar{B}_{jq} + A_{qp}^T B_{jq}^j) + \eta_p \sum_{\substack{j=p \\ j \neq q}}^{m} (B_{pj}^{pT} A_{pq} + \bar{B}_{pj}^T A_{jq}) \eta_j$$

$$\left. + \eta_q \sum_{\substack{j=p \\ j \neq q}}^{m} \eta_j (\bar{B}_{jp}^T A_{jq} + B_{jq}^{jT}) A_{pq} \right],$$

and $\lambda_M(\sigma_{pp})$ is maximal eigenvalue of matrices σ_{pp}, $[\lambda_M(\sigma_{pq}^T \sigma_{pq})]^{1/2}$ is a norm of matrices $(\sigma_{pq}^T \sigma_{pq})$.

Theorem 2.2.2. *Assume that*

(1) *the system (2.2.20) underwent first level (2.2.21) and second level (2.2.22) decompositions;*

(2) *the matrix-valued function (2.2.10) is constructed from elements (2.2.26) and (2.2.28) and is positive definite, i.e.*

(2.2.30) $v(x, \eta) = \eta^T U(x) \eta > 0, \quad \eta \in R_+^m, \quad \eta > 0;$

(3) *the matrix S with elements*

$$s_{pq} = \begin{cases} \lambda_M(\sigma_{pp}), & p = 1, 2, \ldots, m, \\ [\lambda_M(\sigma_{pq}^T \sigma_{pq})]^{1/2}, & p < q, \\ s_{qp}, & p > q, \quad \text{for all} \quad (p, q) = 1, 2, \ldots, m, \end{cases}$$

(a) *negative semi-definite;*

(b) *negative definite.*

Then, *equilibrium state* $x = 0$ *of the system (2.2.20) is*

(a) *uniformly stable in the whole;*

(b) *uniformly asymptotically stable in the whole.*

Proof. For the function (2.2.30) and vector $u = (\|x_1\|, \ldots, \|x_m\|)^T$ it is easy to obtain estimate

$$Dv(x, \eta) = \sum_{i=1}^{m} x_i^T \sigma_{ii} x_i + 2 \sum_{i=1}^{m} \sum_{j=i+1}^{s} x_i^T \sigma_{ij} x_j$$

$$\leq \sum_{i=1}^{m} \lambda_M(\sigma_{ii}) \|x_m\|^2 + 2 \sum_{i=1}^{m} \sum_{j=i+1}^{m} [\lambda_M(\sigma_{ij}^T \sigma_{ij})]^{1/2} \|x_i\| \, \|x_j\| = u^T S u$$

$$\text{for all} \quad x_i \in R^{n_i}, \quad i = 1, 2, \ldots, m.$$

Hence, according to Theorem 2.2.1, the proof of Theorem 2.2.2 is complete.

2.2.6 Examples of third order systems

We consider linear system

$$(2.2.31) \qquad \frac{dx}{dt} = Px, \quad x \in R^3,$$

with matrix

$$(2.2.32) \qquad P = \begin{pmatrix} -3 & -2 & 2 \\ 3 & -4 & 1 \\ 3 & 3 & -4 \end{pmatrix}.$$

For independent first level decomposition subsystems of the form

$$(2.2.33) \qquad \begin{aligned} \frac{dx_1}{dt} &= -3x_1, \\[4pt] \frac{dx_2}{dt} &= -4x_2, \\[4pt] \frac{dx_3}{dt} &= -4x_3, \end{aligned}$$

we consider the following auxiliary functions

$$(2.2.34) \qquad v_{11} = x_1^2, \quad v_{22} = x_2^2, \quad v_{33} = x_3^2.$$

For independent second level decomposition subsystems of the form

$$(2.2.35) \qquad \frac{dx_{ij}}{dt} = P_{ij}x_{ij}$$

with $x_{ij} = (x_i, x_j)^{\mathrm{T}}$, $i, j = 1, 2, 3$ $(i \neq j)$;

$$P_{12} = \begin{pmatrix} -3 & -2 \\ 3 & -4 \end{pmatrix}, \quad P_{13} = \begin{pmatrix} -3 & 2 \\ 3 & 4 \end{pmatrix}, \quad P_{23} = \begin{pmatrix} -4 & 1 \\ 3 & -4 \end{pmatrix}$$

we consider auxiliary functions

$$v_{12} = v_{21} = x_{12}^{\mathrm{T}} B_{12} x_{12},$$
$$(2.2.36) \qquad v_{13} = v_{31} = x_{13}^{\mathrm{T}} B_{13} x_{13},$$
$$v_{23} = v_{32} = x_{23}^{\mathrm{T}} B_{23} x_{23},$$

where B_{ij} $(i, j = 1, 2, 3, i \neq j)$ are determined from the following Liapunov equations

$$P_{12}^{\mathrm{T}} B_{12} + B_{12} P_{12} = -G_{12},$$
$$(2.2.37) \qquad P_{13}^{\mathrm{T}} B_{13} + B_{13} P_{12} = -G_{13},$$
$$P_{23}^{\mathrm{T}} B_{23} + B_{23} P_{23} = -G_{23},$$

with

$$G_{12} = 252 \, \mathrm{diag}\,(1,1), \quad G_{13} = 84 \, \mathrm{diag}\,(1,1), \quad G_{23} = 104 \, \mathrm{diag}\,(1,1).$$

Then we have

$$B_{12} = \begin{pmatrix} 43 & 1 \\ 1 & 31 \end{pmatrix}, \quad B_{13} = \begin{pmatrix} 31 & 17 \\ 17 & 19 \end{pmatrix}, \quad B_{23} = \begin{pmatrix} 19 & 8 \\ 8 & 15 \end{pmatrix}.$$

The function

$$v(x, \eta) = \eta^{\mathrm{T}} U_1(x) \eta, \quad \eta = (1, 1, 1)^{\mathrm{T}} \in R_+^3$$

with elements (2.2.34) and (2.2.36) is positive definite. Calculating all constants from Assumptions 2.2.4 and 2.2.5 applied to the system (2.2.31) we obtain the estimates (2.2.17) in the form

(2.2.38) $D^+v(x,\eta) = \eta^T D^+ U_1(x)\eta = x^T A_1 x,$

with

$$A_1 = \begin{pmatrix} -678 & 142 & 314 \\ 142 & -720 & 116 \\ 314 & 116 & -384 \end{pmatrix}.$$

One can easily verify that the matrix A_1 is negative definite, which implies the validity of all conditions of Theorem 2.2.2 providing the uniform asymptotic stability in the whole of solution $x = 0$ of the system (2.2.31) with the matrix (2.2.32).

Omitting in the matrix-valued function $U_1(x)$ all nondiagonal elements (which correspond to a vector approach (see Šiljak [2])) we obtain in place of equation (2.2.38) the next relation

$$D^+v(x) = x^T \tilde{A}_1 x,$$

with

$$\tilde{A}_1 = \begin{pmatrix} -6 & 1 & 5 \\ 1 & -8 & 4 \\ 5 & 4 & -8 \end{pmatrix}.$$

The matrix $\tilde{A}_1$ is negative definite. Therefore, the vector function (2.2.39) also determines the stability of the system (2.2.31).

Let us consider system (2.2.31) with the matrix

(2.2.39) $$P = \begin{pmatrix} 0 & -2 & -2 \\ 3 & -4 & 1 \\ 3 & 3 & -4 \end{pmatrix}.$$

For independent first level decomposition subsystems of the form

(2.2.40)
$$\frac{dx_1}{dt} = 0,$$
$$\frac{dx_2}{dt} = -4x_2,$$
$$\frac{dx_3}{dt} = -4x_3$$

we shall consider the functions (2.2.34).

For independent second level decomposition subsystems (2.2.35) with

$$P_{12} = \begin{pmatrix} 0 & -2 \\ 3 & -4 \end{pmatrix}, \quad P_{13} = \begin{pmatrix} 0 & -2 \\ 3 & -4 \end{pmatrix}, \quad P_{23} = \begin{pmatrix} -4 & 1 \\ 3 & -4 \end{pmatrix}$$

we shall construct functions v_{ij} $(i \neq j = 1, 2, 3)$ in the form of (2.2.36) using the Liapunov equations (2.2.37) with matrices:

$$G_{12} = 6 \operatorname{diag}(1, 1), \quad G_{13} = 6 \operatorname{diag}(1, 1), \quad G_{23} = 13 \operatorname{diag}(1, 1).$$

As a result we obtain:

$$B_{12} = \begin{pmatrix} 3,625 & -1 \\ -1 & 1,25 \end{pmatrix}, \quad B_{12} = B_{13}, \quad B_{23} = \begin{pmatrix} 2,375 & 1 \\ 1 & 1,875 \end{pmatrix}.$$

For the function $v(x, \eta)$ we have

(2.2.41) $$D^+ v(x, \eta) = \eta^T D^+ U_2(x) \eta = x^T A_2 x,$$

where

$$A_2 = \begin{pmatrix} -24 & -13,75 & 1,75 \\ -13,75 & -46 & 26 \\ 1,75 & 26 & -46 \end{pmatrix}.$$

The matrix A_2 in equation (2.2.41) is negative definite, and since the function $U_2(x)$ is a hierarchical Liapunov function, all the conditions of Theorem 2.2.2 hold. Therefore the solution $x = 0$ of the system (2.2.31) with matrix (2.2.39) is uniformly asymptotically stable in the whole.

Omitting nondiagonal elements in the matrix $U_2(x)$ we obtain the estimates

(2.2.42) $$D^+ v(x) \leq \varphi^T(\|x\|) \tilde{A}_2 \varphi(\|x\|),$$

where

$$\tilde{A}_2 = \begin{pmatrix} 0 & 1 & 1 \\ 1 & -8 & 2 \\ 1 & 2 & -8 \end{pmatrix}.$$

The matrix $\tilde{A}_2$ in equality (2.2.42) is not negative definite, and therefore, the vector function (2.2.17) does not determine the stability of solution $x = 0$ of the system (2.2.31) with matrix (2.2.39).

Consider the system (2.2.31) with the matrix

(2.2.43)
$$P = \begin{pmatrix} 1 & -4 & -2 \\ 3 & -4 & -1 \\ 3 & 3 & -4 \end{pmatrix}.$$

For independent first level decomposition subsystems of the form

(2.2.44)
$$\frac{dx_1}{dt} = x_1,$$
$$\frac{dx_2}{dt} = -4x_2,$$
$$\frac{dx_3}{dt} = -4x_3$$

we shall consider the functions (2.2.34).

For the second level decomposition subsystems (2.2.36) with

$$P_{12} = \begin{pmatrix} 1 & -4 \\ 3 & -4 \end{pmatrix}, \quad P_{13} = \begin{pmatrix} 1 & -2 \\ 3 & -4 \end{pmatrix}, \quad P_{23} = \begin{pmatrix} -4 & -1 \\ 3 & -4 \end{pmatrix}$$

we shall consider the functions v_{ij} $(i \neq j) = 1,2,3$ in the form of (2.2.36). As before, we determine matrices B_{ij} $(i \neq j) = 1,2,3$ from equations (2.2.37) with

$$G_{12} = 6\,\mathrm{diag}\,(1,1), \quad G_{13} = 2\,\mathrm{diag}\,(1,1), \quad G_{23} = 38\,\mathrm{diag}\,(1,1).$$

As a result we have

$$B_{12} = \begin{pmatrix} 5,125 & -2,375 \\ -2,375 & 3,125 \end{pmatrix}, \quad B_{13} = \begin{pmatrix} 3,5 & -1,5 \\ -1,5 & 1 \end{pmatrix},$$

$$B_{23} = \begin{pmatrix} 5,5 & 1 \\ 1 & 4,5 \end{pmatrix}.$$

We have the following expression for the derivative of the function $v(x,\eta)$

$$D^+v(x,\eta) = \eta^T D^+ U_3(x)\eta = x^T A_3 x,$$

where

$$A_3 = \begin{pmatrix} -14 & 1 & 23,25 \\ 1 & -96 & 23,25 \\ 23,25 & 23,25 & -88 \end{pmatrix}.$$

One can easily see that the matrix A_3 is negative definite. Therefore, system (2.2.31) with matrix (2.2.43) satisfies all conditions of Theorem 2.2.2 and solution $x = 0$ of the system (2.2.31) with matrix (2.2.43) is uniformly asymptotically stable in the whole.

Omitting nondiagonal elements in the matrix-valued function $U_3(x)$ we obtain the estimates

$$D^+v(x) \leq \varphi^T(\|x\|)\tilde{A}_3\varphi(\|x\|) \quad \text{for all} \quad x \in R^3$$

with

$$\tilde{A}_3 = \begin{pmatrix} 2 & -1 & 1 \\ -1 & -8 & 2 \\ 1 & 2 & -8 \end{pmatrix}.$$

The matrix $\tilde{A}_3$ is not negative definite, which precludes the use of vector function (2.2.34) for the stability analysis of solution $x = 0$ of the system (2.2.31) with matrix (2.2.43).

2.3 Dynamics of the Systems with Regular Hierarchy Subsystems

2.3.1 Ikeda-Šiljak hierarchical decomposition

Unlike the decomposition of the large scale system presented in Section 2.2 we shall discuss here the decomposition of the system according to Ikeda and Šiljak [1]. However, in contrast to the hierarchical Liapunov function proposed by Ikeda and Šiljak [2] we suggest a more general technique of construction of hierarchical matrix-valued Liapunov functions.

Consider dynamical system (2.2.1) which admits mathematical (or physical) decomposition into s subsystems (2.2.2) which link functions $h_i(t, x_1, \ldots, x_m)$. Assume that $x_i = 0$, $i = 1, 2, \ldots, m$ is the only equilibrium state of subsystems (2.2.4), and for $h_i(t, x) = 0$, $i = 1, 2, \ldots, m$ subsystems (2.2.2) are disconnected, and

(2.3.1) $$R^n = R^{n_1} \times R^{n_2} \times \ldots \times R^{n_m},$$

where R^n and R^{n_i} are state spaces of system (2.2.1) and (2.2.2) respectively.

The construction of system (2.2.4) in terms of system (2.2.1) under conditions (2.3.1) is called *"regular" first level decomposition*. Actually, this step in system (2.2.1) transformation coincides with the first stage considered in Section 2.2.1.

Further each of subsystems (2.2.2) is decomposed to the following:

$$(2.3.2) \qquad \frac{dx_{ij}}{dt} = f_{ij}(t, x_{ij}) + h_{ij}(t, x_i), \qquad j = 1, 2, \ldots, M_i.$$

Here $x_{ij}(t) \in R^{n_{ij}}$ is a state of subsystems (2.3.2) at time $t \in R$, $f_{ij} \in C(R \times R^{n_{ij}}, R^{n_{ij}})$ and $h_{ij} \in C(R \times R^{n_i}, R^{n_{ij}})$, for all $i = 1, 2, \ldots, m$. It is supposed that $x_{ij} = 0$ are only equilibrium states of components M_i:

$$(2.3.3) \qquad \frac{dx_{ij}}{dt} = f_{ij}(t, x_{ij}), \qquad j \in M_i,$$

which are separated and

$$(2.3.4) \qquad R^{n_i} = R^{n_{i1}} \times R^{n_{i2}} \times \ldots \times R^{n_i M_i}.$$

Construction of system (2.3.3) in terms of system (2.2.1) under condition (2.3.1) is called *"regular" second level decomposition.*

Remark 2.3.1. Proceeding with the process and fulfilling third level decomposition, and so on, one can arrive at the subsystem equations of the prescribed order up to the first one.

2.3.2 Hierarchical Liapunov's matrix-valued function

A key idea of the new method of Liapunov function construction for system (2.2.1) is the application of matrix-valued function

$$(2.3.5) \qquad U(t, x) = [U_{ik}(t, \cdot)], \qquad U_{ik} = U_{ki},$$

the elements U_{ik}, $i \neq k$ of which are constructed in terms of matrix-valued functions of smaller dimensions. Here $U_{ii} \in C(R_+ \times R^{n_i}, R)$ for all $i = 1, 2, \ldots, m$ and $U_{ik} \in C(R_+ \times R^{n_i} \times R^{n_k}, R)$. The functions U_{ii} are constructed for subsystems (2.3.3), and functions U_{ik} take into account link functions h_i between subsystems (2.2.2) of first level decomposition.

The explicit form of the functions U_{ii} is

$$(2.3.6) \qquad U_{ii}(t, x_i) = \xi_i^{\mathrm{T}} B_i(t, \cdot) \xi_i, \qquad i = 1, 2, \ldots, m,$$

where $\xi_i \in R_+^{n_i}$, $\xi_i > 0$ and submatrix functions

$$B_i(t, \cdot) = [u_{pq}^{(i)}(t, \cdot)], \quad p = 1, 2, \ldots, M_i, \quad q = 1, 2, \ldots, M_i$$

with elements $u_{pp}^{(i)} \in C(R_+ \times R^{n_{ip}}, R)$, $u_{pq}^{(i)} = u_{qp}^{(i)}$

$$u_{pq}^{(i)} \in C(R_+ \times R^{n_{ip}} \times R^{n_{iq}}, R), \quad i = 1, 2, \ldots, m.$$

Besides, functions $u_{pp}^{(i)}$ are constructed for subsystems (2.3.3) and functions $u_{pq}^{(i)}$ are constructed in terms of link functions between subsystems (2.3.3).

Thus, the block structure of matrix-valued function (2.3.6) allows on each hierarchical level certain elements of the matrix-valued function to correspond to subsystems and link functions. This technique involves a Liapunov function that takes into account both dynamical properties of subsystems and hierarchical structure of a large scale system.

Under appropriate assumptions the function

$$(2.3.7) \qquad\qquad v(t, x, \eta) = \eta^{\mathrm{T}} U(t, x) \eta,$$

where $U \in C(R_+ \times R^n, R^{m \times m})$, $\eta \in R_+^m$, $\eta > 0$ is *hierarchical Liapunov function* for the whole system (2.2.1).

Assumption 2.3.1. *There exist*

(1) *open connected nieghborhoods* $\mathcal{N}_{ip} \subseteq R^{n_{ip}}$ *of the states* $x_{ip} = 0$, $i = 1, 2, \ldots, m$, $p = 1, 2, \ldots, M_i$;

(2) *functions* $\varphi_{ip}, \overline{\varphi}_{ip}$, $\varphi_{ip} \colon \mathcal{N}_{ip} \to R_+$; $\overline{\varphi}_{ip} \colon \mathcal{N}_{ip} \to R_+$, $(\varphi_{ip}, \overline{\varphi}_{ip}) \in K(KR)$, $i = 1, 2, \ldots, m$, $p = 1, 2, \ldots, M_i$;

(3) *constants* $\underline{\alpha}_{pp}^{(i)} > 0$, $\overline{\alpha}_{pp}^{(i)} > 0$, $\alpha_{pq}^{(i)} = \alpha_{qp}^{(i)}$, $q = 1, 2, \ldots, M_i$;

(4) *matrix-valued functions* B_i, *the elements of which satisfy the estimates*

    (a) $\underline{\alpha}_{pp}^{(i)} \varphi_{ip}^2(\|x_{ip}\|) \le u_{pp}^{(i)}(t, x_{ip}) \le \overline{\alpha}_{pp}^{(i)} \varphi_{ip}^2(\|x_{ip}\|)$
       for all $(t, x_{ip}) \in R_+ \times \mathcal{N}_{ip}$;

    (b) $\underline{\alpha}_{pq}^{(i)} \varphi_{ip}(\|x_{ip}\|) \varphi_{iq}(\|x_{iq}\|) \le u_{pq}^{(i)}(t, x_{ip}, x_{iq})$
       $\le \overline{\alpha}_{pq}^{(i)} \overline{\varphi}_{ip}(\|x_{ip}\|) \overline{\varphi}_{iq}(\|x_{iq}\|)$
       for all $(t, x_{ip}, x_{iq}) \in R_+ \times \mathcal{N}_{ip} \times \mathcal{N}_{iq}$.

Proposition 2.3.1. *If all conditions of Assumption 2.3.1 are satisfied for the functions* $U_{ii}(t, x_i)$ *the following bilateral estimates are valid*

$$(2.3.8) \qquad W_i^{\mathrm{T}} \Phi_i^{\mathrm{T}} A_{ii} \Phi_i W_i \le U_{ii}(t, x_i) \le \overline{W}_i^{\mathrm{T}} \Phi_i^{\mathrm{T}} B_{ii} \Phi_i \overline{W}_i,$$

where

$$W_i^{\mathrm{T}} = (\varphi_{i1}(\|x_{i1}\|), \ldots, \varphi_{iM_i}(\|x_{iM_i}\|));$$

$$\overline{W}_i^{\mathrm{T}} = (\overline{\varphi}_{i1}(\|x_{i1}\|), \ldots, \overline{\varphi}_{iM_i}(\|x_{iM_i}\|));$$

$$\Phi_i^{\mathrm{T}} = \Phi_i = \mathrm{diag}\,[\xi_{i1}, \ldots, \xi_{iM_i}],$$

$$A_{ii} = [\underline{\alpha}_{pq}^{(i)}], \qquad B_{ii} = [\overline{\alpha}_{pq}^{(i)}].$$

Proof. Proposition 2.3.1 is proved by direct substitution by estimates (a,b) of (2.1.4) from Assumption 2.3.1 into expression of function (2.3.7).

Assumption 2.3.2. *There exist*

(1) *open connected nieghborhoods* $\mathcal{N}_i \subseteq R^{n_i}$ *of the states* $x_i = 0$, $i = 1, 2, \ldots, m;$

(2) *functions* $\varphi_i, \overline{\varphi}_i,\ \varphi_i \colon \mathcal{N}_i \to R_+;\ \overline{\varphi}_i \colon \mathcal{N}_i \to R_+ (\varphi_i, \overline{\varphi}_i) \in K(KR);$

(3) *the constants* $\beta_{ik}, \overline{\beta}_{ik}, \beta_{ik} = \beta_{ki}, \overline{\beta}_{ik} = \overline{\beta}_{ki}$ *for all* $(k \neq i)$, *and estimates*

$$(2.3.9) \qquad \beta_{ik}\|W_i\|\,\|W_k\| \le U_{ik}(t, x_i, x_k) \le \overline{\beta}_{ik}\|\overline{W}_i\|\,\|\overline{W}_k\|$$

are satisfied for all $(t, x_i, x_k) \in R_+ \times \mathcal{N}_i \times \mathcal{N}_k$ *for all* $(i \neq k) = 1, 2, \ldots, m.$

Proposition 2.3.2. *If all conditions of Assumption 2.3.1 and 2.3.2 are satisfied, then for the function* $v(t, x, \eta)$ *the bilateral inequality*

$$(2.3.10) \qquad W^{\mathrm{T}}H^{\mathrm{T}}AHW \le v(t, x, \eta) \le \overline{W}^{\mathrm{T}}H^{\mathrm{T}}AH\overline{W}$$

is satisfied for all $(t, x) \in R_+ \times \mathcal{N}$. *Here*

$$W^{\mathrm{T}} = (W_1, \ldots, W_s), \qquad \overline{W}^{\mathrm{T}} = (\overline{W}_1, \ldots, \overline{W}_s),$$

$$H^{\mathrm{T}} = H = \mathrm{diag}\,[\eta_1, \ldots, \eta_s], \quad A = [\beta_{ik}], \quad B = [\overline{\beta}_{ik}],$$

$$\beta_{ii} = \lambda_m(\Phi_i^{\mathrm{T}}A_{ii}\Phi_i), \qquad \overline{\beta}_{ii} = \lambda_M(\Phi_i^{\mathrm{T}}A_{ii}\Phi_i).$$

Proof. Estimates (2.3.10) are obtained by direct substitution by estimates (2.3.8) and (2.3.9) into expression (2.3.7) of the function $v(t, x, \eta)$.

Assumption 2.3.3. *There exist*

(1) *open connected nieghborhoods* $\mathcal{N}_{ip}$, $\mathcal{N}_{ip} \subseteq R^{n_{ip}}$ *of the states* $x_{ip} = 0$, $i = 1, 2, \ldots, m$, $p = 1, 2, \ldots, M_i$;

(2) *functions* $u_{pq}^{(i)} \in C(R_+ \times R^{n_{ip}} \times R^{n_{iq}}, R)$, $q = 1, 2, \ldots, m$;

(3) *functions* β_{ip}, $\beta_{ip} : \mathcal{N}_{ip} \to R_+$, $\beta_{ip} \in K(KR)$;

(4) *real numbers* $p_p^{(i)}$, $\mu_p^{(i)}$, $\mu_{pq}^{(i)}$ *such that*

(a) $\displaystyle\sum_{p=1}^{M_i} \xi_{ip}^2 \{D_t^+ u_{pp}^{(i)} + (D_{x_{ip}}^+ u_{pp}^{(i)})^{\mathrm{T}} f_{ip}(t, x_{ip})\} \leq \sum_{p=1}^{M_i} p_p^{(i)} \beta_{ip}^{(i)}(\|x_{ip}\|)$

for all $(t, x_{ip}) \in R_+ \times \mathcal{N}_{ip}$;

(b) $\displaystyle\sum_{p=1}^{M_i} \xi_{ip}^2 \{(D_{x_{ip}}^+ u_{pp}^{(i)})^{\mathrm{T}} h_{ip}(t, x_i)\} + 2 \sum_{p=1}^{M_i-1} \sum_{q=p+1}^{M_i} \xi_{ip} \xi_{iq}$

$\times \{D_t^+ u_{pq}^{(i)} + (D_{x_{ip}}^+ u_{pq}^{(i)})^{\mathrm{T}} f_{ip}(t, x_{ip}) + (D_{x_{ip}}^+ u_{pq}^{(i)})^{\mathrm{T}} h_{ip}(t, x_i)$

$+ (D_{x_{iq}}^+ u_{pq}^{(i)})^{\mathrm{T}} f_{ip}(t, x_{iq}) + (D_{x_{iq}}^+ u_{pq}^{(i)})^{\mathrm{T}} h_{iq}(t, x_i)\}$

$\leq \displaystyle\sum_{p=1}^{M_i} \mu_p^{(i)} \beta_{ip}^2(\|x_{ip}\|) + 2 \sum_{p=1}^{M_i-1} \sum_{q=p+1}^{M_i} \mu_{pq} \beta_{ip}^{(i)}(\|x_{ip}\|) \beta_{iq}^{(i)}(\|x_{iq}\|)$

for all $(t, x_{ip}, x_{iq}) \in R_+ \times \mathcal{N}_{ip} \times \mathcal{N}_{iq}$.

Proposition 2.3.3. *If all conditions of Assumption 2.3.2 are satisfied, then for the total derivative of function* U_{ii} *along solutions of system (2.3.2) estimate*

$$(2.3.11) \qquad D^+ U_{ii}(t, x_i) \leq \lambda_M(S_{ii}) \|z_i\|^2$$

is valid, where $z_i^{\mathrm{T}} = (\beta_{i1}(\|x_{i1}\|), \ldots, \beta_{iM_i}(\|x_{iM_i}\|))$, $\lambda_M(S_{ii})$ *are the maximal eigenvalues of matrices* S_{ii} *with elements*

$$\sigma_{pp}^{(i)} = p_p^{(i)} + \mu_p^{(i)},$$

$$\sigma_{pq}^{(i)} = \sigma_{qp}^{(i)} = \mu_{pq}^{(i)},$$

$$p \neq q \quad \text{for all} \quad (p, q) = 1, 2, \ldots, M_i, \quad i = 1, 2, \ldots, m.$$

Remark 2.3.1. *If in conditions of Assumption 2.3.2 estimates (4)(a) and (4)(b) are satisfied with the inequality sign "$\geq$" for constants* $\tilde{p}_p^{(i)}$, $\tilde{\mu}_p^{(i)}$, $\tilde{\mu}_{pq}^{(i)}$, *then estimate (2.3.11) becomes*

$$(2.3.12) \qquad D^+ U_{ii}(t, x_i) \geq \lambda_m(\tilde{S}_{ii}) \|z_i\|^2,$$

where $\tilde{S}_{ii}$ *are matrices with elements*

$$\tilde{\sigma}_{pp}^{(i)} = \tilde{p}_p^{(i)} + \tilde{\mu}_p^{(i)},$$

$$\tilde{\sigma}_{pq}^{(i)} = \tilde{\sigma}_{qp}^{(i)} = \tilde{\mu}_{pq}^{(i)}, \quad (p, q) = 1, 2, \ldots, M_i.$$

Here $\lambda_m(\tilde{S}_{ii})$ *is a minimal eigenvalue of matrices* $\tilde{S}_{ii}$, $i = 1, 2, \ldots, m$.

Assumption 2.3.4. *There exist*

(1) *open connected nieghborhoods* $\mathcal{N}_i$, $\mathcal{N}_i \subseteq R^{n_i}$ *of the states* $x_i = 0$, $i = 1, 2, \ldots, m$;

(2) *functions* β_{ip}, $i = 1, 2, \ldots, m$, $p = 1, 2, \ldots, M_i$, $\beta_{ip} \in K(KR)$;

(3) *functions* $U_{ik} \in C(R_+ \times \mathcal{N}_i \times \mathcal{N}_k, R)$ *and real numbers* θ_{ik}, $i, k = 1, 2, \ldots, m$ *are such that*

$$\sum_{i=1}^{m} \eta_i^2 (D_{x_i}^+ U_{ii}(t, x_i))^{\mathrm{T}} h_i(t, x_i) + 2 \sum_{i=1}^{m-1} \sum_{k=i+1}^{m} \eta_i \eta_k$$

$$\times \{ D_t^+ U_{ik} + (D_{x_i}^+ U_{ik}(t, x_i, x_k))^{\mathrm{T}} f_i(t, x_i) + (D_{x_i}^+ U_{ik}(t, x_i, x_k))^{\mathrm{T}} h_i(t, x)$$

$$+ (D_{x_k}^+ U_{ik}(t, x_i, x_k))^{\mathrm{T}} f_k(t, x_k) + (D_{x_k}^+ U_{ik}(t, x_i, x_k))^{\mathrm{T}} h_k(t, x) \}$$

$$\leq \sum_{i=1}^{m} \theta_{ii} \|z_i\|^2 + 2 \sum_{i=1}^{m-1} \sum_{k=i+1}^{m} \theta_{ik} \|z_i\| \, \|z_k\|$$

for all $(t, x_i, x_k) \in R_+ \times \mathcal{N}_i \times \mathcal{N}_k$.

Proposition 2.3.4. *If all conditions of Assumption 2.3.3 are satisfied, then*

$$(2.3.13) \qquad D^+ v(t, x, \eta) \leq z^{\mathrm{T}} S z \quad \text{for all} \quad (t, x) \in R_+ \times \mathcal{N},$$

where $\mathcal{N} = \mathcal{N}_1 \times \ldots \times \mathcal{N}_m$, $z^{\mathrm{T}} = (z_1 \| \cdot \|, \ldots, z_m \| \cdot \|)$, S *is the matrix* $m \times m$ *with elements*

$$c_{pp} = \eta_p^2 \lambda_M(s_{pp}) + \theta_{pp},$$

$$c_{pk} = c_{kp} = \theta_{pk} \quad \text{for all} \quad (p \neq k) 1, 2, \ldots, m.$$

Remark 2.3.2. If in conditions of Assumption 2.3.4 estimate (3) is satisfied with inequality sign "$\geq$" for constants $\tilde{\theta}_{ik}$, $i, k = 1, 2, \ldots, m$ and conditions of Remark 2.3.1 are fulfilled, then estimate (2.3.13) reads

$$(2.3.14) \qquad D^+ v(t, x, \eta) \geq z^{\mathrm{T}} \tilde{S} z \quad \text{for all} \quad (t, x) \in R_+ \times \mathcal{N},$$

where $\tilde{S}$ *is the matrix* $m \times m$ *with elements*

$$\tilde{c}_{pp} = \eta_p^2 \lambda_m(\tilde{s}_{pp}) + \tilde{\theta}_{pp},$$

$$\tilde{c}_{pk} = \tilde{c}_{kp} = \tilde{\theta}_{pk} \quad \text{for all} \quad (p \neq k) = 1, 2, \ldots, m.$$

Conditions contained in Assumptions 2.3.1–2.3.3 define functions and estimates of its total derivatives, for which hierarchical Liapunov function can be constructed of elements $u_{pq}^{(i)}(t, \cdot)$, $i = 1, 2, \ldots, m$, $(p, q) = 1, 2, \ldots, M_i$. This function has the form of (2.3.8) with estimates of total derivative along the solutions of system (2.2.1) in the form (2.3.13) or (2.3.14).

2.3.3 Stability and instability conditions

Estimates (2.3.10) together with inequalities (2.3.13) and (2.3.14) allow us to establish sufficient stability and instability conditions for state $x = 0$ of system (2.2.1).

Theorem 2.3.1. *Let vector function f in system (2.2.1) be continuous on $R \times \mathcal{N}$ (on $R \times R^n$) and*

(1) *all conditions of Assumptions 2.3.1–2.3.3 be satisfied;*
(2) *there exists a positive number δ (or $\delta = +\infty$) such that the set $\{v_\zeta(t)\}$ is asymptotically contractive for every $\zeta \in (0, \delta)$;*
(3) *in estimates (2.3.10), (2.3.13) and (2.3.14)*
   (a) *the matrix A is positive definite;*
   (b) *the matrix B is positive definite;*
   (c) *the matrix S is negative semi-definite or equal to zero;*
   (d) *the matrix S is negative definite;*
   (e) *the matrix $\tilde{S}$ is positive definite.*

Then, respectively

(a) *conditions (1)–(3a) and (3b) are sufficient for the state $x = 0$ of system (2.2.1) to be stable;*
(b) *conditions (1)–(3a), (3b) and (3c) are sufficient for the state $x = 0$ of system (2.2.1) to be uniformly stable;*
(c) *conditions (1)–(3a), (3b) and (3d) are sufficient for the state $x = 0$ of system (2.2.1) to be uniformly asymptotically stable;*
(d) *conditions (1)–(3a), (3e) and conditions of Remark 2.3.2 are sufficient for the state $x = 0$ of system (2.2.1) to be unstable.*

Proof. We start with assertion (a) of Theorem 2.3.1. Under conditions (1) and (3a) of Theorem 2.3.1 function $v(t, x, \eta)$ is positive definite. The set $\{v_\zeta(t)\}$ is asymptotically contractive for all $\zeta \in (0, \delta)$, where $\delta = \min\{W^T H^T B H \overline{W}$ for all $x \in N\}$, $\overline{W}^T = (\overline{W}_1, \ldots, \overline{W}_m)$.

Condition (3c) implies that $D^+v(t, x, \eta)$ is negative semi-definite or equal to zero. Moreover, all conditions of Theorem 1.8.1 and Theorem 1.6.1 are satisfied. This proves assertion (a) of Theorem 2.3.1. Assertions (c)–(d) of Theorem 2.3.1 are proved similarly in view of Theorems 1.6.4 and 1.8.6.

Example 2.3.1. We consider the nonlinear system

$$\frac{dx_{11}}{dt} = -3x_{11}^3 - 3x_{11}x_{12}^2 - x_{12}^3 - x_{11}^2 x_{12}$$
$$+ (0.01)[x_{12}^3 + x_{21}x_{22}^2 + x_{21}^2 x_{22} + x_{22}^3];$$

$$\frac{dx_{12}}{dt} = -4x_{12}^3 - 4x_{11}^2 x_{12} - 2x_{11}^3 - 2x_{11}x_{12}^2$$
$$+ (0.1)[x_{21}^3 + x_{21}x_{22}^2] - (0.02)[x_{22}^3 + x_{21}^2 x_{22}];$$

(2.3.15)

$$\frac{dx_{21}}{dt} = -4x_{21}^3 - 4x_{21}x_{22}^2 - x_{22}^3 - x_{22}x_{21}^2$$
$$+ (0.01)[x_{11}^3 + x_{11}x_{12}^2 + x_{11}^2 x_{12} + x_{12}^3];$$

$$\frac{dx_{22}}{dt} = -2x_{22}^3 - 2x_{21}^2 x_{22} + x_{21}^3 + x_{21}x_{22}^2$$
$$+ (0.1)[x_{11}^3 + x_{11}x_{12}^2] - (0.02)[x_{12}^3 + x_{11}^2 x_{12}].$$

We introduce the matrices $A_1, \ldots, A_4$ and vectors $x_1 = (x_{11}, x_{12})^{\mathrm{T}}$, $x_2 = (x_{21}, x_{22})^{\mathrm{T}}$. The system (2.3.15) can be transformed into the form

(2.3.16)
$$\frac{dx_1}{dt} = A_1 x_1^3 + A_2 x_2^3,$$
$$\frac{dx_2}{dt} = A_3 x_2^3 + A_4 x_1^3,$$

where

$$A_1 = \begin{pmatrix} -3 & -1 \\ -2 & -4 \end{pmatrix}, \qquad A_2 = \begin{pmatrix} 0.01 & 0.01 \\ 0.01 & -0.02 \end{pmatrix}$$

$$A_3 = \begin{pmatrix} -4 & -1 \\ 1 & -2 \end{pmatrix}, \qquad A_4 = A_2.$$

At the first level decomposition we consider the independent subsystems

(2.3.17)
$$\frac{dx_1}{dt} = A_1 x_1^3,$$
$$\frac{dx_2}{dt} = A_3 x_2^3$$

and the functions connecting them

(2.3.18) $h_1(x_2) = A_2 x_2^3, \quad h_2(x_1) = A_4 x_1^3.$

The decomposition of subsystems (2.3.15) leads to independent equations of the first order

(2.3.19)

$$\frac{dx_{11}}{dt} = -3x_{11}^3 \triangleq f_{11}(x_{11}), \quad \frac{dx_{12}}{dt} = -4x_{12}^3 \triangleq f_{12}(x_{12}),$$

$$\frac{dx_{21}}{dt} = -4x_{21}^3 \triangleq f_{21}(x_{21}), \quad \frac{dx_{22}}{dt} = -2x_{22}^3 \triangleq f_{22}(x_{22}),$$

with connecting functions

(2.3.20)

$$h_{11}(x_1) = -3x_{11}x_{12}^2 - x_{12}^3 - x_{11}^2 x_{12},$$

$$h_{12}(x_1) = -4x_{11}^2 x_{12} - 2x_{11}^3 - 2x_{11}x_{12}^2,$$

$$h_{21}(x_2) = -4x_{21}x_{22}^2 - x_{22}^3 - x_{22}x_{21}^2,$$

$$h_{22}(x_2) = -2x_{21}^2 x_{22} + 2x_{21}^3 + x_{21}x_{22}^2.$$

As elements of the matrix $B_1(\cdot)$ we take the functions

(2.3.21) $u_{11}^{(1)} = 2x_{11}^2, \quad u_{12}^{(1)} = u_{21}^{(1)} = (0.1)x_{11}x_{12}, \quad u_{22}^{(1)} = x_{12}^2$

and for the matrix $B_2(\cdot)$

(2.3.22) $u_{11}^{(2)} = x_{21}^2, \quad u_{21}^{(2)} = u_{12}^{(2)} = (0.1)x_{21}x_{22}, \quad u_{22}^{(2)} = x_{22}^2$

respectively.

It is easy to see that for the functions $U_{11}(x_1)$ and $U_{22}(x_2)$, the matrix-valued functions B_1 and B_2 have elements (2.3.21) and (2.3.22) and all conditions of Assumption 2.3.1 are fulfilled with constants

$$\underline{\alpha}_{11} = 2, \quad \underline{\alpha}_{12} = 0.1, \quad \underline{\alpha}_{22} = 1, \quad \overline{\alpha}_{11} = 2, \quad \overline{\alpha}_{12} = 0.1, \quad \overline{\alpha}_{22} = 1,$$

$$\underline{\beta}_{11} = 1, \quad \underline{\beta}_{12} = -0.1, \quad \underline{\beta}_{22} = 1, \quad \overline{\beta}_{11} = 1, \quad \overline{\beta}_{12} = 0.1, \quad \overline{\beta}_{22} = 1,$$

and functions

$$\varphi_{11}(\|x_{11}\|) = |x_{11}|, \quad \varphi_{12}(\|x_{12}\|) = |x_{12}|,$$

$$\varphi_{21}(\|x_{21}\|) = |x_{21}|, \quad \varphi_{22}(\|x_{22}\|) = |x_{22}|.$$

For the function

$$(2.3.23) \qquad U_{12}(x) = U_{21}(x) = x_1^{\mathrm{T}} \operatorname{diag}[0.01; 0.01] x_2$$

the estimates of Assumption 2.3.2 are true with the constants

$$\beta_{12} = -1 \times 10^{-2}; \qquad \overline{\beta}_{12} = 1 \times 10^{-2}.$$

The conditions (4a) and (4b) of the Assumption 2.3.2 for the system (2.3.16) are equivalent to the next:

(a) $(D_{x_{11}}^{+} u_{11}^{(1)}(x_{11}))^{\mathrm{T}} f_{11}(x_{11}) \le p_{11}\beta_{11}^{2}(\|x_{11}\|)$ for all $x_{11} \in \mathcal{N}_{11}$;

(b) $(D_{x_{11}}^{+} u_{11}^{(1)}(x_{11}))^{\mathrm{T}} h_{11}(x_1) \le p_{12}\beta_{11}^{2}(\|x_{11}\|) + p_{13}\beta_{11}(\|x_{11}\|)\beta_{12}(\|x_{12}\|)$ for all $(x_{11}, x_{12}) \in \mathcal{N}_{11} \times \mathcal{N}_{12}$;

(c) $(D_{x_{12}}^{+} u_{22}^{(1)}(x_{12}))^{\mathrm{T}} f_{12}(x_{12}) \le p_{21}\beta_{12}^{2}(\|x_{12}\|)$ for all $x_{12} \in \mathcal{N}_{12}$;

(d) $(D_{x_{12}}^{+} u_{22}^{(1)}(x_{12}))^{\mathrm{T}} h_{12}(x_1) \le p_{22}\beta_{12}^{2}(\|x_{12}\|) + p_{23}\beta_{11}(\|x_{11}\|)\beta_{12}(\|x_{12}\|)$ for all $(x_{11}, x_{12}) \in \mathcal{N}_{11} \times \mathcal{N}_{12}$;

(e) $(D_{x_{11}}^{+} u_{12}^{(1)}(x_1))^{\mathrm{T}} f_{11}(x_{11}) \le p_{31}\beta_{11}^{2}(\|x_{11}\|) + p_{32}\beta_{11}(\|x_{11}\|)\beta_{12}(\|x_{12}\|)$ for all $x_1 \in \mathcal{N}_1$;

(f) $(D_{x_{11}}^{+} u_{12}^{(1)}(x_1))^{\mathrm{T}} h_n(t, x_1) \le p_{33}\beta_{11}^{2}(\|x_{11}\|) + p_{34}\beta_{11}(\|x_{11}\|)\beta_{12}(\|x_{12}\|)$ $+ p_{35}\beta_{12}^{2}(\|x_{12}\|)$ for all $x_1 \in \mathcal{N}_1$;

(g) $(D_{x_{12}}^{+} u_{12}^{(1)}(x_1))^{\mathrm{T}} f_{12}(x_{12}) \le p_{41}\beta_{12}^{2}(\|x_{12}\|) + p_{42}\beta_{11}(\|x_{11}\|)\beta_{12}(\|x_{12}\|)$ for all $x_1 \in \mathcal{N}_1$;

(h) $(D_{x_{12}}^{+} u_{12}^{(1)}(x_1))^{\mathrm{T}} h_{12}(x_1) \le p_{43}\beta_{11}^{2}(\|x_{11}\|) + p_{44}\beta_{11}(\|x_{11}\|)\beta_{12}(\|x_{12}\|)$ $+ p_{45}\beta_{12}^{2}(\|x_{12}\|)$ for all $x_1 \in \mathcal{N}_1$;

(i) $(D_{x_{21}}^{+} u_{11}^{(2)}(x_{21}))^{\mathrm{T}} f_{21}(x_{21}) \le p_{51}\beta_{21}^{2}(\|x_{21}\|)$ for all $x_{21} \in \mathcal{N}_{21}$;

(j) $(D_{x_{21}}^{+} u_{21}^{(2)}(x_{21}))^{\mathrm{T}} h_{21}(x_2) \le p_{52}\beta_{21}^{2}(\|x_{21}\|) + p_{53}\beta_{21}(\|x_{21}\|)\beta_{22}(\|x_{22}\|)$ for all $(x_{21}, x_{22}) \in \mathcal{N}_{21} \times \mathcal{N}_{22}$;

(k) $(D_{x_{22}}^{+} u_{22}^{(2)}(x_{22}))^{\mathrm{T}} f_{22}(x_{22}) \le p_{61}\beta_{22}^{2}(\|x_{22}\|)$ for all $x_{22} \in \mathcal{N}_{22}$;

(l) $(D_{x_{22}}^{+} u_{22}^{(2)}(x_{22}))^{\mathrm{T}} h_{22}(x_2) \le p_{62}\beta_{22}^{2}(\|x_{22}\|) + p_{63}\beta_{21}(\|x_{21}\|)\beta_{22}(\|x_{22}\|)$ for all $(x_{21}, x_{22}) \in \mathcal{N}_{21} \times \mathcal{N}_{22}$;

(m) $(D_{x_{21}}^{+} u_{12}^{(2)}(x_{21}))^{\mathrm{T}} f_{21}(x_{21}) \le p_{71}\beta_{21}^{2}(\|x_{21}\|) + p_{72}\beta_{21}(\|x_{21}\|)\beta_{22}(\|x_{22}\|)$ for all $(x_{21}, x_{22}) \in \mathcal{N}_{21} \times \mathcal{N}_{22}$;

(n) $(D_{x_{21}}^{+} u_{12}^{(2)}(x_2))^{\mathrm{T}} h_{21}(x_2) \le p_{73}\beta_{21}^{2}(\|x_{21}\|) + p_{74}\beta_{21}(\|x_{21}\|)\beta_{22}(\|x_{22}\|)$ $+ p_{75}\beta_{22}^{2}(\|x_{22}\|)$ for all $(x_{21}, x_{22}) \in \mathcal{N}_{21} \times \mathcal{N}_{22}$;

(o) $(D^+_{x_{22}} u^{(2)}_{12}(x_2))^T f_{22}(x_{22}) \leq p_{81} \beta^2_{22}(\|x_{22}\|) + p_{82} \beta_{21}(\|x_{21}\|)\beta_{22}(\|x_{22}\|)$ for all $(x_{21}, x_{22}) \in \mathcal{N}_{21} \times \mathcal{N}_{22}$;

(p) $(D^+_{x_{22}} u^{(2)}_{12}(x_2))^T h_{22}(x_2) \leq p_{83} \beta^2_{21}(\|x_{21}\|) + p_{84} \beta_{21}(\|x_{21}\|)\beta_{22}(\|x_{22}\|)$ $+ p_{85} \beta^2_{22}(\|x_{22}\|)$ for all $(x_{21}, x_{22}) \in \mathcal{N}_{21} \times \mathcal{N}_{22}$.

The conditions (a) – (p) are fulfilled with the constants:

$$p_{11} = 12, \quad p_{12} = p_{14} = 2, \quad p_{13} = 4,$$

$$p_{21} = -8, \quad p_{22} = p_{24} = 2, \quad p_{23} = 4,$$

$$p_{31} = p_{32} = p_{34} = 0.15, \quad p_{35} = 0.05, \quad p_{33} = 0,$$

$$p_{41} = p_{42} = p_{44} = 0.2, \quad p_{43} = p_{45} = 0,$$

$$p_{51} = -8, \quad p_{52} = p_{54} = 1, \quad p_{53} = 2,$$

$$p_{61} = -4, \quad p_{62} = p_{64} = 1, \quad p_{63} = 2,$$

$$p_{71} = p_{72} = p_{74} = 0.2, \quad p_{75} = 0.1, \quad p_{73} = 0,$$

$$p_{81} = p_{82} = 0.1, \quad p_{83} = p_{84} = 0.2, \quad p_{85} = 0.$$

For the functions U_{11}, U_{22} and U_{12} the conditions of Assumption 2.3.3 are fulfilled with the constants

$$\theta_{11} = \theta_{22} = 0, \quad \theta_{12} = \theta_{13} = 0.02619, \quad \theta_{32} = 0.042,$$

$$\theta_{31} = 0.0513, \quad \theta_{21} = \theta_{23} = 0.01433, \quad \theta_{33} = 0.0938.$$

In the estimate (2.3.10), the matrices A and B are

$$A = \begin{pmatrix} 0.99 & -0.01 \\ -0.01 & 0.9 \end{pmatrix}, \quad B = \begin{pmatrix} 2 & 0.01 \\ 0.01 & 1.1 \end{pmatrix}$$

and the matrix S in the estimate (2.3.13) is

$$S = \begin{pmatrix} -3.34 & 0.11 \\ 0.11 & -1.44 \end{pmatrix}.$$

All calculations have been made for $\xi_1 = (1,1)^T$, $\xi_2 = (1,1)^T$ and $\eta = (1,1)^T$. It is easy to see that for the system (2.3.15), conditions (1), (2), (3a) and (3b) of Theorem 2.3.1 are fulfilled and therefore the solution $(x = 0) \in R^4$ is uniformly asymptotically stable.

Finally we note that the matrix-valued Liapunov function methodology leads to more adequate scalar Liapunov functions for nonlinear systems and simplifies their construction then via the vector Liapunov function concept.

2.3.4 Linear nonautonomous system

Consider the *linear nonautonomous system*

$$(2.3.24) \qquad \frac{dx}{dt} = A(t)x, \quad x(t_0) = x_0,$$

where $x \in R^n$ and the matrix $A(t)$ has the form

$$A(t) = \begin{pmatrix} A_1(t) & A_1^*(t) \\ A_2^*(t) & A_2(t) \end{pmatrix}.$$

Here $A_1(t)$ is $n_1 \times n_1$, $A_2(t)$ is $n_2 \times n_2$, $A_1^*(t)$ is $n_2 \times n_1$, $A_2^*(t)$ is $n_1 \times n_2$ matrices with elements continuous on every finite interval $I \subset R_+$.

Assume that the system (2.3.24) is decomposed into two interconnected subsystems

$$(2.3.25) \qquad \frac{dx_i}{dt} = A_i(t)x_i + A_i^*(t)x_j, \quad i,j = 1,2, \quad i \neq j,$$

where $x_i \in R^{n_i}$, $n_1 + n_2 = n$,

$$A_i(t) = \begin{pmatrix} A_{i1}(t) & A_{i1}^*(t) \\ A_{i2}^*(t) & A_{i2}(t) \end{pmatrix}, \qquad A_i^*(t) = \begin{pmatrix} A_{i3}(t) & A_{i3}^*(t) \\ A_{i4}^*(t) & A_{i4}(t) \end{pmatrix}$$

and each independent subsystem corresponding to (2.3.24) consists of two interconnected subsystems

$$(2.3.26) \qquad \begin{aligned} \frac{dx_{i1}}{dt} &= A_{i1}(t)x_{i1} + A_{i1}^*(t)x_{i2}, \\ \frac{dx_{i2}}{dt} &= A_{i2}(t)x_{i1} + A_{i2}^*(t)x_{i2}, \quad i = 1,2, \end{aligned}$$

where $x_{i1} \in R^{n_{i1}}$, $x_{i2} \in R^{n_{i2}}$, $n_{i1} + n_{i2} = n_i$, $i = 1,2$. Alongside the system (2.3.24) we consider the matrix-valued function

$$U(t,x) = \begin{pmatrix} U_1(t,x_1) & U_3(t,x) \\ U_3(t,x) & U_2(t,x_2) \end{pmatrix},$$

where $U_i(t,x_i) = \xi_i^T B_i(t,\cdot)\xi_i$, $U_3(t,x) = x_1^T P_1(t)x_2$ and $\xi_i \in R_+^2$, $i=1,2$ is an arbitrary continuous matrix. In addition

$$(2.3.27) \qquad u_{11}^{(i)}(t,x_{i1}) = x_{i1}^T P_{i1}(t)x_{i1}, \qquad u_{22}^{(i)}(t,x_{i2}) = x_{i2}^T P_{i2}(t)x_{i2},$$

$$(2.3.28) \qquad u_{12}^{(i)}(t,x_i) = x_{i1}^T P_{i3}(t)x_{i2}, \qquad i=1,2,$$

where $P_{i1}(t)$ and $P_{i2}(t)$ are symmetric continuous and positive definite matrices for $t \in R_+$, $P_{i3}(t)$ are arbitrary continuous matrices and matrices $P_{i1}(t)$, $P_{i2}(t)$ and $P_{i3}(t)$ are continuously differentiable on R_+. Functions $U_i(t,x_i)$ correspond to subsystems (2.3.26) and function $U_3(t,x)$ shows the connections in system (2.3.25). It can be easily verified that for functions (2.3.27) and (2.3.28) the following estimates are satisfied

(i) $\lambda_m(P_{i1})\|x_{i1}\|^2 \le u_{11}^{(i)}(t,x_{i1}) \le \lambda_M(P_{i1})\|x_{i1}\|^2$,

(ii) $\lambda_m(P_{i2})\|x_{i2}\|^2 \le u_{i2}^{(i)}(t,x_{i2}) \le \lambda_M(P_{i2})\|x_{i2}\|^2$,

(iii) $-\lambda_M^{1/2}(P_{i3}P_{i3}^T)\|x_{i1}\|\,\|x_{i2}\| \le u_{12}^{(i)}(t,x_i) \le \lambda_M^{1/2}(P_{i3}P_{i3}^T)\|x_{i1}\|\,\|x_{i2}\|$,

where $\lambda_m(P_{ij})$ and $\lambda_M(P_{ij})$ are minimal and maximal eigenvalues of the matrices $P_{ij}(t)$, $i,j = 1,2$ and $\lambda_M^{1/2}(P_{i3}P_{i3}^T)$ are the norms of matrices $P_{i3}P_{i3}^T$.

Proposition 2.3.5. *Provided estimates* (i)−(iii) *hold, for functions* $U_i(t,x_i)$ *the estimates*

$$(2.3.29) \qquad w_i^T C_i w_i \le U_i(t,x_i) \le w_i^T D_i w_i$$

are satisfied, where $w_i^T = (\|x_{i1}\|^T, \|x_{i2}\|^T)$, $i=1,2$,

$$C_i = \begin{pmatrix} \xi_{i1}^2 \lambda_m(P_{i1}) & -\xi_{i1}\xi_{i2}\lambda_M^{1/2}(P_{i3}P_{i3}^T) \\ -\xi_{i1}\xi_{i2}\lambda_M^{1/2}(P_{i3}P_{i3}^T) & \xi_{i1}^2 \lambda_m(P_{i2}) \end{pmatrix},$$

$$D_i = \begin{pmatrix} \xi_{i1}^2 \lambda_M(P_{i1}) & \xi_{i1}\xi_{i2}\lambda_M^{1/2}(P_{i3}P_{i3}^T) \\ \xi_{i1}\xi_{i2}\lambda_M^{1/2}(P_{i3}P_{i3}^T) & \xi_{i2}^2 \lambda_M(P_{i2}) \end{pmatrix}.$$

It can be easily shown that for the function $U_3(t,x)$ the estimate

$$(2.3.30) \qquad -\lambda_M^{1/2}(P_1 P_1^T)\|w_1\|\,\|w_2\| \le U_3(t,x) \le \lambda_M^{1/2}(P_1 P_1^T)\|w_1\|\,\|w_2\|$$

takes place.

By means of the matrix-valued function $U(t,x)$ and the vector $\eta \in R_+^2$ we introduce the function

$$(2.3.31) \qquad v(t,x,\eta) = \eta^T U(t,x)\eta.$$

Proposition 2.3.6. *If for the elements of the matrix-valued function* $U(t, x)$ *estimates (2.3.29) and (2.3.30) take place, then for function (2.3.31) the bilateral inequality*

$$W^{\mathrm{T}} C W \leq v(t, x, \eta) \leq W^{\mathrm{T}} D W$$

is satisfied, where $W^{\mathrm{T}} = (\|w_1\|^{\mathrm{T}}, \|w_2\|^{\mathrm{T}})$ *and*

$$C = \begin{pmatrix} \eta_1^2 \lambda_m(C_1) & -\eta_1 \eta_2 \lambda_M^{1/2}(P_1 P_1^{\mathrm{T}}) \\ -\eta_1 \eta_2 \lambda_M^{1/2}(P_1 P_1^{\mathrm{T}}) & \eta_2^2 \lambda_m(C_2) \end{pmatrix},$$

$$D = \begin{pmatrix} \eta_1^2 \lambda_M(D_1) & \eta_1 \eta_2 \lambda_M^{1/2}(P_1 P_1^{\mathrm{T}}) \\ \eta_1 \eta_2 \lambda_M^{1/2}(P_1 P_1^{\mathrm{T}}) & \eta_2^2 \lambda_M(D_2) \end{pmatrix}.$$

Alongside the function (2.3.31) we consider its total derivative

(2.3.32) $$Dv(t, x, \eta) = \eta^{\mathrm{T}} DU(t, x)\eta,$$

where $DU = Du_{ij}(t, \cdot)$ and D is the Euler derivative.

We compute the derivative of function $U_1(t, \cdot)$, $U_2(t, \cdot)$, $U_3(t, \cdot)$ by virtue of system (2.3.26)

(i) $D_t u_{i1}^{(i)}(t, x_{i1}) + (D_{x_{i1}} u_{i1}^{(i)})^{\mathrm{T}} A_{i1} x_{i1} \leq p_{i1}(t) \|x_{i1}\|^2$;

(ii) $D_t u_{i2}^{(i)}(t, x_{i2}) + (D_{x_{i2}} u_{i2}^{(i)})^{\mathrm{T}} A_{i2} x_{i2} \leq p_{i2}(t) \|x_{i2}\|^2$;

(iii) $(D_{x_{i1}} u_{i1}^{(i)})^{\mathrm{T}} A_{i1}^* x_{i2} \leq p_{i3}(t) \|x_{i1}\| \|x_{i2}\|$;

(iv) $(D_{x_{i2}} u_{i2}^{(i)})^{\mathrm{T}} A_{i2}^* x_{i1} \leq p_{i4}(t) \|x_{i1}\| \|x_{i2}\|$;

(v) $D_t u_{12}^{(i)}(t, x_i) + (D_{x_i} u_{12}^{(i)})^{\mathrm{T}} A_i x_i \leq p_{i5}(t) \|x_{i1}\|^2 + p_{i6}(t) \|x_{i2}\|^2$
    $+ p_{i7}(t) \|x_{i1}\| \|x_{i2}\|$,

where $D_\alpha = \partial / \partial \alpha$, $p_{i1}(t)$, $p_{i2}(t)$, $p_{i5}(t)$ and $p_{i6}(t)$ are maximal eigenvalues of the matrices

$$A_{i1}^{\mathrm{T}} P_{i1} + P_{i1} A_{i1} + \frac{dP_{i1}(t)}{dt}, \qquad A_{i2}^{\mathrm{T}} P_{i2} + P_{i2} A_{i2} + \frac{dP_{i2}(t)}{dt},$$

respectively; $p_{i3}(t)$, $p_{i4}(t)$ and $p_{i7}(t)$ are norms of the matrices $2P_{i1} A_{i1}^*$, $2A_{i2}^{*T} P_{i2}$ and

$$A_{i1}^{\mathrm{T}} P_{i3} + P_{i3} A_{i2} + \frac{dP_{i3}(t)}{dt},$$

respectively.

Proposition 2.3.7. *If estimates* (i) – (v) *are satisfied, the total derivative of function* $U_i(t, x_i)$ *by virtue of subsystem* (2.3.26) *is estimated by inequality*

$$(2.3.33) \qquad DU_i(t, x_i) \le \lambda_M(\bar{C}_i) \|\omega_i\|^2,$$

where $\omega_i^{\mathrm{T}} = (\|x_{i1}\|, \|x_{i2}\|)$ *and* $\lambda_M(\bar{C}_i)$ *are maximal eigenvalues of matrix* $\bar{C}_i$ *with the elements*

$$\sigma_{11}^i = \xi_{i1}^2 p_{i1}(t) + 2\xi_{i1}\xi_{i2}p_{i5}(t),$$

$$\sigma_{12}^i = \sigma_{21}^i = \xi_{i1}\xi_{i2}p_{i7}(t) + \frac{1}{2}(\xi_{i1}p_{i3}(t) + \xi_{i2}p_{i4}(t)),$$

$$\sigma_{22}^i = \xi_{i2}^2 p_{i2}(t) + 2\xi_{i1}\xi_{i2}p_{i6}(t).$$

The estimations of derivatives are easily obtained

$$(2.3.34)$$
$$(DU_i(t, x_i))^{\mathrm{T}}A_i^* x_j \le \theta_{ii}(t)\|x_1\|\|x_2\|, \quad i, j = 1, 2;$$

$$(DU_3(t, x))^{\mathrm{T}}A(t)x \le \theta_{12}(t)\|x_1\|^2 + \theta_{13}(t)\|x_2\|^2 + \theta_{14}(t)\|x_1\|\|x_2\|,$$

where $\theta_{12}(t)$ and $\theta_{13}(t)$ are maximal eigenvalues of matrices $P_1 A_2^*$ and $A_1^* P_1$, respectively; θ_{ii} and θ_{i4} are norms of matrices $A_i^{*T} P_i^* + P_i^* A_i^*$, $A_1^{\mathrm{T}} P_1 + P_1 A_2$, respectively.

In addition

$$P_i^* = \begin{pmatrix} P_{i1} & P_{i3} \\ P_{i3}^{\mathrm{T}} & P_{i2} \end{pmatrix}, \quad i = 1, 2.$$

Proposition 2.3.8. *Provided inequalities* (2.3.33) *and* (2.3.34) *are satisfied the total derivative of the function* (2.3.31) *along solutions of system* (2.3.24) *is estimated by the inequality*

$$Dv(t, x, \eta) \le W^{\mathrm{T}}SW,$$

where $W^{\mathrm{T}} = (\omega_1, \omega_2)$ and matrix S has the elements

$$s_{11} = \eta_1^2 \lambda_M(\bar{C}_1) + 2\eta_1\eta_2\theta_{12}(t),$$

$$s_{12} = s_{21} = \eta_1\eta_2\theta_{14}(t) + \frac{1}{2}(\eta_1^2\theta_{11}(t) + \eta_2^2\theta_{22}(t)),$$

$$s_{22} = \eta_2^2 \lambda_M(\bar{C}_2) + 2\eta_1\eta_2\theta_{13}(t).$$

Theorem 2.3.2. *Assume that for the linear nonautonomous system (2.3.24)*

(i) *there exists a matrix-valued function* $U : R_+ \times R^n \to R^{2 \times 2}$;
(ii) *there exists a vector* $\eta \in R_+^2$;
(iii) *matrix C is positive definite;*
(iv) *matrix S is negative semi-definite or equal to zero;*
(v) *matrix S is negative definite.*

Then, respectively

(i) *conditions (i) – (iv) are sufficient for the state $x = 0$ of system (2.3.24) to be stable in the whole;*
(ii) *conditions (i) – (iii) and (v) are sufficient for the state $x = 0$ of system (2.3.24) to be asymptotically stable in the whole.*

We consider some system equations of perturbed motion to illustrate how the proposed technique works.

Example 2.3.2. Consider linear homogeneous system

$$(2.3.35) \qquad \frac{dx}{dt} = A(t)x, \quad x \in R^4,$$

matrix $A(t)$ is of the form

$$A(t) = \begin{pmatrix} -\frac{8-\sin 2t}{2(1+\cos^2 t)} & -\frac{\sin t}{1+\cos^2 t} & \frac{1}{1+\cos^2 t} & 0 \\ 0 & -6 & 0 & 0 \\ 0 & 0 & -4 & 0 \\ 0 & \frac{1}{1+\cos^2 t} & \frac{2}{1+\cos^2 t} & -\frac{10-0.5\sin 2t}{1+\cos^2 t} \end{pmatrix}.$$

First level decomposition results in two subsystems

$$(2.3.36) \qquad \begin{aligned} \frac{dx_1}{dt} &= A_1 x_1 + A_1^* x_2, \\ \frac{dx_2}{dt} &= A_2 x_2 + A_2^* x_1, \end{aligned}$$

where $x_1 \in R^2$, $x_2 \in R^2$ and

$$A_1 = \begin{pmatrix} a_{11} & a_{12} \\ 0 & a_{22} \end{pmatrix}, \qquad A_1^* = \begin{pmatrix} a_{13} & 0 \\ 0 & 0 \end{pmatrix},$$

$$A_2 = \begin{pmatrix} a_{33} & 0 \\ a_{34} & a_{44} \end{pmatrix}, \qquad A_2^* = \begin{pmatrix} 0 & 0 \\ 0 & a_{24} \end{pmatrix},$$

$$a_{11} = -\frac{8 - \sin 2t}{2(1 + \cos^2 t)}, \quad a_{12} = -\frac{\sin t}{1 + \cos^2 t}, \quad a_{22} = -6, \quad a_{13} = \frac{1}{1 + \cos^2 t},$$

$$a_{33} = -4, \quad a_{34} = -\frac{2}{1 + \cos^2 t}, \quad a_{41} = -\frac{10 - 0.5 \sin 2t}{1 + \cos^2 t}, a_{24} = \frac{2}{1 + \cos^2 t}.$$

The independent subsystems

(2.3.37)

$$\frac{dx_1}{dt} = A_1 x_1,$$

$$\frac{dx_2}{dt} = A_2 x_2$$

of system (2.3.36) are connected by interconnection functions $h_1 = A_1^* x_2$ and $h_2 = A_2^* x_1$.

On the second level decomposition of system (2.3.37) we get independent equations

(2.3.38)

$$\frac{dx_{11}}{dt} = -\frac{8 - \sin 2t}{2(1 + \cos^2 t)} x_{11},$$

$$\frac{dx_{12}}{dt} = -6x_{12},$$

$$\frac{dx_{21}}{dt} = -4x_{21};$$

$$\frac{dx_{22}}{dt} = -\frac{10 - 0.5 \sin 2t}{1 + \cos^2 t} x_{22},$$

and interconnection functions

$$h_{11} = -\frac{\sin t}{1 + \cos^2 t} x_{12}, \quad h_{12} = h_{21} = 0, \quad h_{22} = \frac{2}{1 + \cos^2 t} x_{21}.$$

In the matrix $B_1(t, \cdot)$ the elements $u_{ij}^{(1)}(t, \cdot)$ are taken in the form

(2.3.39)

$$u_{11}^{(1)} = (1 + \cos^2 t)x_{11}^2, \quad u_{22}^{(1)} = x_{12}^2,$$

$$u_{12}^{(1)} = u_{21}^{(1)} = 2 \cdot 10^{-3}(1 + \cos^2 t)x_{11}x_{12}$$

and in the matrix $B_2(t, \cdot)$

(2.3.40)

$$u_{11}^{(2)} = x_{21}^2, \quad u_{22}^{(2)} = (1 + \cos^2 t)x_{22}^2,$$

$$u_{12}^{(2)} = u_{21}^{(2)} = 10^{-3}(1 + \cos^2 t)x_{21}x_{22}.$$

For the functions $U_1(t, x_1)$ and $U_2(t, x_2)$, whose matrix-valued functions B_1 and B_2 have elements (2.3.36) and (2.3.40), Assumption 2.3.1 holds, where

$$\underline{\alpha}_{11}^{(1)} = 1, \quad \underline{\alpha}_{12}^{(1)} = \underline{\alpha}_{21}^{(1)} = 1 \cdot 10^{-3}, \quad \underline{\alpha}_{22}^{(1)} = 1,$$

$$\overline{\alpha}_{11}^{(1)} = 2, \quad \overline{\alpha}_{12}^{(1)} = \overline{\alpha}_{21}^{(1)} = 2 \cdot 10^{-3}, \quad \overline{\alpha}_{22}^{(1)} = 1,$$

$$\underline{\alpha}_{11}^{(2)} = 1, \quad \underline{\alpha}_{12}^{(2)} = \underline{\alpha}_{21}^{(2)} = 1 \cdot 10^{-3}, \quad \underline{\alpha}_{22}^{(2)} = 1,$$

$$\overline{\alpha}_{11}^{(2)} = 1, \quad \overline{\alpha}_{12}^{(2)} = \overline{\alpha}_{21}^{(2)} = 2 \cdot 10^{-3}, \quad \overline{\alpha}_{22}^{(2)} = 2,$$

and

$$\phi_{11}(\|x_{11}\|) = |x_{11}|, \quad \phi_{12}(\|x_{12}\|) = |x_{12}|,$$

$$\phi_{21}(\|x_{21}\|) = |x_{21}|, \quad \phi_{22}(\|x_{22}\|) = |x_{22}|.$$

For the function

$$U_3(t, x) = x_1^{\mathrm{T}} \operatorname{diag} [10^{-3}(1 + \cos^2 t), 10^{-3}(1 + \cos^2 t)]x_2$$

the estimates in Assumption 2.3.2 are satisfied with constants

$$\beta_{12} = -2 \cdot 10^{-3}, \quad \overline{\beta}_{12} = 2 \cdot 10^{-3}.$$

Assumption 2.3.3 holds with the constants

$$p_{11} = -8, \quad p_{13} = 2, \quad p_{21} = -12, \quad p_{32} = 4.5 \cdot 10^{-3},$$

$$p_{35} = 10^{-3}, \quad p_{43} = 12 \cdot 10^{-3}, \quad p_{51} = -8, \quad p_{61} = -20,$$

$$p_{63} = 4, \quad p_{72} = 9 \cdot 10^{-3}, \quad p_{82} = 10.5 \cdot 10^{-3}, \quad p_{83} = 2 \cdot 10^{-3},$$

$$p_{12} = p_{22} = p_{32} = p_{31} = p_{33} = p_{34} = p_{41} = p_{43} = p_{44} = p_{45} = 0,$$

$$p_{52} = p_{53} = p_{62} = p_{71} = p_{73} = p_{74} = p_{75} = p_{81} = p_{84} = p_{85} = 0$$

and the functions

$$\beta_{11}(\|x_{11}\|) = |x_{11}|, \quad \beta_{12}(\|x_{12}\|) = |x_{12}|,$$

$$\beta_{21}(\|x_{21}\|) = |x_{21}|, \quad \beta_{22}(\|x_{22}\|) = |x_{22}|.$$

Assumption 2.3.4 holds for the functions U_1, U_2 and U_3 with the constants

$$\theta_{11} = \theta_{12} = \theta_{21} = \theta_{22} = 0, \quad \theta_{13} = \theta_{23} = 2.002,$$
$$\theta_{31} = \theta_{32} = 2 \cdot 10^{-3}, \quad \theta_{33} = 26.5 \cdot 10^{-3}.$$

Matrices A and B in (2.3.10) and matrix S in (2.3.13) are

$$A = \begin{pmatrix} 0.999 & -2 \cdot 10^{-3} \\ -2 \cdot 10^{-3} & 0.999 \end{pmatrix}, \quad B = \begin{pmatrix} 2 & 2 \cdot 10^{-3} \\ 2 \cdot 10^{-3} & 2 \end{pmatrix},$$
$$S = \begin{pmatrix} -7.976 & 2.0285 \\ 2.0285 & -7.642 \end{pmatrix},$$

respectively. Herewith, it is supposed that $\xi_1 = (1,1)^T$, $\xi_2 = (1,1)^T$ and $\eta = (1,1)^T$. It is easily verified that the matrices A and B are positive definite and the matrix S is negative definite. Applying Theorem 2.3.2 and taking into consideration that system (2.3.35) is linear we find that its solution $x = 0$ is uniformly asymptotically stable in the whole.

Example 2.3.3. We consider the linear nonautonomous system

(2.3.41)
$$\frac{dx}{dt} = A(t)x,$$

where $x \in R^6$ and matrix $A(t)$ is

$$A(t) = \begin{pmatrix} a_{11}(t) & a_{12}(t) & a_{13}(t) & a_{14}(t) & a_{15}(t) & a_{16}(t) \\ 0 & -3 & a_{23}(t) & 0 & 0 & 1 \\ 0 & a_{32}(t) & -3 & 0 & a_{35}(t) & a_{36}(t) \\ a_{41}(t) & 0 & a_{43}(t) & a_{44}(t) & 0 & a_{46}(t) \\ a_{51}(t) & 0 & a_{53}(t) & 0 & a_{55}(t) & a_{56}(t) \\ 0 & -1 & 0 & a_{64}(t) & 0 & -4 \end{pmatrix},$$

where

$$a_{11}(t) = -\frac{10 + \sin 2t}{2(1 + \sin^2 t)}, \quad a_{12}(t) = -\frac{\cos t}{1 + \sin^2 t}, \quad a_{13}(t) = -\frac{\sin t}{1 + \sin^2 t},$$
$$a_{14}(t) = \frac{\cos t}{1 + \sin^2 t}, \quad a_{15}(t) = -\frac{\cos t}{1 + \sin^2 t}, \quad a_{16}(t) = \frac{1}{1 + \sin^2 t},$$

$$a_{23}(t) = \frac{1}{1 + \sin^2 t}, \qquad a_{32}(t) = -\frac{\cos t}{1 + \cos^2 t}, \qquad a_{35}(t) = \frac{\cos t}{1 + \cos^2 t},$$

$$a_{36}(t) = \frac{\sin t}{1 + \cos^2 t}, \qquad a_{41}(t) = -\frac{\cos t}{1 + \cos^2 t}, \qquad a_{43}(t) = \frac{2 \cos t}{1 + \cos^2 t},$$

$$a_{44}(t) = \frac{6 - 0.5 \sin 2t}{1 + \cos^2 t}, \qquad a_{46}(t) = \frac{1}{1 + \cos^2 t}, \qquad a_{51}(t) = \frac{\cos t}{1 + \sin^2 t},$$

$$a_{53}(t) = \frac{1}{1 + \sin^2 t}, \qquad a_{55}(t) = -\frac{8 + 0.5 \sin 2t}{1 + \sin^2 t}, \qquad a_{56}(t) = \frac{\cos t}{1 + \sin^2 t},$$

$$a_{64}(t) = -\frac{1}{1 + \sin^2 t}.$$

The first level decomposition results in three subsystems

(2.3.42)
$$\frac{dx_1}{dt} = A_{11}x_1 + A_{12}x_2 + A_{13}x_3,$$
$$\frac{dx_2}{dt} = A_{22}x_2 + A_{21}x_1 + A_{23}x_3,$$
$$\frac{dx_3}{dt} = A_{33}x_3 + A_{31}x_1 + A_{32}x_2,$$

where $x_i \in R^2$, $i = 1, 2, 3$ and matrices A_{ij}, $i, j = 1, 2, 3$ are of the form

$$A_{11} = \begin{pmatrix} a_{11} & a_{12} \\ 0 & -3 \end{pmatrix}, \quad A_{12} = \begin{pmatrix} a_{13} & a_{14} \\ a_{23} & 0 \end{pmatrix}, \quad A_{13} = \begin{pmatrix} a_{15} & a_{16} \\ 0 & 1 \end{pmatrix},$$

$$A_{21} = \begin{pmatrix} 0 & a_{32} \\ a_{41} & 0 \end{pmatrix}, \quad A_{22} = \begin{pmatrix} -3 & 0 \\ a_{43} & a_{44} \end{pmatrix}, \quad A_{23} = \begin{pmatrix} a_{35} & a_{36} \\ 0 & a_{46} \end{pmatrix},$$

$$A_{31} = \begin{pmatrix} a_{51} & 0 \\ 0 & -1 \end{pmatrix}, \quad A_{32} = \begin{pmatrix} a_{53} & 0 \\ 0 & a_{64} \end{pmatrix}, \quad A_{33} = \begin{pmatrix} a_{55} & a_{56} \\ 0 & -4 \end{pmatrix}.$$

In system (2.3.42) the independent subsystems

(2.3.43)
$$\frac{dx_1}{dt} = A_{11}x_1, \quad x_1 \in R^2,$$
$$\frac{dx_2}{dt} = A_{22}x_2, \quad x_2 \in R^2,$$
$$\frac{dx_3}{dt} = A_{33}x_3, \quad x_3 \in R^2$$

are connected by the link functions

$$h_1 = A_{12}x_2 + A_{13}x_3, \quad h_2 = A_{21}x_1 + A_{23}x_3, \quad h_3 = A_{31}x_1 + A_{32}x_2.$$

The second level decomposition results in independent equations

$$\frac{dx_{11}}{dt} = -\frac{10 + \sin 2t}{2(1 + \sin^2 t)} x_{11}, \quad \frac{dx_{12}}{dt} = -3x_{12},$$

$$\frac{dx_{21}}{dt} = -3x_{21}, \quad \frac{dx_{22}}{dt} = -\frac{6 - 0.5 \sin 2t}{1 + \cos^2 t} x_{22},$$

$$\frac{dx_{31}}{dt} = -\frac{8 - 0.5 \sin 2t}{1 + \sin^2 t} x_{31}, \quad \frac{dx_{32}}{dt} = -4x_{32}$$

and interconnections functions

$$h_{11} = -\frac{\cos t}{1 + \sin^2 t} x_{12}, \quad h_{12} = 0, \quad h_{21} = 0, \quad h_{22} = \frac{2 \cos t}{1 + \cos^2 t} x_{21},$$

$$h_{31} = -\frac{\cos t}{1 + \sin^2 t} x_{32}, \quad h_{32} = 0.$$

In the matrices $B_i(t, \cdot)$, $i = 1, 2, 3$ we take the elements in the form

(2.3.44)
$$u_{11}^{(1)} = (1 + \sin^2 t)x_{11}^2, \quad u_{22}^{(1)} = x_{12}^2,$$
$$u_{12}^{(1)} = u_{21}^{(1)} = 10^{-3} \cdot (1 + \sin^2 t)x_{11}x_{12},$$

(2.3.45)
$$u_{11}^{(2)} = x_{12}^2(1 + \sin^2 t)x_{11}^2, \quad u_{22}^{(2)} = (1 + \cos^2 t)x_{22}^2,$$
$$u_{12}^{(2)} = u_{21}^{(2)} = 10^{-3} \cdot (1 + \cos^2 t)x_{21}x_{22},$$

(2.3.46)
$$u_{11}^{(3)} = (1 + \sin^2 t)x_{31}^2, \quad u_{22}^{(3)} = x_{32}^2,$$
$$u_{12}^{(3)} = u_{21}^{(3)} = 10^{-3} \cdot (1 + \sin^2 t)x_{31}x_{32}.$$

For the functions $U_{ii}(t, x_i)$, $i = 1, 2, 3$ with matrices B_1, B_2 and B_3 (with elements (2.3.44)–(2.3.46)) all conditions of Assumption 2.3.2 are satisfied with the constants

$$\underline{\alpha}_{11}^{(1)} = \underline{\alpha}_{22}^{(1)} = \underline{\alpha}_{11}^{(2)} = \underline{\alpha}_{11}^{(3)} = \underline{\alpha}_{22}^{(3)} = 1, \quad \underline{\alpha}_{12}^{(2)} = \underline{\alpha}_{12}^{(3)} = -2 \cdot 10^{-3},$$

$$\overline{\alpha}_{11}^{(1)} = \overline{\alpha}_{22}^{(2)} = \overline{\alpha}_{11}^{(3)} = 2, \quad \overline{\alpha}_{11}^{(1)} = \overline{\alpha}_{11}^{(2)} = \overline{\alpha}_{22}^{(3)} = 1,$$

$$\overline{\alpha}_{12}^{(1)} = \overline{\alpha}_{12}^{(2)} = \overline{\alpha}_{12}^{(3)} = -2 \cdot 10^{-3}$$

and the functions

$$\phi_{11}(\|x_{11}\|) = |x_{11}|, \quad \phi_{21}(\|x_{21}\|) = |x_{21}|, \quad \phi_{31}(\|x_{31}\|) = |x_{31}|,$$
$$\overline{\phi}_{11}(\|x_{11}\|) = |x_{11}|, \quad \overline{\phi}_{21}(\|x_{21}\|) = |x_{21}|, \quad \overline{\phi}_{31}(\|x_{31}\|) = |x_{31}|,$$
$$\phi_{12}(\|x_{12}\|) = |x_{12}|, \quad \phi_{22}(\|x_{22}\|) = |x_{22}|, \quad \phi_{32}(\|x_{32}\|) = |x_{32}|,$$
$$\overline{\phi}_{12}(\|x_{12}\|) = |x_{12}|, \quad \overline{\phi}_{22}(\|x_{22}\|) = |x_{22}|, \quad \overline{\phi}_{32}(\|x_{32}\|) = |x_{32}|.$$

Functions U_{12}, U_{13} and U_{23} are taken as

$$U_{12} = x_1^T \text{diag} \left[10^{-3} \cdot (1 + \sin^2 t), \; 10^{-3} \cdot (1 + \cos^2 t)\right]x_2,$$
$$U_{13} = x_1^T \text{diag} \left[10^{-3} \cdot (1 + \sin^2 t), \; 10^{-3} \cdot (1 + \cos^2 t)\right]x_3,$$
$$U_{23} = x_2^T \text{diag} \left[10^{-3} \cdot (1 + \cos^2 t), \; 10^{-3} \cdot (1 + \cos^2 t)\right]x_3.$$

Estimates in Assumption 2.3.1 are satisfied with the constants

$$\beta_{11} = -2 \cdot 10^{-3}, \quad \beta_{13} = -2 \cdot 10^{-3}, \quad \beta_{23} = -2 \cdot 10^{-3},$$
$$\overline{\beta}_{11} = -2 \cdot 10^{-3}, \quad \overline{\beta}_{13} = -2 \cdot 10^{-3}, \quad \overline{\beta}_{23} = -2 \cdot 10^{-3}.$$

Conditions of Assumption 2.3.2 are satisfied with the constants

$$p_1^{(1)} = -10, \qquad p_2^{(1)} = -6, \qquad \mu_1^{(1)} = 0, \qquad \mu_2^{(1)} = 2 \cdot 10^{-3},$$
$$\mu_{12}^{(1)} = 1.0115, \qquad p_1^{(2)} = -6, \qquad p_2^{(2)} = -12, \qquad \mu_1^{(2)} = 4 \cdot 10^{-3},$$
$$\mu_2^{(2)} = 0, \qquad \mu_{12}^{(2)} = 2.0125, \qquad p_1^{(3)} = -16, \qquad p_2^{(3)} = -8,$$
$$\mu_1^{(3)} = 0, \qquad \mu_2^{(3)} = 2 \cdot 10^{-3}, \qquad \mu_{12}^{(3)} = 1.0165$$

and the functions

$$\beta_{11}(\|x_{11}\|) = |x_{11}|, \quad \beta_{21}(\|x_{21}\|) = |x_{21}|, \quad \beta_{31}(\|x_{31}\|) = |x_{31}|,$$
$$\beta_{12}(\|x_{12}\|) = |x_{12}|, \quad \beta_{22}(\|x_{22}\|) = |x_{22}|, \quad \beta_{32}(\|x_{32}\|) = |x_{32}|.$$

Conditions of Assumption 2.3.3 for the functions U_{12}, U_{13} and U_{23} are satisfied with the constants

$$\theta_{11} = 2.74 \cdot 10^{-3}, \quad \theta_{22} = 2.74 \cdot 10^{-3}, \quad \theta_{33} = 3.73 \cdot 10^{-3},$$
$$\theta_{12} = 8.25 \cdot 10^{-3}, \quad \theta_{13} = 1 \cdot 10^{-2}, \quad \theta_{23} = 16.5 \cdot 10^{-3}.$$

Matrices A and B in estimate (2.3.10) and matrix S in estimate (2.3.13) are of the form

$$A = \begin{pmatrix} 0.998 & -2 \cdot 10^{-3} & -2 \cdot 10^{-3} \\ -2 \cdot 10^{-3} & 0.998 & -2 \cdot 10^{-3} \\ -2 \cdot 10^{-3} & -2 \cdot 10^{-3} & 0.998 \end{pmatrix},$$

$$B = \begin{pmatrix} 2 & 2 \cdot 10^{-3} & -2 \cdot 10^{-3} \\ 2 \cdot 10^{-3} & 2 & -2 \cdot 10^{-3} \\ -2 \cdot 10^{-3} & -2 \cdot 10^{-3} & 2 \end{pmatrix},$$

$$S = \begin{pmatrix} -5.75226 & 0.00825 & 0.01 \\ 0.00825 & 0.0165 & 0.0165 \\ 0.01 & -5.36924 & -7.87127 \end{pmatrix},$$

respectively. Here we assume $\xi_1 = \xi_2 = \xi_3 = \eta = (1, 1, 1)^{\mathrm{T}}$. It is easily to check that matrices A and B are positive definite and matrix S is negative definite.

Applying Theorem 2.3.2 we find that the equilibrium state $x = 0$ of system (2.3.41) is uniformly asymptotically stable in the whole.

2.4 Stability Analysis of Large Scale Systems

The aim of this section is to present a new method of constructing the matrix-valued function and then to obtain efficient stability conditions for one class of large scale systems admitting one-level decomposition.

2.4.1 A class of large scale systems

We consider a system with a finite number of degrees of freedom whose motion is described by the equations

$$(2.4.1) \qquad \frac{dx_i}{dt} = f_i(x_i) + g_i(t, x_1, \ldots, x_m), \quad i = 1, 2, \ldots, m,$$

where $x_i \in R^{n_i}$, $t \in T_\tau$, $T_\tau = [\tau, +\infty)$, $f_i \in C(R^{n_i}, R^{n_i})$, $g_i \in C(T_\tau \times R^{n_1} \times \cdots \times R^{n_m}, R^{n_i})$.

Suppose that the functions $f_i(x_i) + g_i(t, x_1, \ldots, x_m)$ are smooth enough to guarantee existence, uniqueness and continuous dependence of solutions $x_i(t) = x_i(t; t_0, x_{i0})$ of (2.4.1).

Introduce the designation

$$(2.4.2) \qquad G_i(t,x) = g_i(t, x_1, \ldots, x_m) - \sum_{j=1, j \neq i}^{m} g_{ij}(t, x_i, x_j),$$

where $g_{ij}(t, x_i, x_j) = g_i(t, 0, \ldots, x_i, \ldots, x_j, \ldots, 0)$ for all $i \neq j$; $i, j = 1, 2, \ldots, m$. Taking into consideration (2.4.2) system (2.4.1) is rewritten as

$$(2.4.3) \qquad \frac{dx_i}{dt} = f_i(x_i) + \sum_{j=1, j \neq i}^{m} g_{ij}(t, x_i, x_j) + G_i(t, x).$$

Actually equations (2.4.3) describe the class of large scale nonlinear nonautonomously connected systems. It is of interest to extend the method of matrix Liapunov functions to this class of equations in view of the new method of construction of nondiagonal elements of matrix-valued functions.

2.4.2 Construction of nondiagonal elements of matrix-valued function

In order to extend the method of matrix Liapunov functions to systems (2.4.3) it is necessary to estimate variation of matrix-valued function elements and their total derivatives along solutions of the corresponding systems. Such estimates are provided by the assumptions below.

Assumption 2.4.1. *There exist open connected neighborhoods $\mathcal{N}_i \subseteq R^{n_i}$ of the equilibrium states $x_i = 0$, functions $v_{ii} \in C^1(R^{n_i}, R_+)$, the comparison functions φ_{i1}, φ_{i2} and ψ_i of class $K(KR)$ and real numbers $\underline{c}_{ii} > 0$, $\bar{c}_{ii} > 0$ and γ_{ii} such that*

(1) $v_{ii}(x_i) = 0$ *for all* $(x_i = 0) \in \mathcal{N}_i$;
(2) $\underline{c}_{ii} \varphi_{i1}^2(\|x_i\|) \leq v_{ii}(x_i) \leq \bar{c}_{ii} \varphi_{i2}^2(\|x_i\|)$;
(3) $(D_{x_i} v_{ii}(x_i))^T f_i(x_i) \leq \gamma_{ii} \psi_i^2(\|x_i\|)$ *for all* $x_i \in \mathcal{N}_i$,
    $i = 1, 2, \ldots, m$.

It is clear that under conditions of Assumption 2.4.1 the equilibrium states $x_i = 0$ of nonlinear isolated subsystems

$$(2.4.4) \qquad \frac{dx_i}{dt} = f_i(x_i), \quad i = 1, 2, \ldots, m$$

are

(a) *uniformly asymptotically stable in the whole*, if $\gamma_{ii} < 0$ and $(\varphi_{i1}, \varphi_{i2}, \psi_i) \in KR$-class;

(b) *stable*, if $\gamma_{ii} = 0$ and $(\varphi_{i1}, \varphi_{i2}) \in K$-class;

(c) *unstable*, if $\gamma_{ii} > 0$ and $(\varphi_{i1}, \varphi_{i2}, \psi_i) \in K$-class.

The approach proposed in this section takes large scale systems (2.4.3) into consideration, subsystems (2.4.4) having various dynamical properties specified by conditions of Assumption 2.4.1.

Assumption 2.4.2. *There exist open connected neighborhoods* $\mathcal{N}_i \subseteq R^{n_i}$ *of the equilibrium states* $x_i = 0$, *functions* $v_{ij} \in C^{1,1,1}(\mathcal{T}_\tau \times R^{n_i} \times R^{n_j}, R)$, *comparison functions* $\varphi_{i1}, \varphi_{i2} \in K(KR)$, *positive constants* $(\eta_1, \ldots, \eta_m)^T \in R^m$, $\eta_i > 0$ *and arbitrary constants* $\underline{c}_{ij}, \bar{c}_{ij}$, $i, j = 1, 2, \ldots, m$, $i \neq j$ *such that*

(1) $v_{ij}(t, x_i, x_j) = 0$ *for all* $(x_i, x_j) = 0 \in \mathcal{N}_i \times \mathcal{N}_j$, $t \in \mathcal{T}_\tau$, $i, j = 1, 2, \ldots, m$, $(i \neq j)$;

(2) $\underline{c}_{ij}\varphi_{i1}(\|x_i\|)\varphi_{j1}(\|x_j\|) \leq v_{ij}(t, x_i, x_j) \leq \bar{c}_{ij}\varphi_{i2}(\|x_i\|)\varphi_{j2}(\|x_j\|)$ *for all* $(t, x_i, x_j) \in \mathcal{T}_\tau \times \mathcal{N}_i \times \mathcal{N}_j$, $i \neq j$;

(3) $D_t v_{ij}(t, x_i, x_j) + (D_{x_i} v_{ij}(t, x_i, x_j))^T f_i(x_i)$
$+(D_{x_j} v_{ij}(t, x_i, x_j))^T f_j(x_j) + \frac{\eta_i}{2\eta_j}(D_{x_i} v_{ii}(x_i))^T g_{ij}(t, x_i, x_j)$
$+\frac{\eta_j}{2\eta_i}(D_{x_j} v_{jj}(x_j))^T g_{ji}(t, x_i, x_j) = 0;$ (2.4.5)

It is easy to notice that first order partial equations (2.4.5) are a somewhat generalization of the classical Liapunov equation proposed in (see Liapunov [1]) for determination of auxiliary function in the theory of his direct method of motion stability investigation. In a particular case these equations are transformed into the systems of algebraic equations whose solutions can be constructed analytically.

Assumption 2.4.3. *There exist open connected neighborhoods* $\mathcal{N}_i \subseteq R^{n_i}$ *of the equilibrium states* $x_i = 0$, *comparison functions* $\psi \in K(KR)$, $i = 1, 2, \ldots, m$, *real numbers* $\alpha_{ij}^1, \alpha_{ij}^2, \alpha_{ij}^3, \nu_{ki}^1, \nu_{kij}^1, \mu_{kij}^1$ *and* μ_{kij}^2, $i, j, k = 1, 2, \ldots, m$, *such that*

(1) $(D_{x_i} v_{ii}(x_i))^T G_i(t, x) \leq \psi_i(\|x_i\|) \sum_{k=1}^{m} \nu_{ki}^1 \psi(\|x_k\|) + R_1(\psi)$ *for all* $(t, x_i, x_j) \in \mathcal{T}_\tau \times \mathcal{N}_i \times \mathcal{N}_j$;

(2) $(D_{x_i} v_{ij}(t, \cdot))^T g_{ij}(t, x_i, x_j) \leq \alpha_{ij}^1 \psi_i^2(\|x_i\|) + \alpha_{ij}^2 \psi_i(\|x_i\|)\psi_j(\|x_j\|) + \alpha_{ij}^3 \psi_j^2(\|x_j\|) + R_2(\psi)$ *for all* $(t, x_i, x_j) \in \mathcal{T}_\tau \times \mathcal{N}_i \times \mathcal{N}_j$;

(3) $(D_{x_i} v_{ij}(t,\cdot))^{\mathrm{T}} G_i(t,x) \leq \psi_j(\|x_j\|) \sum\limits_{k=1}^{m} \nu_{ijk}^2 \psi_k(\|x_k\|) + R_3(\psi)$

for all $(t, x_i, x_j) \in \mathcal{T}_\tau \times \mathcal{N}_i \times \mathcal{N}_j;$

(4) $(D_{x_i} v_{ij}(t,\cdot))^{\mathrm{T}} g_{ik}(t, x_i, x_k) \leq \psi_j(\|x_j\|)(\mu_{ijk}^1 \psi_k(\|x_k\|) + \mu_{ijk}^2 \psi_i(\|x_i\|))$

$+ R_4(\psi)$ for all $(t, x_i, x_j) \in \mathcal{T}_\tau \times \mathcal{N}_i \times \mathcal{N}_j$, and $k \neq j.$

Here $R_s(\psi)$ are polynomials in $\psi = (\psi_1(\|x_1\|), \ldots, \psi_m(\|x_m\|))$ in a power higher than three, $R_s(0) = 0$, $s = 1, \ldots, 4$.

Under conditions (2) of Assumptions 2.4.1 and 2.4.2 it is easy to establish for function

(2.4.6) $\qquad v(t, x, \eta) = \eta^{\mathrm{T}} U(t,x)\eta = \sum\limits_{i,j=1}^{m} v_{ij}(t,\cdot)\eta_i \eta_j$

the bilateral estimate

(2.4.7) $\qquad u_1^{\mathrm{T}} H^{\mathrm{T}} \underline{C} H u_1 \leq v(t, x, \eta) \leq u_2^{\mathrm{T}} H^{\mathrm{T}} \bar{C} H u_2,$

where

$$u_1 = (\varphi_{11}(\|x_1\|), \ldots, \varphi_{m1}(\|x_m\|))^{\mathrm{T}},$$

$$u_2 = (\varphi_{12}(\|x_1\|), \ldots, \varphi_{m2}(\|x_m\|))^{\mathrm{T}}$$

which holds true for all $(t, x) \in \mathcal{T}_\tau \times \mathcal{N}$, $\mathcal{N} = \mathcal{N}_1 \times \cdots \times \mathcal{N}_m$.

Based on conditions (3) of Assumptions 2.4.1, 2.4.2 and conditions (1) – (4) of Assumption 2.4.3 it is easy to establish the inequality estimating the auxiliary function variation along solutions of system (2.4.3). This estimate reads

(2.4.8) $\qquad Dv(t, x, \eta)|_{(2.4.1)} \leq u_3^{\mathrm{T}} M u_3,$

where $u_3 = (\psi_1(\|x_1\|), \ldots, \psi_m(\|x_m\|))$ and holds for all $(t,x) \in \mathcal{T}_\tau \times \mathcal{N}$.

Elements σ_{ij} of matrix M in the inequality (2.4.8) have the following structure

$$\sigma_{ii} = \eta_i^2 \gamma_{ii} + \eta_i^2 \nu_{ii} + \sum\limits_{k=1, k\neq i}^{m} (\eta_k \eta_i \nu_{kii}^2 + \eta_i^2 \nu_{kii}^2) + 2 \sum\limits_{j=1, j\neq i}^{m} \eta_i \eta_j (\alpha_{ij}^1 + \alpha_{ji}^3);$$

$$\sigma_{ij} = \frac{1}{2}(\eta_i^2 \nu_{ji}^1 + \eta_j^2 \nu_{ij}^1) + \sum\limits_{k=1, k\neq j}^{m} \eta_k \eta_j \nu_{kij}^2 + \sum\limits_{k=1, k\neq i}^{m} \eta_i \eta_j \nu_{kij}^2$$

$$+ \eta_i \eta_j (\alpha_{ij}^2 + \alpha_{ji}^2) + \sum\limits_{\substack{k=1, k\neq i, \\ k\neq j}}^{m} (\eta_k \eta_j \mu_{kji}^1 + \eta_i \eta_j \mu_{ijk}^2 + \eta_i \eta_k \mu_{kij}^1 + \eta_i \eta_j \mu_{jik}^2),$$

$$i = 1, 2, \ldots, m, \quad i \neq j.$$

2.4.3 Test for stability analysis

Sufficient criteria of various types of stability of the equilibrium state $x = 0$ of system (2.4.3) are formulated in terms of the sign definiteness of matrices C, $\bar{C}$ and M from estimates (2.4.7), (2.4.8). We shall show that the following assertion is valid.

Theorem 2.4.1. *Assume that the perturbed motion equations are such that all conditions of Assumptions 2.4.1 – 2.4.3 are fulfilled and moreover*

(1) *matrices C and $\bar{C}$ in estimate (2.4.7) are positive definite;*
(2) *matrix M in inequality (2.4.8) is negative semi-definite (negative definite).*

Then the equilibrium state $x = 0$ of system (2.4.1) is uniformly stable (uniformly asymptotically stable).

If, additionally, in conditions of Assumptions 2.4.1 – 2.4.3 all estimates are satisfied for $\mathcal{N}_i = R^{n_i}$ and comparison functions $(\varphi_{i1}, \varphi_{i2}) \in KR$-class, then the equilibrium state of system (2.4.1) is uniformly stable in the whole (uniformly asymptotically stable in the whole).

Proof. If all conditions of Assumptions 2.4.1 – 2.4.2 are satisfied, then it is possible for system (2.4.1) to construct function $v(t, x, \eta)$ which together with total derivative $Dv(t, x, \eta)$ satisfies the inequalities (2.4.7), and (2.4.8). Condition (1) of Theorem 2.4.1 implies that function $v(t, x, \eta)$ is positive definite and decreasing for all $t \in \mathcal{T}_\tau$. Under condition (2) of Theorem 2.4.1 function $Dv(t, x, \eta)$ is negative semi-definite (definite). Therefore all conditions of Theorems 1.8.1, 1.8.3 are fulfilled. The proof of the second part of Theorem 2.4.1 is based on Theorem 1.8.4.

2.4.4. Linear large scale system

Assume that in the system

$$\frac{dx_1}{dt} = A_{11}x_1 + A_{12}x_2 + A_{13}x_3,$$

(2.4.9)
$$\frac{dx_2}{dt} = A_{21}x_1 + A_{22}x_2 + A_{23}x_3,$$

$$\frac{dx_3}{dt} = A_{31}x_1 + A_{32}x_2 + A_{33}x_3,$$

the state vectors $x_i \in R^{n_i}$, $i = 1, 2, 3$, and $A_{ij} \in R^{n_i \times n_j}$ are constant matrices for all $i, j = 1, 2, 3$.

For the independent systems

$$(2.4.10) \qquad \frac{dx_i}{dt} = A_{ii}x_i, \quad i = 1, 2, 3$$

we construct auxiliary functions $v_{ii}(x_i)$ as the quadratic forms

$$(2.4.11) \qquad v_{ii}(x_i) = x_i^{\mathrm{T}} P_{ii} x_i, \quad i = 1, 2, 3$$

whose matrices P_{ii} are determined by the algebraic Liapunov equations

$$(2.4.12) \qquad A_{ii}^{\mathrm{T}} P_{ii} + P_{ii} A_{ii} = -G_{ii}, \quad i = 1, 2, 3,$$

where G_{ii} are prescribed matrices of definite sign. In order to construct nondiagonal elements $v_{ij}(x_i, x_j)$ of matrix-valued function $U(x)$ we employ equation (2.4.5) from Assumption 2.4.2. Note that for system (2.4.9)

$$f_i(x_i) = A_{ii}x_i, \quad f_j(x_j) = A_{jj}x_j,$$

$$g_{ij}(x_i, x_j) = A_{ij}x_j, \quad G_i(t, x) = 0, \quad i = 1, 2, 3.$$

Since for the bilinear forms

$$(2.4.13) \qquad v_{ij}(x_i, x_j) = v_{ji}(x_j, x_i) = x_i^{\mathrm{T}} P_{ij} x_j,$$

the correlations

$$D_{x_i} v_{ij}(x_i, x_j) = x_j^{\mathrm{T}} P_{ij}^{\mathrm{T}}, \quad D_{x_j} v_{ij}(x_i, x_j) = x_i^{\mathrm{T}} P_{ij},$$

are true, and equation (2.4.5) becomes

$$x_i^{\mathrm{T}} \left(A_{ii}^{\mathrm{T}} P_{ij} + P_{ij} A_{jj} + \frac{\eta_i}{\eta_j} P_{ii} A_{ij} + \frac{\eta_j}{\eta_i} A_{ji}^{\mathrm{T}} P_{ii} \right) x_j = 0.$$

From this correlation for determining matrices P_{ij} we get the system of algebraic equations

$$(2.4.14) \qquad A_{ii} P_{ij} + P_{ij} A_{jj} = -\frac{\eta_i}{\eta_j} P_{ii} A_{ij} - \frac{\eta_j}{\eta_i} A_{ji}^{\mathrm{T}} P_{ii},$$

$$i \neq j, \quad i, j = 1, 2, 3.$$

Since for (2.4.11), and (2.4.13) the estimates

$$v_{ii}(x_i) \geq \lambda_m(P_{ii})\|x_i\|^2, \quad x_i \in R^{n_i},$$

$$v_{ij}(x_i, x_j) \geq -\lambda_M^{1/2}(P_{ij}P_{ij}^T)\|x_i\|\|x_j\|, \quad (x_i, x_j) \in R^{n_i} \times R^{n_j}$$

hold true, for function $v(x, \eta) = \eta^T U(x)\eta$ the inequality

$$(2.4.15) \qquad\qquad w^T H^T C H w \leq v(x, \eta)$$

is satisfied for all $x \in R^n$, where $w = (\|x_1\|, \|x_2\|, \|x_3\|)^T$ and the matrix

$$C = \begin{pmatrix} \lambda_m(P_{11}) & -\lambda_M^{1/2}(P_{12}P_{12}^T) & -\lambda_M^{1/2}(P_{13}P_{13}^T) \\ -\lambda_M^{1/2}(P_{12}P_{12}^T) & \lambda_m(P_{22}) & -\lambda_M^{1/2}(P_{23}P_{23}^T) \\ -\lambda_M^{1/2}(P_{13}P_{13}^T) & -\lambda_M^{1/2}(P_{23}P_{23}^T) & \lambda_m(P_{33}) \end{pmatrix}.$$

Here $\lambda_m(\cdot)$ are minimal eigenvalues of matrices P_{11}, P_{22}, P_{33}, and $\lambda_M^{1/2}(\cdot)$ is the norm of matrices $(\cdot)$.

For system (2.4.9) the constants from Assumption 2.4.3 are:

$$\alpha_{ij}^1 = \alpha_{ij}^2 = 0; \quad \alpha_{ij}^3 = \lambda_M(A_{ij}^T P_{ij} + P_{ij}^T A_{ij}),$$

$$\nu_{ki}^1 = \nu_{ijk}^2 = 0; \nu_{ijk}^1 = \lambda_M^{1/2}[(P_{ij}^T A_{ik})(P_{ij}^T A_{ik})], \quad \mu_{ijk}^2 = 0.$$

Therefore the elements σ_{ij} of matrix M in (2.4.8) for system (2.4.9) have the structure

$$\sigma_{ii} = -\eta_i^2 \lambda_m(G_{ii}) + 2 \sum_{j=1, j \neq i}^{3} \eta_i \eta_j \alpha_{ij}^3, \quad i = 1, 2, 3,$$

$$\sigma_{ij} = \sum_{\substack{k=1, k \neq i, \\ k \neq j}}^{3} (\eta_k \eta_j \nu_{ijk}^1 + \eta_i \eta_k \nu_{kij}^1), \quad i, j = 1, 2, 3, \quad i \neq j.$$

Consequently, the function $Dv(x, \eta)$ variation along solutions of system (2.4.9) is estimated by the inequality

$$(2.4.16) \qquad\qquad Dv(x, \eta)\big|_{(2.4.9)} \leq w^T M w$$

for all $(x_1, x_2, x_3) \in R^{n_1} \times R^{n_2} \times R^{n_3}$.

We summarize our presentation as follows.

Corollary 2.4.1. *Assume for system (2.4.9) the following conditions are satisfied:*

(1) *algebraic equations (2.4.12) have the sign-definite matrices P_{ii}, $i = 1, 2, 3$ as their solutions;*

(2) *algebraic equations (2.4.14) have constant matrices P_{ij}, for all $i, j = 1, 2, 3$, $i \neq j$ as their solutions;*

(3) *matrix C in (2.4.15) is positive definite;*

(4) *matrix M in (2.4.16) is negative semi-definite (negative definite).*

Then the equilibrium state $x = 0$ of system (2.4.9) is uniformly stable (uniformly asymptotically stable).

This corollary follows from Theorem 2.4.1 and hence its proof is obvious.

2.4.5 Discussion and numerical example

To start to illustrate the possibilities of the proposed method of Liapunov function construction we consider a system of two connected equations that was studied earlier by the Bellman-Bailey approach (see Bailey [1], Piontkovskii and Rutkovskaya [1], etc.).

Partial case of system (2.4.9) is the system

(2.4.17)
$$\frac{dx_1}{dt} = Ax_1 + C_{12}x_2$$
$$\frac{dx_2}{dt} = Bx_2 + C_{21}x_1,$$

where $x_1 \in R^{n_1}$, $x_2 \in R^{n_2}$, and A, B, C_{12} and C_{21} are constant matrices of corresponding dimensions. For independent subsystems

(2.4.18)
$$\frac{dx_1}{dt} = Ax_1,$$
$$\frac{dx_2}{dt} = Bx_2$$

the functions $v_{11}(x_1)$ and $v_{22}(x_2)$ are constructed as the quadratic forms

(2.4.19)
$$v_{11} = x_1^{\mathrm{T}} P_{11} x_1, \quad v_{22} = x_2^{\mathrm{T}} P_{22} x_2,$$

where P_{11} and P_{22} are sign-definite matrices.

Function $v_{12} = v_{21}$ is searched for as a bilinear form $v_{12} = x_1^{\mathrm{T}} P_{12} x_2$ whose matrix is determined by the equation

$$(2.4.20) \quad A^{\mathrm{T}} P_{12} + P_{12} B = -\frac{\eta_1}{\eta_2} P_{11} C_{12} - \frac{\eta_2}{\eta_1} C_{21}^{\mathrm{T}} P_{22}, \quad \eta_1 > 0, \quad \eta_2 > 0.$$

According to Lankaster [1] equation (2.4.20) has a unique solution, provided that matrices A and $-B$ have no common eigenvalues.

Matrix C in (2.4.15) for system (2.4.17) reads

$$(2.4.21) \qquad C = \begin{pmatrix} \lambda_m(P_{11}) & -\lambda_M^{1/2}(P_{12} P_{12}^{\mathrm{T}}) \\ -\lambda_M^{1/2}(P_{12} P_{12}^{\mathrm{T}}) & \lambda_m(P_{22}) \end{pmatrix}.$$

Estimate (2.4.16) for function $Dv(x, \eta)$ by virtue of system (2.4.17) is

$$(2.4.22) \qquad Dv(x, \eta) \mid_{(2.4.17)} \leq w^{\mathrm{T}} \Xi w,$$

where $w = (\|x_1\|, \|x_2\|)^{\mathrm{T}}$, $\Xi = [\sigma_{ij}]$, $i, j = 1, 2$;

$$\sigma_{11} = \lambda_1 \eta_1^2 + \eta_1 \eta_2 \alpha_{22},$$

$$\sigma_{22} = \lambda_2 \eta_2^2 + \eta_1 \eta_2 \beta_{22},$$

$$\sigma_{12} = \sigma_{21} = 0.$$

The notations are

$$\lambda_1 = \lambda_M(A^{\mathrm{T}} P_{11} + P_{11} A),$$

$$\lambda_2 = \lambda_M(B^{\mathrm{T}} P_{22} + P_{22} B),$$

$$\alpha_{22} = \lambda_M(C_{12}^{\mathrm{T}} P_{12} + P_{12}^{\mathrm{T}} C_{12}),$$

$$\beta_{22} = \lambda_M(C_{21}^{\mathrm{T}} P_{12}^{\mathrm{T}} + P_{12} C_{21}),$$

$\lambda(\cdot)$ is maximal eigenvalue of matrix $(\cdot)$. Partial case of Assumption 2.4.1 is as follows.

Corollary 2.4.2. *For system (2.4.17) let functions $v_{ij}(\cdot)$, $i, j = 1, 2$ be constructed so that matrix C for system (2.4.17) is positive definite and matrix Ξ in inequality (2.4.22) is negative definite. Then the equilibrium $x = 0$ of system (2.4.17) is uniformly asymptotically stable.*

We consider the numerical example. Let the matrices from system (2.4.17) be of the form

$$(2.4.23) \qquad A = \begin{pmatrix} -2 & 1 \\ 3 & -2 \end{pmatrix}, \qquad B = \begin{pmatrix} -4 & 1 \\ 2 & -1 \end{pmatrix},$$

$$(2.4.24) \quad C_{12} = \begin{pmatrix} -0.5 & -0.5 \\ 0.8 & -0.7 \end{pmatrix}, \qquad C_{21} = \begin{pmatrix} 1.1 & 0.5 \\ -0.6 & -0.3 \end{pmatrix}.$$

Functions v_{ii} for subsystems

$$\dot{x} = Ax, \quad x = (x_1, x_2)^{\mathrm{T}},$$
$$\dot{y} = By, \quad y = (y_1, y_2)^{\mathrm{T}}$$

are taken as the quadratic forms

(2.4.25)
$$v_{11} = 1.75x_1^2 + x_1 x_2 + 1.5x_2^2,$$
$$v_{22} = 0.35y_1^2 + 0.9y_1 y_2 + 0.95y_2^2.$$

Let $\eta = (1,1)^{\mathrm{T}}$. Then $\lambda_1 = \lambda_2 = -1$,

$$P_{12} = \begin{pmatrix} -0.011 & 0.021 \\ -0.05 & -0.022 \end{pmatrix},$$

$$\alpha_{22} = 0.03, \quad \beta_{22} = -0.002.$$

It is easy to verify that $\sigma_{11} < 0$, and $\sigma_{22} < 0$, and hence all conditions of Corollary 2.4.2, are fulfilled in view that

$$\lambda_M^{1/2}(P_{12}P_{12}^{\mathrm{T}}) \le (\lambda_m(P_{11})\lambda_m(P_{22}))^{1/2},$$

for the values of $\lambda_M^{1/2}(P_{12}P_{12}^{\mathrm{T}}) = 0.06$, $\lambda_m(P_{11}) = 1.08$, $\lambda_m(P_{22}) = 0.115$. This implies uniform asymptotic stability in the whole of the equilibrium state of system (2.4.17) with matrices (2.4.23), and (2.4.24).

Let us show now that stability of system (2.4.17) with matrices (2.4.23), and (2.4.24) can not be studied in terms of the Bailey [1] theorem.

We recall that in this theorem the conditions of exponential stability of the equilibrium state are

(1) for subsystems (2.4.18) there must exist functions (2.4.19) satisfying estimates
   (a) $c_{i1}\|x_i\|^2 \le v_i(t, x_i) \le c_{i2}\|x_i\|^2$,
   (b) $Dv_i(t, x_i) \le -c_{i3}\|x_i\|^2$,
   (c) $\|\partial v_i/\partial x_i\| \le c_{i4}\|x_i\|$ for $x_i \in R^{n_i}$,
   where c_{ij} are some positive constants, $i = 1, 2$, $j = 1, 2, 3, 4$;
(2) the norms of matrices C_{ij} in system (2.4.17) must satisfy the inequality (see Abdullin, Anapolskii, et al. [1, p. 106])

(2.4.26)
$$\|C_{12}\|\|C_{21}\| < \left(\frac{c_{11}c_{21}}{c_{12}c_{22}}\right)^{1/2} \left(\frac{c_{13}c_{23}}{c_{14}c_{24}}\right),$$

where $\left(\frac{c_{13}c_{23}}{c_{14}c_{24}}\right) \le 1$.

We note that this inequality is refined as compared with the one obtained firstly by Bailey [1].

The constants $c_{11}, \ldots, c_{24}$ for functions (2.4.25) and system (2.4.17) with matrices (2.4.23), and (2.4.24) take the values

$$c_{11} = 1.08, \quad c_{21} = 0.115, \quad c_{12} = 2.14,$$

$$c_{22} = 2.14, \quad c_{22} = 1.135, \quad c_{13} = c_{23} = 1, \quad c_{14} = 4.83, \quad c_{24} = 2.4.$$

Condition (2.4.26) requires that $\|C_{12}\|\|C_{21}\| < 0.0184$ whereas for system (2.4.17), (2.4.23), and (2.4.24) we have

$$\|C_{12}\|\|C_{21}\| = 1.75.$$

Thus, the Bailey theorem turns out to be nonapplicable to this system and the condition (2.4.28) is "too sufficient" for the property of stability.

2.5 Overlapping Decomposition and Matrix-Valued Function Construction

The purpose of this section is to adopt the method of matrix-valued function construction presented in the previous section to the case of extension (via overlapping decomposition) of the dynamical system.

2.5.1 Dynamical system extension

Consider the dynamical system

(2.5.1)
$$\frac{dx}{dt} = Ax, \quad x(t_0) = x_0,$$

where $x(t) \in R^n$, $t \in R_+$, A is $n \times n$-constant matrix. Vector x is divided into three subvectors x_i, $i = 1, 2, 3$, so that $x = (x_1^T, x_2^T, x_3^T)^T \in R^n$ and $x_i \in R^{n_i}$, $n = n_1 + n_2 + n_3$.

Matrix A of system (2.5.1) is represented in the block form

(2.5.2)
$$A = \begin{pmatrix} A_{11} & A_{12} & A_{13} \\ A_{21} & A_{22} & A_{23} \\ A_{31} & A_{32} & A_{33} \end{pmatrix},$$

where submatrices $A_{ij} \in R^{n_i \times n_j}$ for all $i, j = 1, 2, 3$. Further three components of vector x, are transformed into two components of vector y according to the rule: $y_1 = (x_1^T, x_2^T)^T$, and $y_2 = (x_2^T, x_3^T)^T$. Besides $y = (y_1^T, y_2^T)^T \in R^{\tilde{n}}$, where $\tilde{n} = n_1 + 2n_2 + n_3$.

By means of linear nondegenerate transform

$$(2.5.3) \qquad\qquad y = Tx,$$

where T is $\tilde{n} \times n$-matrix of the form

$$T = \begin{pmatrix} I_1 & 0 & 0 \\ 0 & I_2 & 0 \\ 0 & I_2 & 0 \\ 0 & 0 & I_3 \end{pmatrix},$$

I_1, I_2, and I_3 are identity matrices whose dimensions correspond to the dimensions of subvectors x_1, x_2, and x_3 of vector x. We reduce system (2.5.1) to the form (see Ikeda and Šiljak [1])

$$(2.5.4) \qquad \begin{aligned} \frac{dy_1}{dt} &= \tilde{A}_{11} y_1 + \tilde{A}_{12} y_2, \\ \frac{dy_2}{dt} &= \tilde{A}_{21} y_1 + \tilde{A}_{22} y_2. \end{aligned}$$

Here

$$\tilde{A}_{22} = \begin{pmatrix} A_{11} & A_{12} \\ A_{21} & A_{22} \end{pmatrix}, \quad \tilde{A}_{12} = \begin{pmatrix} 0 & A_{13} \\ 0 & A_{23} \end{pmatrix},$$

$$\tilde{A}_{22} = \begin{pmatrix} A_{22} & A_{23} \\ A_{32} & A_{33} \end{pmatrix}, \quad \tilde{A}_{21} = \begin{pmatrix} A_{21} & 0 \\ A_{31} & 0 \end{pmatrix}.$$

Designate $\tilde{A} = (\tilde{A}_{ij})$; $i, j = 1, 2$ and note that

$$(2.5.5) \qquad\qquad \tilde{A} = TAT^I + M,$$

where
(2.5.6)

$$T^I = \begin{pmatrix} I_1 & 0 & 0 & 0 \\ 0 & \frac{1}{2}I_2 & \frac{1}{2}I_2 & 0 \\ 0 & 0 & 0 & I_3 \end{pmatrix}, \quad M = \begin{pmatrix} 0 & \frac{1}{2}A_{12} & -\frac{1}{2}A_{12} & 0 \\ 0 & \frac{1}{2}A_{22} & -\frac{1}{2}A_{22} & 0 \\ 0 & -\frac{1}{2}A_{22} & \frac{1}{2}A_{22} & 0 \\ 0 & -\frac{1}{2}A_{32} & \frac{1}{2}A_{32} & 0 \end{pmatrix}.$$

The notion of the extended system used in conjunction with the Liapunov vector function was first introduced by Ikeda and Šiljak [1].

Definition 2.5.1. System (2.5.4) is an *extension of system* (2.5.1) if there exists a linear transformation of maximal rank (2.5.3) such that for $y_0 = Tx_0$

$$(2.5.7) \qquad x(t, x_0) = T^I y(t, y_0), \quad t \geq t_0,$$

where $y(t, y_0)$ is solution of system (2.5.4).

It is proved (see Ikeda and Šiljak [1]), that system (2.5.4) is an extension of (2.5.1) in the sense of Definition 2.5.1 iff one of conditions $MT = 0$ or $T^I M = 0$ is satisfied.

Remark 2.5.1. It is noted by Ikeda and Šiljak [1], and Šiljak [2] that the extension procedure of system (2.5.1) or of a general nonlinear system can result in the applicability of the vector Liapunov function while its immediate application to the initial system is difficult or impossible. Below we cite an example showing that this is not a common case, i.e. there exist systems of (2.5.1) type to which it is impossible to apply the method of the vector Liapunov function even after their extension in the sense of Definition 2.5.1 to the form (2.5.4).

2.5.2 Liapunov matrix-valued function construction

We select out of the extended system (2.5.4) two independent subsystems

$$(2.5.8) \qquad \begin{aligned} \frac{dz_1}{dt} &= \tilde{A}_{11} z_1, \quad z_1(t_0) = z_{10}, \\ \frac{dz_2}{dt} &= \tilde{A}_{22} z_2, \quad z_2(t_0) = z_{20} \end{aligned}$$

and assume that for the given sign-definite matrices G_{11} and G_{22} the algebraic Liapunov equations

$$(2.5.9) \qquad \tilde{A}_{11}^{\mathrm{T}} P_{11} + P_{11} \tilde{A}_{11} = G_{11},$$

$$(2.5.10) \qquad \tilde{A}_{22}^{\mathrm{T}} P_{22} + P_{22} \tilde{A}_{22} = G_{22}$$

have the solutions in the form of sign-definite symmetric matrices P_{11} and P_{22} respectively.

Let matrices P_{11} and P_{22} be determined by equations (2.5.9) and (2.5.10). Assume that there exist constants $\eta_1, \eta_2 > 0$ such that the algebraic equation

$$(2.5.11) \qquad \tilde{A}_{11}^{\mathrm{T}} P_{12} + P_{12} \tilde{A}_{22} = -\frac{\eta_1}{\eta_2} P_{11} \tilde{A}_{12} - \frac{\eta_2}{\eta_1} \tilde{A}_{21}^{\mathrm{T}} P_{22}$$

has bounded matrix P_{12} as its solution.

We shall dwell on some comments on the equations (2.5.9) and (2.5.11). It is known (see Barbashin [2]), that equations (2.5.9), and (2.5.10) allow constructing Liapunov functions for independent subsystems (2.5.8) as quadratic forms with respect to a given derivative.

Let $\lambda_i(\tilde{A}_{11})$ and $\lambda_j(\tilde{A}_{22})$ be the roots of characteristic equations

$$(2.5.12) \qquad \det(\tilde{A}_{11} - \lambda E_1) = 0,$$

$$(2.5.13) \qquad \det(\tilde{A}_{22} - \lambda E_2) = 0.$$

Proposition 2.5.1. *If the roots of characteristic equation (2.5.12) are such that the expressions $\lambda_i(\tilde{A}_{11}) + \lambda_k(\tilde{A}_{11})$ do not vanish for any values of i, and k, then there exist a unique matrix P_{11} satisfying equation (2.5.9) whatever the matrix G_{11}.*

A similar assertion takes place for equation (2.5.10) as well. As applied to equation (2.5.9) the well-known classical Liapunov theorems yields the following result.

Proposition 2.5.2. *If all roots of characteristic equation (2.5.9) have negative real parts, then, given matrix G_{11} of definite sign and the matrix P_{11}, solving equation (2.5.9) is positive definite.*

A similar assertion is true for equation (2.5.10) as well. We recall also one more useful assumption on subsystem (2.5.8) of the extended system (2.5.1).

For the proofs of these assertions see Barbashin [2], Demidovich [1], Hahn [2], and Liapunov [1] in context with the extended system (2.5.4).

Proposition 2.5.3. *If the matrices $\tilde{A}_{11}$ and $-\tilde{A}_{22}$ do not have common eigenvalues, then the algebraic equation (2.5.11) has a unique solution in the form of bounded matrix P_{12}.*

Proposition 2.5.3 follows from Theorem 85.1 of Lankaster [1].

Equations (2.5.9) – (2.5.11) make the basis of the proposed a new method of Liapunov matrix-valued function construction.

Now we construct a two-index system of functions

$$(2.5.14) \qquad U(y) = [v_{ij}(y_1, y_2)], \quad i, j = 1, 2$$

with elements

$$(2.5.15) \qquad v_{11}(y_1) = y_1^{\mathrm{T}} P_{11} y_1, \quad v_{22}(y_2) = y_2^{\mathrm{T}} P_{22} y_2,$$

$$(2.5.16) \qquad v_{12}(y_1, y_2) = v_{21}(y_1, y_2) = y_1^{\mathrm{T}} P_{12} y_2.$$

Here matrices P_{11}, P_{22}, and P_{12} are determined by the equations (2.5.9) – (2.5.11).

For quadratic forms (2.5.15) and bilinear form (2.5.16) the estimates

$$v_{11}(y_1) \geq \lambda_m(P_{11})\|y_1\|^2,$$

(2.5.17)
$$v_{22}(y_2) \geq \lambda_m(P_{22})\|y_2\|^2,$$

$$v_{12}(y_1, y_2) \geq -\lambda_M^{1/2}(P_{12}P_{12}^T)\|y_1\|\|y_2\|,$$

are true, where $\lambda_m(P_{11})$, and $\lambda_m(P_{22})$ are minimal eigenvalues of matrices P_{11} and P_{22}, and $\lambda_M^{1/2}(\cdot)$ is the norm of matrix $P_{12}P_{12}^T$, coordinated with the vector norm.

It is easy to show that the function

(2.5.18) $v(y, \eta) = \eta^T U(y)\eta, \quad \eta \in R_+^2, \quad \eta > 0$

satisfies for all $y \in R^{\tilde{n}}$ the estimate

(2.5.19) $v(y, \eta) \geq u^T H^T C H u,$

where $u = (\|y_1\|, \|y_2\|)^T$, $H = \text{diag}(\eta_1, \eta_2)$, and

$$C = \begin{pmatrix} \lambda_m(P_{11}) & -\lambda_M^{1/2}(P_{12}P_{12}^T) \\ -\lambda_M^{1/2}(P_{12}P_{12}^T) & \lambda_m(P_{22}) \end{pmatrix}.$$

The variation of the total derivative of function (2.5.18) along solutions of system (2.5.4)

$$Dv(y, \eta) = y_1^T(\tilde{A}_{11}^T P_{11} + P_{11}\tilde{A}_{11})y_1\eta_1^2$$

$$+ 2y_1^T[(\tilde{A}_{11}^T P_{12} + P_{12}\tilde{A}_{22})\eta_1\eta_2 + \eta_1^2 P_{11}\tilde{A}_{11} + \eta_2^2\tilde{A}_{22}P_{22}]y_2$$

(2.5.20)
$$+ y_2^T(\tilde{A}_{22}^T P_{22} + P_{22}\tilde{A}_{22})y_2 + \eta_1\eta_2 y_1^T(P_{12}\tilde{A}_{21} + \tilde{A}_{21}^T P_{12}^T)y_1$$

$$+ \eta_1\eta_2 y_2^T(\tilde{A}_{12}^T P_{12} + P_{12}^T\tilde{A}_{12})y_2$$

is estimated in view of equation (2.5.11). Denote by

(2.5.21)
$$\lambda = \lambda_M(\tilde{A}_{11}^T P_{11} + P_{11}\tilde{A}_{11}); \quad \beta = \lambda_M(\tilde{A}_{22}^T P_{22} + P_{22}\tilde{A}_{22});$$

$$\varkappa = \lambda_M(P_{12}\tilde{A}_{21} + \tilde{A}_{21}^T P_{12}^T); \quad \chi = \lambda_M(\tilde{A}_{12}^T P_{12} + P_{12}^T\tilde{A}_{12})$$

the maximal eigenvalues of the corresponding matrices. In view of designations (2.5.21) for all $(y_1, y_2) \in R^{n_1} \times R^{n_2}$

$$Dv(y, \eta)\big|_{(2.5.4)} \leq \lambda \eta_1^2 \|y_1\|^2 + \varkappa \eta_1 \eta_2 \|y_1\|^2 + \beta \eta_2^2 \|y_2\|^2 + \chi \eta_1 \eta_2 \|y_2\|.$$

Hence it follows that for all $(y_1, y_2) \in R^{n_1} \times R^{n_2}$ the inequality

$$(2.5.22) \qquad\qquad Dv(y, \eta) \leq u^T S u$$

holds true, where matrix $S = [\sigma_{ij}]$, $i, j = 1, 2$, has the elements

$$\sigma_{11} = \lambda \eta_1^2 + \varkappa \eta_1 \eta_2, \quad \sigma_{22} = \beta \eta_2^2 + \chi \eta_1 \eta_2, \quad \sigma_{12} = \sigma_{21} = 0.$$

2.5.3 Test for stability of system (2.5.1)

Estimates (2.5.19) of function (2.5.18) and its total derivative (2.5.22) enable us to establish new stability test for system (2.5.1) as follows.

Theorem 2.5.1. *Assume that the perturbed motion equations (2.5.1) are such that the following conditions are satisfied:*

(1) *system (2.5.4) is the extension of system (2.5.1) in the sense of Definition 2.5.1;*

(2) *there exist solutions to algebraic equations (2.5.9) – (2.5.11);*

(3) *in estimate (2.5.19) matrix C is positive definite;*

(4) *in estimate (2.5.22) matrix S is negative semi-definite (negative definite).*

Then the equilibrium state $x = 0$ of system (2.5.1) is stable (asymptotically stable).

Proof. Condition (1) of Theorem 2.5.1 and Theorem 2.11 from Ikeda and Šiljak [1] imply that stability (asymptotic stability) of system (2.5.4) yields the corresponding type of stability of system (2.5.1). Therefore it suffices to study stability of the equilibrium state $y = 0$ of system (2.5.4).

Under condition (2) of Theorem 2.5.1 one can construct the elements (2.5.15), and (2.5.16) of the matrix-valued function (2.5.14), which satisfies estimate (2.5.18). Together with condition (3) of Theorem 2.5.1 this means that (2.5.18) is positive definite. Because of condition (4) of Theorem 2.5.1 the total derivative of function (2.5.18) is negative semi-definite (negative definite). Thus, all conditions of Theorem 2.11 by Ikeda and Šiljak [1] and Theorems 25.1 and 25.2 by Hahn [2] are fulfilled and the equilibrium state $y = 0$ of the extended system possesses a certain type of stability. As noted before this is sufficient for stability of the equilibrium state $x = 0$ of system (2.5.1).

Theorem 2.5.2. *Assume that conditions (1) – (3) of Theorem 2.5.1 are satisfied and there exists a vector $\eta = (\eta_1, \eta_2) > 0$ such that instead of condition (4) of Theorem 2.5.1 the following inequalities are fulfilled*

$$(2.5.23) \qquad \eta_1^2 \left(\tilde{A}_{11}^{\mathrm{T}} P_{11} + P_{11} \tilde{A}_{11} \right) + \eta_1 \eta_2 \left(P_{12} \tilde{A}_{21} + \tilde{A}_{21}^{\mathrm{T}} P_{12}^{\mathrm{T}} \right) < 0,$$

$$(2.5.24) \qquad \eta_2^2 \left(\tilde{A}_{22}^{\mathrm{T}} P_{22} + P_{22} \tilde{A}_{22} \right) + \eta_1 \eta_2 \left(\tilde{A}_{12}^{\mathrm{T}} P_{12} + P_{12}^{\mathrm{T}} \tilde{A}_{12} \right) < 0.$$

Then the equilibrium state $x = 0$ of system (2.5.1) is asymptotically stable.

Proof. Together with conditions (1) – (3) of Theorem 2.5.1 inequalities (2.5.23), and (2.5.24) ensure satisfaction of all conditions of Theorem 2.11 by Ikeda and Šiljak [1] and Theorem 25.2 by Hahn [2]. Hence the assertion of Theorem 2.5.2.

2.5.4 Numerical example

The proposed technique of Liapunov matrix-valued function construction for the extended system is illustrated by the example of the third order system

$$(2.5.25) \qquad \frac{dx}{dt} = \begin{pmatrix} -3 & 4 & 6 \\ 4 & -5 & 4 \\ -10 & -4 & -3 \end{pmatrix} x,$$

where $x = (x_1, x_2, x_3)^{\mathrm{T}}$. Diagonal blocks of the matrix are taken as the matrices of coefficients of independent subsystems of the extended system, i.e. system (2.5.25) is extended to

$$(2.5.26) \qquad \frac{dy}{dt} = \begin{pmatrix} -3 & 4 & 0 & 6 \\ 4 & -5 & 0 & 4 \\ 4 & 0 & -5 & 4 \\ -10 & 0 & -4 & -3 \end{pmatrix} y,$$

where $y = (y_1, y_2)^{\mathrm{T}}$, $y_1 = (x_1, x_2)^{\mathrm{T}}$, $y_2 = (x_2, x_3)^{\mathrm{T}}$ are the state vectors of two second order subsystems

$$(2.5.27) \qquad \frac{dy_1}{dt} = \begin{pmatrix} -3 & 4 \\ 4 & -5 \end{pmatrix} y_1 + \begin{pmatrix} 0 & 6 \\ 0 & 4 \end{pmatrix} y_2,$$

$$(2.5.28) \qquad \frac{dy_2}{dt} = \begin{pmatrix} -5 & 4 \\ -4 & -3 \end{pmatrix} y_2 + \begin{pmatrix} 4 & 0 \\ -10 & 0 \end{pmatrix} y_1.$$

According to the adopted notation we have

$$\tilde{A}_{11} = \begin{pmatrix} -3 & 4 \\ 4 & -5 \end{pmatrix}, \quad \tilde{A}_{22} = \begin{pmatrix} -5 & 4 \\ -4 & -3 \end{pmatrix},$$

$$\tilde{A}_{12} = \begin{pmatrix} 0 & 6 \\ 0 & 4 \end{pmatrix}, \quad \tilde{A}_{21} = \begin{pmatrix} 4 & 0 \\ -10 & 0 \end{pmatrix}.$$

Assume that $P_{11} = P_{22} = E$, $\eta = (1,1)^{\mathrm{T}}$ and take

$$v_{11}(y_1) = y_1^{\mathrm{T}} y_1, \quad v_{22}(y_2) = y_2^{\mathrm{T}} y_2,$$

$$v_{12}(y_1, y_2) = v_{21}(y_1, y_2) = y_1^{\mathrm{T}} P_{12} y_2$$

as the elements of matrix-valued function (2.5.14). Here matrix P_{12} is determined by the equation

$$\begin{pmatrix} -3 & 4 \\ 4 & -5 \end{pmatrix} P_{12} + P_{12} \begin{pmatrix} -5 & 4 \\ -4 & -3 \end{pmatrix} = \begin{pmatrix} -4 & 4 \\ 0 & -4 \end{pmatrix}$$

corresponding to equation (2.5.11). It is easy to verify that $P_{12} = \frac{1}{2}E$, where E is 2×2 identity matrix. Since $v_{11} = \|y_1\|^2$, $v_{22} = \|y_2\|^2$, $v_{12} \geq -\frac{1}{2}\|y_1\|\|y_2\|$, the matrix C in estimate (2.5.22) reads

$$C = \begin{pmatrix} 1 & -1/2 \\ -1/2 & 1 \end{pmatrix}$$

and is positive definite. It is easy to see that conditions (1)–(3) of Theorem 2.5.2 are satisfied and conditions (2.5.23), and (2.5.24) of this theorem have the form

$$\begin{pmatrix} -6 & 8 \\ 8 & -10 \end{pmatrix} + \frac{1}{2} \begin{pmatrix} 8 & -10 \\ -10 & 0 \end{pmatrix} = \begin{pmatrix} -2 & 3 \\ 3 & -10 \end{pmatrix} = S_1,$$

$$\begin{pmatrix} -10 & 0 \\ 0 & -6 \end{pmatrix} + \frac{1}{2} \begin{pmatrix} 0 & 6 \\ 6 & 8 \end{pmatrix} = \begin{pmatrix} -10 & 3 \\ 3 & -2 \end{pmatrix} = S_2.$$

One can easily check that the matrices S_1 and S_2 are negative definite. Consequently, the equilibrium state $y = 0$ of system (2.5.26) is asymptotically stable. Since all conditions of Theorem 2.5.2, are fulfilled, the equilibrium state $x = 0$ of system (2.5.1) possesses the same type of stability.

Remark 2.5.2. Example (2.5.25) with the extension (2.5.26) is the one to which vector Liapunov function is not applicable. This can be verified easily by the method proposed by Šiljak [2] for the proof of vector Liapunov function nonapplicability to the non-extended system in one case. The method proposed by us for Liapunov matrix-valued function construction in conjunction with the overlapping decomposition method enlarges the area of the direct Liapunov method in nonlinear dynamics of systems.

2.6 Exponential Polystability Analysis of Separable Motions

One of applications of the method of the matrix-valued function is the problem on polystability of nonlinear systems with separable motions. In this Section this problem is studied with the aim of establishing various sufficiency conditions for the corresponding motions. Some results are illustrated by examples.

2.6.1 Statement of the problem

Consider a system of differential equations of perturbed motion

$$(2.6.1) \qquad \frac{dx}{dt} = f(t, x), \quad x(t_0) = x_0,$$

where $x \in R^n$, $f \in C(R_+ \times D, R^n)$, $D \subseteq R^n$, and, hence, $f(t, x) = 0$ for all $t \in R_+$ iff $x = 0$. We assume that this equilibrium state is unique for system (2.6.1).

Let us decompose a vector $x \in R^n$ into two subvectors $x_i \in R^{n_i}$, $i = 1, 2$, $n_1 + n_2 = n$, and rewrite system (2.6.1) as follows:

$$(2.6.2) \qquad \frac{dx_i}{dt} = f_i(t, x_1, x_2), \quad x_i(t_0) = x_{i0},$$

where $f \in C(R_+ \times R^{n_i}, R^{n_i})$, $i = 1, 2$.

We use the following notation for norms of vectors:

$$\|x_i\| = \left(\sum_{k=1}^{n_i} x_k^2 \right)^{1/2}, \quad \|x\| = \left(\sum_{s=1}^{n} x_s^2 \right)^{1/2} = \left(\sum_{j=1}^{2} x_j^2 \right)^{1/2}, \quad i = 1, 2.$$

Assume that the right-hand side of system (2.6.1) is continuous in the region $R_+ \times D$, where $D = \{x : \|x_1\| + \|x_2\| \leq H < +\infty\}$, and the right-hand side of system (2.6.2) is continuous in $R_+ \times \mathcal{D}^*$, where $\mathcal{D}^* = \{x : \|x_1\| \leq H,\ 0 < \|x_2\| < +\infty\}$.

If system (2.6.2) is considered in the region $R_+ \times \mathcal{D}^*$, we assume that its solution $x(t; t_0, x_0)$ is x_2-extendable.

Below, we present some definitions, taking into account the results of He and Wang [1], and Martynyuk [13, 14, 18].

Definition 2.6.1. The *equilibrium state* $x = 0$ of system (2.6.1) is called *exponentially x_1-stable (in small)*, if there exists $\lambda > 0$ and, for any $\varepsilon > 0$, one can find $\delta(\varepsilon) > 0$ such that

$$(2.6.3) \qquad \|x_1(t; t_0, x_0)\| \leq \varepsilon \exp[-\lambda(t - t_0)] \quad \text{for all} \quad t \geq t_0$$

if $\|x_0\| < \delta(\varepsilon)$.

Definition 2.6.2. The *equilibrium state* $x = 0$ of system (2.6.1) is called *globally exponentially x_1-stable*, if there exists $\lambda > 0$ and, for any Δ, $0 < \Delta < +\infty$, one can find $K(\Delta) > 0$ such that

$$\|x_1(t; t_0, x_0)\| \leq \varepsilon \exp[-\lambda(t - t_0)] \quad \text{for all} \quad t \geq t_0$$

if $\|x_0\| < \Delta$.

Definition 2.6.3. The *equilibrium state* $x = 0$ of system (2.6.1) is called *exponentially polystable (in small)*, if for positive constants r_1 and r_2 and any $\varepsilon > 0$, there exists $\lambda > 0$ and $\Delta(\varepsilon)$, such that

$$(2.6.4) \qquad \begin{aligned} \|x_1(t; t_0, x_0)\|^{2r_1} + \|x_2(t; t_0, x_0)\|^{2r_2} &\leq \varepsilon \exp[-\lambda(t - t_0)] \\ \text{for all} \quad t &\geq t_0 \end{aligned}$$

if $\|x_0\| < \Delta$.

Definition 2.6.4. The *equilibrium state* $x = 0$ of system (2.6.1) is called *globally exponentially polystable*, if there exists $\lambda > 0$ and, for any Δ, one can find $R(\Delta) > 0$ such that

$$\|x_1(t; t_0, x_0)\|^{2r_1} + \|x_2(t; t_0, x_0)\|^{2r_2} \leq R(\Delta) \exp[-\lambda(t - t_0)] \quad \text{for all} \quad t \geq t_0$$

if $\|x_0\| < \Delta$, $t_0 > 0$.

We study exponential properties of the solution $x = 0$ in the following cases:

Case 1. We study the exponential stability of the solution $x = 0$ with respect to the vector x_1, i.e. the exponential x_1-stability.

Case 2. We study the exponential polystability of the solution $x = 0$.

Remark 2.6.1. The informative part of the notion of polystability in Definitions 2.6.3 and 2.6.4 is, in fact, the difference between the rates of decrease of components of the solution $x(t; t_0, x_0)$ of system (2.6.2).

2.6.2 A method for the solution of the problem

We investigate the exponential properties of the solution $x = 0$ of system (2.6.1) in Cases 1 and 2 by using scalar and matrix-valued Liapunov functions, respectively.

First, consider Case 1. Suppose that a scalar function $v(t, x) \in C(R_+ \times D^*, R_+)$ is associated with system (2.6.1) and $v(t, x_1, x_2) = 0$ for all $t \in R_+$ if $x_1 = 0$.

Theorem 2.6.1. *Assume that the vector function f in system (2.6.1) is continuous in $R_+ \times D^*$ and there exist*

    (i) *functions $v(t, x) \in C(R_+ \times D^*, R_+)$ and functions $\varphi_1, \varphi_2 \in K$ of the same order of magnitude;*

    (ii) *positive constants c and γ_1 such that*

(2.6.5) $$c\|x_1\|^{\gamma_1} \leq v(t, x_1, x_2) \leq \varphi_1(\|x\|),$$

(2.6.6) $$D^+ v(t, x_1, x_2)\big|_{(2.6.2)} \leq -\varphi_2(\|x\|).$$

Then the equilibrium state $x = 0$ of system (2.6.1) is exponentially x_1-stable in small.

Proof. For functions φ_1 and φ_2 satisfying the conditions of Theorem 2.6.1, there exist constants α_1 and β_1 such that

(2.6.7) $$\alpha_1 \varphi_1(r) \leq \varphi_2(r) \leq \beta_1 \varphi_1(r).$$

In view of (2.6.7), it follows from inequalities (2.6.5) and (2.6.6)

$$D^+ v(t, x_1, x_2)\big|_{(2.6.2)} \leq -\alpha_1 v(t, x_1, x_2)$$

and, further,

$$v(t, x(t)) \leq v(t_0, x_0) \exp[-\alpha_1(t - t_0)] \quad \text{for all} \quad t_0 \geq 0.$$

By using the lower bound for the function $v(t, x)$ and inequality (2.6.5), we obtain

$$\|x_1(t; t_0, x_0)\| \leq c^{-1/\gamma_1} \varphi_1^{1/\gamma_1}(\|x_0\|) \exp\left[-\frac{\alpha_1}{\gamma_1}(t - t_0)\right], \quad t \geq t_0.$$

Denote $\lambda = \alpha_1/\gamma_1$. For any $\varepsilon > 0$, we choose $\delta(\varepsilon) = \varphi_1^{-1}(c\varepsilon^{\gamma_1})$. Then we arrive at estimate (2.6.3) if $\|x_0\| < \delta(\varepsilon)$, $t_0 \geq 0$. The theorem is proved.

Theorem 2.6.2. *Suppose that the vector function f in system (2.6.1) is continuous in $R_+ \times R^n$ and there exist*

   (i) *functions $v(t, x) \in C(R_+ \times R^n, R_+)$ and functions $\varphi_1, \varphi_2 \in KR$ of the same order of magnitude;*

  (ii) *positive constants c and γ_1 such that*

(2.6.8)
$$d\|x_1\|^{\gamma_2} \leq v(t, x_1, x_2) \leq \varphi_1(\|x\|),$$
$$D^+ v(t, x_1, x_2)\big|_{(2.6.2)} \leq -\varphi_2(\|x\|).$$

Then the equilibrium state $x = 0$ of system (2.6.1) is globally exponentially x_1-stable.

Proof. As in the proof of Theorem 2.6.1, we obtain the estimate

$$\|x_1(t; t_0, x_0)\| \leq d^{-1/\gamma_2} \varphi_1^{1/\gamma_2}(\|x_0\|) \exp\left[-\frac{\alpha_2}{\gamma_2}(t - t_0)\right], \quad t \geq t_0.$$

Denote $\lambda = \frac{\alpha_2}{\gamma_2}$. For any $0 < \Delta < +\infty$, we find $K(\Delta) = d^{-1/\gamma_2}\varphi_1^{1/\gamma_2}(\Delta)$. Then

$$\|x(t; t_0, x_0)\| \leq K(\Delta) \exp[-\lambda(t - t_0)], \quad t \geq t_0,$$

for $\|x_0\| < \Delta$, $t_0 \geq 0$.

Consider Case 2. For system (2.6.2), we consider the matrix-valued function

(2.6.9)
$$U(t, x) = [v_{ij}(t, x)], \quad i, j = 1, 2$$

the element $v_{ij}(t, x)$ of which satisfy special conditions.

Assumption 2.6.1. *There exist*

(i) *functions* $\varphi_1, \varphi_2 \in K(KR)$ *of the same order of magnitude;*
(ii) *the matrix function (2.6.9) whose elements satisfy the following estimates:*

(a) $\underline{c}_{11}\|x_1\|^{2r_1} \leq v_{11}(t, x_1) \leq \bar{c}_{11}\varphi_1^2(\|x_1\|)$ *for all* $(t, x) \in R_+ \times D$
(for all $(t, x) \in R_+ \times R^n$ *);*
(b) $\underline{c}_{22}\|x_2\|^{2r_2} \leq v_{22}(t, x_2) \leq \bar{c}_{22}\varphi_2^2(\|x_2\|)$ *for all* $(t, x) \in R_+ \times D$
(for all $(t, x) \in R_+ \times R^n$*), here,* $\underline{c}_{ii} > 0$ *and* $\bar{c}_{ii} > 0$, $i = 1, 2$;
(c) $\underline{c}_{12}\|x_1\|^{r_1}\|x_2\|^{r_2} \leq v_{12}(t, x_1, x_2) \leq \bar{c}_{12}\varphi_1(\|x_1\|)\varphi_2(\|x_2\|)$
(d) $v_{12}(t, x_1, x_2) = v_{21}(t, x_1, x_2)$ *for all* $(t, x) \in R_+ \times D$ *(for all*
$(t, x) \in R_+ \times R^n$*), here,* $\underline{c}_{ij} = \underline{c}_{ji}$, $\bar{c}_{ij} = \bar{c}_{ji}$, $i \neq j$, *and* $r_i > 0$,
$i, j = 1, 2$.

Proposition 2.6.1. *Suppose that all conditions of Assumption 2.6.1 are satisfied. Then the function*

$$v(t, x, \eta) = \eta^T U(t, x)\eta$$

with $\eta \in R_+^2$ *satisfies the bilateral inequality*

$$(2.6.10) \qquad u_1^T A_1 u_1 \leq v(t, x, \eta) \leq u_2^T A_2 u_2$$

for all $(t, x) \in R_+ \times D$ *(for all* $(t, x) \in R_+ \times R^n$*). Here,*

$$u_1^T = (\|x_1\|^{r_1}, \|x_2\|^{r_2}), \quad u_2^T = (\varphi_1(\|x_1\|), \varphi_2(\|x_2\|)),$$

$$A_1 = H^T C_1 H, \quad A_2 = H^T C_2 H, \quad H = \text{diag}(\eta_1, \eta_2),$$

$$C_1 = \begin{pmatrix} \underline{c}_{11} & \underline{c}_{12} \\ \underline{c}_{21} & \underline{c}_{22} \end{pmatrix}, \quad C_2 = \begin{pmatrix} \bar{c}_{11} & \bar{c}_{12} \\ \bar{c}_{21} & \bar{c}_{22} \end{pmatrix}.$$

Proof. By substituting inequalities (a) – (c) from Assumption 2.6.1 into the expression

$$v(t, x, \eta) = \sum_{i,j=1}^{2} \eta_i \eta_j v_{ij}(t, \cdot),$$

we get estimate (2.6.10).

Assumption 2.6.2. *There exist*

(i) *functions* ψ_1, $\psi_2 \in K(KR)$ *of the same order of magnitude*

(ii) *functions* $\mu_{ij}(t)$, $i = 1, 2$, $j = 1, 2, \ldots, 10$, *continuous on any finite interval and such that*

(a) $D_t^+ v_{ii}(t, x_i) + (D_{x_i}^+ v_{ii})^\mathrm{T} f_i(t, x_i) \le \mu_{ij}(t)\psi_i^2(\|x_i\|) + r_{i1}(t, \psi)$ *for all* $(t, x) \in R_+ \times D$ *(for all* $(t, x) \in R_+ \times R^n$ *);*

(b) $(D_{x_i}^+ v_{ii})^\mathrm{T} g_i(t, x_1, x_2) \le \mu_{i2}(t)\psi_i^2(\|x_i\|)$
$+ \mu_{i3}(t)\psi_1(\|x_1\|)\psi_2(\|x_2\|) + \mu_{i4}(t)\psi_2^2(\|x_2\|) + r_{i2}(t, \psi)$
for all $(t, x) \in R_+ \times D$ *(for all* $(t, x) \in R_+ \times R^n$ *);*

(c) $D_t^+ v_{12}(t, x_1, x_2) + (D_{x_i}^+ v_{12})^\mathrm{T} f_i(t, x_i) \le \mu_{i5}(t)\psi_i^2(\|x_i\|)$
$+ \mu_{i6}(t)\psi_1(\|x_1\|)\psi_2(\|x_2\|) + \mu_{i7}(t)\psi_2^2(\|x_2\|) + r_{i3}(t, \psi)$
for all $(t, x) \in R_+ \times D$ *(for all* $(t, x) \in R_+ \times R^n$ *);*

(d) $(D_{x_i}^+ v_{12})^\mathrm{T} g_i(t, x_1, x_2) \le \mu_{i8}(t)\psi_i^2(\|x_i\|)$
$+ \mu_{i9}(t)\psi_1(\|x_1\|)\psi_2(\|x_2\|) + \mu_{i10}(t)\psi_2^2(\|x_2\|) + r_{i4}(t, \psi)$
for all $(t, x) \in R_+ \times D$ *(for all* $(t, x) \in R_+ \times R^n$ *).*

Here, $f_i(t, x_i) = f_i(t, x_i, x_j)$ for $x_j = 0$, $j = 1, 2$, $g_i(t, x_i, x_j) = f_i(t, x_i, x_j)$ $- f_i(t, x_i)$, $i, j = 1, 2$, and $r_{ik}(t, \psi)$, $i = 1, 2$, $k = 1, 2, 3, 4$, are polynomials in ψ_i, $i = 1, 2$, of degree higher than two.

Proposition 2.6.2. *If all conditions of Assumption 2.6.2 are satisfied, then the following estimate is true for the function* $D^+ v(t, x, \eta)$ *for all* $(t, x) \in R_+ \times D$ *(for all* $(t, x) \in R_+ \times R^n$ *):*

(2.6.11) $\qquad \eta^\mathrm{T} D^+ U(t, x)\eta \le u_3^\mathrm{T}(\|x\|)A_3(t)u_3(\|x\|) + R(t, \psi).$

Here, $u_3^\mathrm{T}(\|x\|) = (\psi_1(\|x_1\|), \psi_2(\|x_2\|)$,

$$R(t, \psi) = \eta_1^2(r_{11}(t, \psi) + r_{12}(t, \psi)) + \eta_2^2(r_{21}(t, \psi) + r_{22}(t, \psi))$$
$$+ 2\eta_1\eta_2(r_{13}(t, \psi) + r_{14}(t, \psi) + r_{23}(t, \psi) + r_{24}(t, \psi)),$$

and $A_3(t)$ is a 2×2 matrix continuous on every finite interval with elements defined as follows:

$$a_{11}(t) = \eta_1^2(\mu_{11}(t) + \mu_{12}(t)) + \eta_2^2\mu_{22}(t)$$
$$+ 2\eta_1\eta_2(\mu_{15}(t) + \mu_{18}(t) + \mu_{25}(t) + \mu_{28}(t)),$$

$$a_{22}(t) = \eta_2^2(\mu_{21}(t) + \mu_{22}(t)) + \eta_1^2\mu_{14}(t)$$
$$+ 2\eta_1\eta_2(\mu_{17}(t) + \mu_{110}(t) + \mu_{27}(t) + \mu_{210}(t)),$$

$$a_{12}(t) = \frac{1}{2}(\eta_1^2(\mu_{13}(t) + \eta_2^2\mu_{23}(t))$$
$$+ \eta_1\eta_2(\mu_{16}(t) + \mu_{19}(t) + \mu_{26}(t) + \mu_{29}(t)).$$

Assume that the matrix $A_3(t)$ is negative definite for all $t \in R_+ = [0, +\infty)$. Then, for any $\mu \in (0, 1)$, there exists $H(\mu) > 0$ and $\alpha > 0$ such that, for $x \in \Omega(H) \subseteq \mathcal{D}$ and $\Omega(H) = \{x \colon \|x\| < H(\mu)\}$, the estimate

$$(2.6.12) \qquad u_3^T(\|x\|)^T A_3(t) u_3(\|x\|) + |R(t, \psi)| < -\alpha(1 - \mu)v(t, x, \eta)$$

is true for all $t \in R_+$, and estimate (2.6.11) takes the form

$$(2.6.13) \qquad D^+ v(t, x, \eta) \leq -\alpha(1 - \mu)v(t, x, \eta)$$

in the region $(t, x) \in R_+ \times \Omega$.

Let $\|u\| = (u^T u)^{1/2}$ be the Euclidean norm of a vector u in the cone $K = \{u \colon u \geq 0\}$.

The proof of Proposition 2.6.2 is based on the direct application of the estimates from Assumption 2.6.2 to the function $D^+ v(t, x, \eta)$.

Proposition 2.6.3. *The following estimates are true for the quadratic forms $u_1^T A_1 u_1$ and $u_2^T A_2 u_2$:*

$$(2.6.14) \qquad \lambda_m(A_1) u_1^T u_1 \leq u_1^T A_1 u_1 \leq \lambda_M(A_1) u_1^T u_1,$$

$$(2.6.15) \qquad \lambda_m(A_2) u_{20}^T u_{20} \leq u_{20}^T A_{20} u_{20} \leq \lambda_M(A_2) u_{20}^T u_{20},$$

where $u_{20} = (\varphi_1(\|x_{10}\|), \varphi_2(\|x_{20}\|))$.

Proposition 2.6.3 can be proved by standard methods of theory of quadratic forms.

Theorem 2.6.3. *Suppose that the vector function f of system (2.6.1) is continuous in $R_+ \times \Omega$ and*

(i) *the conditions of Assumptions 2.6.1 and 2.6.2 are satisfied;*
(ii) *the matrices A_1 and A_2 are positive definite;*
(iii) *the matrix $A_3(t)$ is negative definite for all $t \in R_+$.*

Then the equilibrium state $x = 0$ of system (2.6.2) is exponentially stable in small.

Proof. It follows from (2.5.13) that

$$(2.6.16) \qquad v(t, x(t), \eta) \leq v(t_0, x_0, \eta) \exp[-\alpha(1 - \mu)(t - t_0)], \quad t \geq t_0.$$

By virtue of Propositions 2.6.1 and 2.6.3, we have

$$u_1^T(t) A_1 u_1(t) \leq u_{20}^T A_2 u_{20} \exp[-\alpha(1 - \mu)(t - t_0)], \quad t \geq t_0,$$

and, further,

(2.6.17)
$$\lambda_m(A_1)u_1^{\mathrm{T}}(t)u_1(t) \leq \lambda_M(A_2)u_{20}^{\mathrm{T}}u_{20}\exp[-\alpha(1-\mu)(t-t_0)],$$
$$t \geq t_0.$$

Denoting $a = \lambda_m^{-1}(A_1)\lambda_M(A_2)$, we rewrite estimate (2.6.17) as follows:

(2.6.18)
$$\|x_1(t)\|^{2r_1} + \|x_2(t)\|^{2r_2} \leq a(\varphi_1^2(\|x_{10}\|)$$
$$+ \varphi_2^2(\|x_{20}\|))\exp[-\alpha(1-\mu)(t-t_0)], \quad t \geq t_0.$$

Since the functions φ_1, $\varphi_2 \in K$ have the same order of magnitude (see condition (i) in Assumption 2.6.1), there exists a function $\varphi \in K$ such that

(2.5.19)
$$\varphi_1^2(\|x_{10}\|) + \varphi_2^2(\|x_{20}\|) \leq \varphi^2(\|x_0\|).$$

Inequality (2.6.18) holds if the following inequality is satisfied:

(2.6.20)
$$\|x_1(t)\|^{2r_1} + \|x_2(t)\|^{2r_2} \leq a\varphi^2(\|x_0\|)\exp[-\alpha(1-\mu)(t-t_0)],$$
$$t \geq t_0.$$

For any $\varepsilon > 0$, we choose $\delta(\varepsilon) = \min\left(H(\mu), \varphi_1^{-1}(a^{-1/2}\varepsilon^{1/2})\right)$ and denote $\lambda = \alpha(1-\mu)$, $0 < \mu < 1$. Then it follows from inequality (2.6.20) that if $\|x_0\| < \delta(\varepsilon)$, $t_0 \geq 0$, then

$$\|x_1(t)\|^{2r_1} + \|x_2(t)\|^{2r_2} \leq \varepsilon\exp[-\alpha(1-\mu)(t-t_0)], \quad t \geq t_0,$$

i.e., the separable motions of system (2.6.2) are exponentially polystable. The theorem is proved.

Theorem 2.6.4. *Suppose that the vector function f of system (2.6.1) is continuous in $R_+ \times R^n$ and*

   (i) *the conditions of Assumptions 2.6.1 and 2.6.2 with functions φ_1, φ_2 $\in KR$ and ψ_1, $\psi_2 \in KR$, respectively, are satisfied;*
   (ii) *for any $\mu \in (0,1)$, inequality (2.6.12) holds for $(t,x) \in R_+ \times R^n$;*
   (iii) *conditions (ii) and (iii) of Theorem 2.6.3 are satisfied.*

Then the equilibrium state $x = 0$ of system (2.6.2) is globally exponentially stable.

Proof. By analogy with the proof of Theorem 2.6.3, we obtain inequality (2.8.20) with the function $\varphi(\|x_0\|) \in KR$. As above, we denote $\lambda =$

$\alpha(1 - \mu)$, $0 < \mu < 1$, and, for any $0 < \Delta < +\infty$, choose $R(\Delta)$ in the form $R(\Delta) = a\varphi^2(\Delta)$. Then the following estimate is true for $\|x_0\| < \Delta$, $t_0 \geq 0$:

$$\|x_1(t)\|^{2r_1} + \|x_2(t)\|^{2r_2} \leq R(\Delta)\exp[-\alpha(1 - \mu)(t - t_0)], \quad t \geq t_0.$$

Theorem 2.6.4 is proved.

The statement below establishes the relationship between the global exponentially x_1-stability of the solution $x = 0$ and other types of stability of this solution.

Theorem 2.6.5. *The equilibrium state $x = 0$ of system (2.6.2) is globally exponentially x_1-stable if and only if it is exponentially x_1-stable in small and globally uniformly asymptotically x_1-stable.*

Proof. Necessity. If the equilibrium state $x = 0$ of system (2.6.2) is globally exponentially x_1-stable, then it is exponentially x_1-stable in small. Definition 2.6.2 implies that

$$(2.6.21) \qquad \|x_1(t; t_0, x_0)\| < M(\Delta) \quad \text{for all} \quad t \geq t_0 \quad \text{and} \quad \|x_0\| < \Delta,$$

where $M(\Delta) = K(\Delta)\Delta$. Inequality (2.6.21) follows from the fact that the global uniform asymptotic x_1-stability implies the uniform x_1-boundedness of the solution $x = 0$. If $\|x_0\| < \delta(\varepsilon)$ for $t \geq t_0$, where $\delta(\varepsilon) = \varepsilon$, then estimate (2.6.3) yields

$$\|x_1(t; t_0, x_0)\| < \varepsilon \quad \text{for all} \quad t \geq t_0$$

because the equilibrium state $x = 0$ is uniformly x_1-stable.

It is easy to show that, for any $\Delta > 0$, $\varepsilon > 0$, and $t_0 \in r_+$, there exists

$$T(\varepsilon, \Delta) = (1/\lambda)\ln(M(\Delta)/\varepsilon)$$

such that

$$(2.6.22) \qquad \|x_1(t; t_0, x_0)\| < \varepsilon \quad \text{for all} \quad t \geq t_0 + T(\delta, \Delta)$$

whenever $\|x_0\| < \Delta$ and $t_0 \geq 0$. Thus, the equilibrium state $x = 0$ of system (2.6.2) is globally uniformly x_1-stable.

Sufficiency. It follows from the exponential x_1-stability of the solution $x = 0$ in small that one can find $\lambda > 0$ for any $\delta > 0$, $0 < \delta \leq r_0 < 1$, and $a > 0$ such that the condition $\|x_0\| \leq \beta$, $t_0 > 0$ implies the estimate

$$(2.6.23) \qquad \|x_1(t; t_0, x_0)\| \leq a\|x_0\|\exp(-\lambda(t - t_0)) \quad \text{for all} \quad t \geq t_0.$$

For any $\varepsilon > 0$, we choose $\delta(\varepsilon) = \varepsilon/2$. Then, for $\|x_0\| < \delta(\varepsilon)$, inequality (2.6.23) yields

$$\|x_1(t; t_0, x_0)\| \leq \varepsilon \exp(-\lambda(t - t_0)) \quad \text{for all} \quad t \geq t_0.$$

Here, $0 < \delta \leq \varepsilon$, $a \geq 1 \geq \varepsilon$, and $S(r_0) = \{x \colon \|x_1\| < r_0, \ 0 < \|x_2\| < \infty\}$. It follows from the condition of global uniform asymptotic x_1-stability in Theorem 2.6.5 that, for any $\Delta > 0$, there exists $M(\Delta) > 0$ such that

$$(2.6.24) \qquad \|x_1(t; t_0, x_0)\| < M(\Delta) \quad \text{for all} \quad t \geq t_0$$

whenever $\|x_0\| < \Delta$. Furthermore, for any $\Delta > 0$, $\varepsilon > 0$, and $t_0 \in R_+$, one can find $T = T(\varepsilon, \Delta) > 0$ such that the condition $\|x_0\| \leq \Delta$ implies the estimate

$$(2.6.25) \qquad \|x_1(t; t_0, x_0)\| < \delta(\varepsilon) \quad \text{for all} \quad t \geq t_0 + T(\varepsilon, \Delta).$$

Let

$$R(\Delta) = \max(M(\Delta) \exp(\lambda T(\varepsilon, \Delta), a).$$

Let us estimate the solution $x_1(t; t_0, x_0)$ for $t_0 \leq t \leq t_0 + T(\varepsilon, \Delta)$ and $t \leq t_0 + T(\varepsilon, \Delta)$, respectively. Assume that $t_0 \leq t \leq t_0 + T(\varepsilon, \Delta)$. Since

$$R(\Delta) \exp[-\lambda(t - t_0)] \geq R(\Delta) \exp[-\lambda T(\varepsilon, \Delta)] = M(\Delta),$$

we have

$$(2.6.26) \qquad \|x_1(t; t_0, x_0)\| \leq R(\Delta) \exp(-\lambda(t - t_0)), \quad t \geq t_0$$

for $\|x_0\| < \Delta$. Let $t \geq t_0 + T(\varepsilon, \Delta)$. Denote $\tilde{x} = x(t_1; t_0, x_0)$. In this case, we have $\|x_1\| < \delta(\varepsilon)$. Estimate (2.6.23) yields

$$(2.6.27) \qquad \|x_1(t; t_0, x_0)\| \leq \varepsilon \exp(-\lambda(t - t_0)), \quad t \geq t_1.$$

Note that, by virtue of the continuity and uniqueness of solutions of system (2.6.2), the following relation is true:

$$x_1(t; t_0, \tilde{x}) = x_1(t; t_1, x(t_1; t_0, x_0)) = x_1(t; t_1, \tilde{x}), \quad t \geq t_1.$$

It is now easy to show that there exists $\lambda > 0$ and, for any $\beta > 0$, one can find $R(\Delta) > 0$ such that

$$(2.6.28) \qquad \|x_1(t; t_0, x_0)\| \leq R(\Delta) \exp(-\lambda(t - t_0)), \quad t \geq t_0,$$

whenever $\|x_0\| < \Delta$ and $t_0 \geq 0$. For $\|x_0\| \leq r_0$, we have estimate (2.6.26). Hence, it remains to consider the case $r_0 \leq \|x_0\| \leq \Delta < +\infty$. For $\|x_0\|/r_0 \geq 1$, we get $K(\Delta) = R(\Delta)/r_0$, and inequality (2.6.28) implies the following estimate:

$$\|x_1(t; t_0, x_0)\| \leq K(\Delta)\|x_0\| \exp(-\lambda(t - t_0)), \quad t \geq t_0.$$

This completes the proof of the theorem.

2.6.3 Autonomous system

Consider the perturbed motion equation

(2.6.29)
$$\frac{dx_1}{dt} = f_1(x_1) + g_1(x_1, x_2),$$
$$\frac{dx_2}{dt} = f_2(x_2) + g_2(x_1, x_2),$$

where $x_1 \in R^{n_1}$, $x_2 \in R^{n_2}$, $x = (x_1^T, x_2^T)^T \in R^n$, $f_1 \in C(\mathcal{D}_1, R^{n_1})$, $f_2 \in C(\mathcal{D}_2, R^{n_2})$, $g_1 \in C(\mathcal{D}_1 \times \mathcal{D}_2, R^{n_1})$, $g_2 \in C(\mathcal{D}_1 \times \mathcal{D}_2, R^{n_2})$. Here $\mathcal{D}_1 = \{x \in R^{n_1}: 0 < \|x_1\| < h_1\}$, $\mathcal{D}_2 = \{x \in R^{n_2}: 0 < \|x_2\| < h_2\}$, $h_1, h_2 = \mathrm{const} > 0$.

Suppose that system (2.6.29) has a continuous solution $x(t, x_0)$ in open neighborhood $\mathcal{S} \subseteq \mathcal{D}_1 \times \mathcal{D}_2$ of the unique equilibrium state $x = 0$ for any $x_0 \in \mathcal{S}$ and its motions are definite and continuous in $(t, x_0) \in I_0 \times \mathcal{S}$, $I_0 \subseteq R_+$, $I_0 \neq 0$, $I_0 = I_0(x_0)$. We shall establish exponential polystability conditions for system (2.6.29) in the sense of Definition 2.6.1 the method of constructive application of the matrix-valued Liapunov function.

We shall formulate some assumptions which are the basis of the proposed method of analysis of exponential polystability of motion.

Assumption 2.6.3. *There exists*

(1) *open connected neighborhood $\mathcal{S}$ of equilibrium state $x = 0$ of system (2.6.29);*

(2) *matrix-valued function $U(x) = [v_{ij}(\cdot)]$, $i, j = 1, 2$, with elements $v_{ii} \in C(\mathcal{D}_i, R_+)$, $v_{ij} \in C(\mathcal{D}_1 \times \mathcal{D}_2, R)$, $i \neq j$;*

(3) *real constants $\bar{c}_{ii} > 0$, $\underline{c}_{ii} > 0$, $\bar{c}_{12}$, $\underline{c}_{12} \in R$, and*

(4) *comparison functions $\varphi_1, \varphi_2 \in K$ such that*

$$\underline{c}_{11}\|x_1\|^{2r_1} \leq v_{11}(x_1) \leq \bar{c}_{11}\varphi_1^2(\|x_1\|)$$

for all $x_1 \in \mathcal{D}_1$,

$$\underline{c}_{22}\|x_2\|^{2r_2} \le v_{22}(x_2) \le \bar{c}_{22}\varphi_2^2(\|x_2\|)$$

for all $x_2 \in \mathcal{D}_2$,

$$\underline{c}_{12}\|x_1\|^{r_1}\|x_2\|^{r_2} \le v_{12}(x_1,x_2) \le \bar{c}_{12}\varphi_1(\|x_1\|)\varphi_2(\|x_2\|)$$

for all $(x_1,x_2) \in \mathcal{D}_1 \times \mathcal{D}_2$, where r_1 and r_2 are positive constants.

Proposition 2.6.4. *If all conditions of Assumption 2.6.3, are satisfied, then for function*

(2.6.30) $$v(x,\eta) = \eta^{\mathrm{T}}U(x)\eta, \quad ,\eta \in R_+^2, \eta > 0$$

the bilateral inequality

(2.6.31) $$u_1 H^{\mathrm{T}}\underline{C}Hu_1 \le v(x,\eta) \le u_2^{\mathrm{T}}H^{\mathrm{T}}\bar{C}Hu_2$$

holds true for all $(x_1,x_2) \in \mathcal{D}_1 \times \mathcal{D}_2$.

Here

$$u_1 = (\|x_1\|^{r_1},\|x_2\|^{r_2})^{\mathrm{T}}, \qquad u_2 = (\varphi_1(\|x_1\|),\varphi_2(\|x_2\|))^{\mathrm{T}},$$

$$H = \mathrm{diag}\,(\eta_1,\eta_2),$$

$$\underline{C} = \begin{pmatrix} \underline{c}_{11} & \underline{c}_{12} \\ \underline{c}_{21} & \underline{c}_{22} \end{pmatrix}, \quad \bar{C} = \begin{pmatrix} \bar{c}_{11} & \bar{c}_{12} \\ \bar{c}_{21} & \bar{c}_{22,} \end{pmatrix},$$

$$\bar{c}_{12} = \bar{c}_{21}, \quad \underline{c}_{12} = \underline{c}_{21}.$$

Assumption 2.6.4. *Assume that*

(1) *conditions (1), (2) and (4) of Assumption 2.6.3 are satisfied;*
(2) *there exist constants* α_{ij}, $i = 1,2$, $j = 1,2,3$, β_{ij}, $i = 1,2$, $j = 1,\ldots,5$ *such that*
   (a) $(D_{x_1}v_{11})^{\mathrm{T}}f_1(x_1) \le \alpha_{11}\psi_1^2(\|x_1\|)$;
   (b) $(D_{x_1}v_{11})^{\mathrm{T}}g_1(x_1,x_2) \le \alpha_{12}\psi_1(\|x_1\|)\psi_2(\|x_2\|) + \alpha_{13}\psi_1^2(\|x_1\|)$;
   (c) $(D_{x_2}v_{22})^{\mathrm{T}}f_2(x_2) \le \alpha_{21}\psi_2^2(\|x_2\|)$;
   (d) $(D_{x_2}v_{22})^{\mathrm{T}}g_2(x_1,x_2) \le \alpha_{22}\psi_1(\|x_1\|)\psi_2(\|x_2\|) + \alpha_{23}\psi_2^2(\|x_2\|)$;
   (e) $(D_{x_1}v_{12})^{\mathrm{T}}f_1(x_1) \le \beta_{11}\psi_1^2(\|x_1\|) + \beta_{12}\psi_1(\|x_1\|)\psi_2(\|x_2\|)$;
   (f) $(D_{x_1}v_{12})^{\mathrm{T}}g_1(x_1,x_2) \le \beta_{13}\psi_1^2(\|x_1\|) + \beta_{14}\psi_1(\|x_1\|)\psi_2(\|x_2\|) + \beta_{22}\psi_2^2(\|x_2\|)$;
   (g) $(D_{x_1}v_{12})^{\mathrm{T}}f_2(x_2) \le \beta_{21}\psi_1(\|x_1\|)\psi_2(\|x_2\|) + \beta_{22}\psi_2^2(\|x_2\|)$;
   (h) $(D_{x_2}v_{12})^{\mathrm{T}}g_2(x_1,x_2) \le \beta_{23}\psi_1^2(\|x_1\|) + \beta_{24}\psi_1(\|x_1\|)\psi_2(\|x_2\|) + \beta_{25}\psi_2^2(\|x_2\|)$
   for all $(x_1,x_2) \in \mathcal{D}_1 \times \mathcal{D}_2$.

Proposition 2.6.5. *If all conditions of Assumption 2.6.4 are satisfied, then for the total derivative of function (2.6.30) along solutions of system (2.6.29) the inequality*

$$(2.6.32) \qquad Dv(x,\eta)\big|_{(2.6.29)} \le u^{\mathrm{T}} S u$$

holds true for all $x \in \mathcal{D}_1 \times \mathcal{D}_2$.

Here $u = (\psi(\|x_1\|), \psi(\|x_2\|))$, and S is 2×2 matrix with elements

$$\sigma_{11} = \eta_1^2(\alpha_{11} + \alpha_{13}) + 2\eta_1\eta_2(\beta_{11} + \beta_{13} + \beta_{23}),$$

$$\sigma_{22} = \eta_2^2(\alpha_{21} + \alpha_{23}) + 2\eta_1\eta_2(\beta_{15} + \beta_{22} + \beta_{25}),$$

$$\sigma_{12} = \eta_1^2\alpha_{12} + \eta_2^2\alpha_{22} + 2\eta_1\eta_2(\beta_{12} + \beta_{14} + \beta_{21} + \beta_{24}).$$

Proof. We omit the proofs of Propositions 2.6.3 and 2.6.4 because they are similar to the known ones (see Martynyuk and Miladzhanov [1], and Djordjević [2]).

Estimates (2.6.31), (2.6.32) are sufficient for formulation of a new test for the presence of exponential polystability of separable motion in system (2.6.29).

Theorem 2.6.6. *Assume that differential equations of perturbed motion (2.6.29) are such that all conditions of Assumptions 2.6.3 and 2.6.4 are satisfied and moreover:*

(1) *$\inf \frac{\psi_1^2(r)+\psi_2^2(r)}{\varphi_1^2(r)+\varphi_2^2(r)} = \alpha > 0$ for all $r \in [0,a)$;*

(2) *in estimate (2.6.31) matrices C and $\bar{C}$ are positive definite;*

(3) *in inequality (2.6.32) matrix S is negative definite.*

Then the equilibrium state $x = 0$ of system (2.6.29) is exponentially polystable in small.

Proof. Designate $\lambda_1 = \lambda_m(H^{\mathrm{T}}\bar{C}H)$, $\lambda_2 = \lambda_M(H^{\mathrm{T}}CH)$ and $\gamma = \lambda_M(S)$, $\gamma < 0$. By condition (1) of Theorem 2.6.6 $\|u\|^2 \ge \|u_2\|^2$ and therefore the sequence of inequalities

$$(2.6.33) \qquad -\|u\|^2 \le -\alpha\|u_2\|^2 \le -\frac{\alpha}{\lambda_1} v(x,\eta)$$

is satisfied. According to (2.6.33) inequality (2.6.32) becomes

$$(2.6.34) \qquad Dv(x,\eta) \le -\Delta v(x,\eta),$$

where $\Delta = -\frac{\alpha\gamma}{\lambda_1} > 0$. From inequality (2.6.34) it is easy to find

$$
\begin{aligned}
\lambda_2\|u_1\|^2 &= \lambda_2(\|x_1\|^{2r_1} + \|x_2\|^{2r_2}) \\
&\leq v(x_0, \eta)\exp[-\Delta(t - t_0)] \\
&\leq \lambda_1(\varphi_1^2(\|x_{10}\|) + \varphi_2^2(\|x_{20}\|))\exp[-\Delta(t - t_0)].
\end{aligned}
$$

(2.6.35)

Since the functions $(\varphi_1, \varphi_2) \in K$, the fact that $\|x_{10}\| \leq \|x_0\|$ and $\|x_{20}\| \leq \|x_0\|$ implies $\varphi_1(\|x_{10}\|) \leq \varphi_1(\|x_0\|)$ and $\varphi_1(\|x_{10}\|) \leq \varphi_1(\|x_0\|)$. Consequently we get from (2.6.35)

$$
\begin{aligned}
\|x_1(t, x_0)\|^{2r_1} + \|x_2(t, x_0)\|^{2r_2} &\leq \frac{\lambda_1}{\lambda_2}(\varphi_1^2(\|x_{10}\|) + \varphi_2^2(\|x_{20}\|)) \\
\times \exp[-\Delta(t - t_0)] &\leq \frac{\lambda_1}{\lambda_2}(\varphi_1^2(\|x_0\|) + \varphi_2^2(\|x_0\|))\exp[-\Delta(t - t_0)]
\end{aligned}
$$

(2.6.36)

for all $t \geq t_0$.

For arbitrary $\varepsilon > 0$ we take $\delta = \delta(\varepsilon) > 0$ according to the formula

$$
\delta(\varepsilon) = \min\left\{\varphi_1^{-1}\left[\left(\frac{\varepsilon\lambda_2}{2\lambda_1}\right)^{1/2}\right], \quad \varphi_2^{-1}\left[\left(\frac{\varepsilon\lambda_2}{2\lambda_1}\right)^{1/2}\right]\right\}.
$$

Besides, from (2.6.36) we get the estimate of separable motions

(2.6.37) $\|x_1(t, x_0)\|^{2r_1} + \|x_2(t, x_0)\|^{2r_2} \leq \varepsilon\exp[-\Delta(t - t_0)]$

for all $t \geq t_0$ whenever $\|x_0\| < \delta$. This proves the theorem.

Example 2.6.1. Let perturbed motion equations be

$$
\begin{aligned}
\frac{dx_1}{dt} &= A_1x_1 + B_1x_1\|x_1\|^{-r_1}\|x_2\|^{r_2}, \quad 0 < r_1 < 1, \\
\frac{dx_2}{dt} &= A_2x_2 + B_2x_2\|x_1\|^{r_1}\|x_2\|^{-r_2}, \quad 0 < r_2 < 1,
\end{aligned}
$$

(2.6.38)

where $x_1 \in R^{n_1}$, $x_2 \in R^{n_2}$, A_i, B_i, $i = 1, 2$, are matrices of corresponding dimensions. In order to use Theorem 2.6.6 we construct the matrix-valued function $U(x) = [v_{ij}(\cdot)]$, $i, j = 1, 2$ with elements:

$$
\begin{aligned}
v_{11}(x_1) &= (x_1^{\mathrm{T}}x_1)^{r_1}, \quad v_{22}(x_2) = (x_2^{\mathrm{T}}x_2)^{r_2}, \\
v_{12}(x_1, x_2) &= v_{21}(x_1, x_2) = \alpha(x_1^{\mathrm{T}}x_1)^{\frac{r_1}{2}}(x_2^{\mathrm{T}}x_2)^{\frac{r_2}{2}},
\end{aligned}
$$

(2.6.39)

where $\alpha = \text{const}$, $|\alpha| < 1$.

Denote $\lambda_1 = \lambda_M(A_1 + A_1^{\mathrm{T}})$, $\beta_1 = \lambda_M(B_1 + B_1^{\mathrm{T}})$, and $\lambda_2 = \lambda_M(A_2 + A_2^{\mathrm{T}})$, $\beta_2 = \lambda_M(B_2 + B_2^{\mathrm{T}})$ are maximal eigenvalues of the corresponding matrices.

It is easy to show that in region $D_1 \times D_2$

$$Dv_{11}(x_1)\big|_{(2.6.38)} \leq r_1 \lambda_1 \|x_1\|^{2r_1} + r_2 \beta_1 \|x_1\|^{r_1} \|x_2\|^{r_2},$$

$$Dv_{22}(x_2)\big|_{(2.6.38)} \leq r_2 \lambda_2 \|x_2\|^{2r_2} + r_1 \beta_1 \|x_1\|^{r_1} \|x_2\|^{r_2},$$

(2.6.40)

$$Dv_{12}(x_2)\big|_{(2.6.38)} \leq \frac{\alpha r_1}{2} \lambda_1 \|x_1\|^{r_1} \|x_2\|^{r_2} + \frac{\alpha r_1}{2} \lambda_1 \|x_2\|^{2r_2}$$

$$+ \frac{\alpha r_2}{2} \lambda_2 \|x_1\|^{r_1} \|x_2\|^{r_2} + \frac{\alpha r_2}{2} \lambda_2 \|x_1\|^{2r_1}.$$

Hence it follows that the variation of total derivative of function $v(x, \eta)$ by virtue of system (2.6.38) is estimated by the inequality

(2.6.41) $$Dv(x, \eta)\big|_{(2.6.38)} \leq \psi^{\mathrm{T}} S \psi,$$

where $\psi = (\|x_1\|^{r_1}, \|x_2\|^{r_2})^{\mathrm{T}}$ and 2×2 matrix S has the elements

$$\sigma_{11} = \eta_1^2 r_1 \lambda_1 + \eta_1 \eta_2 \alpha r_2 \beta_2;$$

$$\sigma_{22} = \eta_2^2 r_2 \lambda_2 + \eta_1 \eta_2 \alpha r_1 \beta_1;$$

$$\sigma_{12} = \sigma_{21} = \eta_1^2 r_1 \beta_1 + \eta_2^2 r_2 \beta_2 + \alpha r_1 \eta_1 \eta_2 \lambda_1 + \alpha r_2 \eta_1 \eta_2 \lambda_2.$$

It is easy to check that all conditions of Theorem 2.6.6 are fulfilled for the function $U(x)$ with elements (4.6.39) if

(2.6.42) $$\sigma_{11} < 0, \quad \sigma_{22} < 0, \quad \sigma_{11}\sigma_{22} - \sigma_{12}^2 > 0.$$

Conditions (2.6.42) are sufficient for exponential polystability of system (2.6.35) motions.

2.6.4 Polystability by the first order approximations

System (2.6.2) is represented as two groups of equations

(2.6.43)

$$\frac{dx_1}{dt} = A(t)x_1 + B(t)x_2 + Y(t, x_1, x_2),$$

$$\frac{dx_2}{dt} = C(t)x_1 + D(t)x_2 + Z(t, x_1, x_2).$$

Here A, B, C and D are matrix functions of t continuous for all $t \in R_+$, and the dimensions of which are coordinated with the dimensions of vectors $x_1 \in R^{n_1}$ and $x_2 \in R^{n_2}$, $n_1 + n_2 = n$. Vector functions Y and Z contain variables x_1, and x_2 in power higher than two and together with linear approximation satisfy existence conditions for solutions of system (2.6.43).

Definition 2.6.5. *State* $x = (x_1^T, x_2^T)^T = 0$ of system (2.6.43) is *poly-stable*, if it is uniformly Liapunov stable and (simultaneously) exponentially x_1-stable, i.e. for any $\varepsilon > 0$ and $t_0 > 0$ one can find numbers $\delta(\varepsilon) > 0$ and $\gamma > 0$ such that for $\|x_0\| < \delta$ the inequalities

$$\|x(t; t_0, x_0)\| < \varepsilon, \quad \|x_1(t; t_0, x_0)\| < \varepsilon \exp[-\gamma(t - t_0)]$$

hold true for all $t \geq t_0$.

Theorem 2.6.7. *Suppose that the perturbed motion equations (2.6.43) are such that:*

(1) *the equilibrium state* $x = (x_1^T, x_2^T)$ *of system*

(2.6.44)
$$\frac{dx_1}{dt} = A(t)x_1 + B(t)x_2,$$
$$\frac{dx_2}{dt} = C(t)x_1 + D(t)x_2$$

is polystable in the sense of Definition 2.6.5;

(2) *vector functions* Y *and* Z *satisfy the conditions*

$$Y(t, 0, 0) = Y(t, 0, x_2) = 0, \quad Z(t, 0, 0) = Z(t, 0, x_2) = 0,$$

$$\frac{\|Y(t, x_1 x_2)\| + \|Z(t, x_1, x_2)\|}{\|x_1\|} \to 0$$

for $\|x_1\| + \|x_2\| \to 0$ *uniformly in* t.

Then the equilibrium state $x = (x_1^T, x_2^T)^T = 0$ of system (2.6.43) is polystable in the sense of Definition 2.6.5.

Proof. If condition (1) of Theorem 2.6.7 is satisfied, it is possible to construct for system (2.6.44) the matrix-valued function $U(t, x)$ and to find vector $\eta \in R_+$ such that the function $v(t, x, \eta) = \eta^T U(t, x)\eta$ for all $t \geq 0$, $\|x\| < +\infty$, will satisfy the conditions

(a) $\|x_1\| \leq v(t, x, \eta) \leq M\|x\|$, $M = \text{const} > 0$.

(b) $|v(t, x', \eta) - v(t, x''\eta)| \leq M\|x' - x''\|$,

(c) $Dv(t, x, \eta)\big|_{(2.6.44)} \leq -\alpha v(t, x, \eta)$.

It is easy to see that for the function $Dv(t, x, \eta)\big|_{(2.6.43)}$ the estimate

(2.6.45) $$Dv(t, x, \eta)\big|_{(2.6.43)} \leq -\alpha v(t, x, \eta) + H(t, x),$$

where $H(t,x) = (\mathrm{grad}\ v, X(t,x))$, $X(t,x) = (Y^{\mathrm{T}}, Z^{\mathrm{T}})^{\mathrm{T}}$.

For $H(t,x)$ in estimate (2.6.45)

$$(2.6.46) \qquad\qquad |H(t,x)| \le \varepsilon M v(t,x,\eta),$$

where $\varepsilon \to 0$ as $\|x\| \to 0$, because of conditions (a), (b) imposed on function $v(t,x,\eta)$ and due to condition (2) of Theorem 2.6.7. If inequality (2.6.46) is true, there exists a β $(0 < \beta < d < +\infty)$ such that in the domain $t \ge 0$, $\|x\| \le \beta$ estimate (2.6.45) becomes

$$(2.6.47) \qquad\qquad Dv(t,x,\eta)\big|_{(2.6.44)} \le -\alpha_1 v(t,x,\eta),$$

where $\alpha_1 = \mathrm{const} > 0$. Note that for arbitrary solution $x(t; t_0, x_0)$ of system (2.6.44) with the initial conditions $t \ge 0$, $\|x_0\| \le \delta$ $(0 < \delta < \beta)$ estimate $\|x(t; t_0, x_0)\| \le \beta$ holds true at least on some interval (t_0, t^*). Therefore due to condition (a) imposed on function $v(t,x,\eta)$ we get from inequality (2.6.47)

$$(2.6.48) \quad \|x_1(t; t_0, x_0)\| \le v(t, x(t; t_0, x_0)) \le M\|x_0\| \exp[-\alpha_1(t - t_0)].$$

Condition (2) of Theorem 2.6.7 and inequality (2.6.48) imply that there exists a constant $\alpha_2 = \mathrm{const} \to 0$ as $\|x\| \to 0$ such that

$$(2.6.49) \qquad\qquad \|X(t, x(t; t_0, x_0))\| \le \alpha_2 \|x_0\| \exp[-\alpha_1(t - t_0)],$$

for all $t \in (t_0, t^*)$.

Let $K(t, \tau)$ be the Cauchy matrix of linear system (2.6.44). It is known that solution $x(t; t_0, x_0)$ of system (2.6.43) can be represented as

$$(2.6.50) \qquad x(t; t_0, x_0) = K(t, t_0)x_0 + \int_{t_0}^{t} K(t, \tau) X(\tau, x(\tau; t_0, x_0))\, d\tau,$$

for all $t \ge t_0$. Since the state $x = (x_1^{\mathrm{T}}, x_2^{\mathrm{T}})^{\mathrm{T}} = 0$ of system (2.6.44) is uniformly Liapunov stable, there exists a constant $N > 0$ such that $\|K(t, t_0)\| \le N$ for all $t \ge t_0$, $t_0 \ge 0$. In view of this fact and estimating (2.6.49) from (2.6.50) we get

$$(2.6.51) \qquad\qquad \|x(t; t_0, x_0)\| \le N(1 + \alpha_1^{-1}\alpha_2)\|x_0\|$$

for all $t \in (t_0, t^*)$.

Let ε be arbitrary small, $0 < \varepsilon < \beta$ so that $\delta < \min\{M^{-1}, [N(1 + \alpha_1^{-1}\alpha_2)]^{-1}\}\varepsilon$. Moreover, estimates (2.6.48) and (2.6.45) yield

(2.6.52)
$$\|x_1(t; t_0, x_0)\| \le \varepsilon \exp[-\alpha_1(t - t_0)],$$
$$\|x(t; t_0, x_0)\| < \varepsilon$$

for all $t \in (t_0, t^*)$.

Inequalities (2.6.52) hold for all values of time for which estimate (2.6.51) takes place. According to the choice ε, $\varepsilon < \beta$, estimate (2.6.52) is fulfilled for all $t \ge t_0$. This proves Theorem 2.6.7.

Note that if in Theorem 2.6.7 condition (2) is replaced by

(2)$'$ in domain $t \ge 0$, $\|x\| < d < +\infty$, for given function $v(t, x, \eta)$ the inequality

$$\|Y(t, x_1, x_2)\| + \|Z(t, x_1, x_2)\| \le \gamma v(t, x, \eta),$$

where $\gamma = \text{const} > 0$, sufficiently small, then the Theorem 2.6.7 remains valid.

Theorem 2.6.7 may be extended to systems more general than (2.6.43). In particular, consider the perturbed motion equations in the form

(2.6.53)
$$\frac{dx_1}{dt} = A(t)x_1 + B(t)x_2 + Y(t, x_1, x_2, x_3),$$
$$\frac{dx_2}{dt} = C(t)x_1 + D(t)x_2 + Z(t, x_1, x_2, x_3),$$
$$\frac{dx_3}{dt} = W(t, x_1, x_2, x_3).$$

In domain $\mathcal{D}_2 = \{t \ge t_0, \ \|x\| \le d < +\infty, \ \|x_3\| < +\infty\}$ we assume that the existence and uniqueness conditions are fulfilled for solutions of system (2.6.53) and other solutions of system (2.6.53) for which x_3 is extendable, i.e. definite for all $t \ge 0$ for which $\|x\| \le d$.

Definition 2.6.6. *The equilibrium state* $y = (x_1^{\mathrm{T}}, x_2^{\mathrm{T}}, x_3^{\mathrm{T}})^{\mathrm{T}} = 0$ *of system* (2.6.53) *is polystable with respect to a part of variables, if it is uniformly* $(x_1^{\mathrm{T}}, x_2^{\mathrm{T}})$*-stable and (simultaneously) exponentially* x_1*-stable, i.e. for any values of* ε, $t \ge t_0$, *there exist numbers* $\delta(\varepsilon) > 0$ *and* $\gamma > 0$ *such that for* $\|x_0\| + \|x_{30}\| < \delta$ *the inequalities*

$$\|x(t; t_0, x_0, x_{30})\| < \varepsilon, \quad \|x_1(t; t_0, x_0, x_{30})\| < \varepsilon \exp[-\gamma(t - t_0)]$$

are fulfilled for all $t \ge t_0$.

The following assertion is proved in the same way as Theorem 2.6.7.

Theorem 2.6.8. *Assume that*

(1) *the equilibrium state* $x = (x_1^T, x_2^T)^T = 0$ *of system* (2.6.44) *is uniformly Liapunov stable and (simultaneously) exponentially x_1-stable;*

(2) *in domain $\mathcal{D}_2$ the conditions*

$$Y(t,0,0,0) = Y(t,0,x_2,x_3) = 0,$$

$$Z(t,0,0,0) = Z(t,0,x_2,x_3) = 0,$$

$$W(t,0,0,0) = 0,$$

$$\frac{\|Y(t,x_1,x_2,x_3)\| + \|Z(t,x_1,x_2,x_3)\|}{\|x_1\|} \to 0$$

are fulfilled for $\|x_1\| + \|x_2\| \to 0$.

Then the equilibrium state $y = (x_1^T, x_2^T, x_3^T)^T$ *of system* (2.6.53) *is polystable with respect to a part of variables.*

Example 2.6.2. Angular motion of a solid with respect to the mass center subjected to the linear moments of forces is described by the equation

$$(2.6.54) \qquad \frac{dx}{dt} = L(t)x + X(x),$$

where $x = (x_1, x_2, x_3) \in R^3$, L is a matrix 3×3 whose elements are continuous for all $t \in R_+$ functions characterizing the action of linear moments, dissipative and accelerating forces, and vector $X(x)$ is of the form

$$X(x) = ((I_2 - I_3)I_1^{-1}x_2x_3, (I_3 - I_1)I_2^{-1}x_2x_3, (I_1 - I_2)I_3^{-1}x_2x_3)^T.$$

Here x_1, x_2, and x_3 are projections of the x-angular velocity vector on main central axes of inertia, I_i are main central moments of inertia.

Assume that the equilibrium state $x = (x_1, x_2, x_3) = 0$ of the linear system

$$(2.6.55) \qquad \frac{dx}{dt} = L(t)x, \qquad x \in R^3$$

is uniformly Liapunov stable and (simultaneously) exponentially (x_1, x_2)-stable. It is easy to verify that for the vector function $X(x)$ the condition (2) of Theorem 2.6.7 is satisfied and therefore, the equilibrium state $x = (x_1, x_2, x_3)^T = 0$ of system (2.6.54) possesses the same properties the linear approximation (2.6.55).

2.7 Integral and Lipschitz Stability

Classical properties of stability and instability of motion in the Liapunov sense were complemented by the consideration of other dynamical properties of solutions of systems of equations of perturbed motion. The concepts of integral stability and uniform Lipschitz stability enlarged the collection of dynamical properties of solutions that can be investigated by the direct Liapunov method. The purpose of this section is to obtain new conditions of stability and uniform Lipschitz stability based on the use of the principle of comparison with a matrix-valued Liapunov function.

2.7.1 Definitions

We shall consider the system

$$(2.7.1) \qquad \frac{dx}{dt} = f(t, x), \quad x(t_0) = x_0, \quad t_0 \in \mathcal{T}_0,$$

where $r \in C(\mathcal{T}_0 \times R^n, R^n)$ and its perturbed system

$$(2.7.2) \qquad \frac{dx}{dt} = f(t, x) + r(t, x), \quad x(t_0) = x_0, \quad t_0 \in \mathcal{T}_0,$$

where $r(t, x) \neq 0$ for $x = 0$.

Definition 2.7.1. The *equilibrium state* $x = 0$ of system (2.7.1) is called

(i) *integrally stable with respect to* $\mathcal{T}_i \subseteq R$, if for any $\varepsilon > 0$ and $t_0 \in \mathcal{T}_i$, there exist positive functions $\delta_1(t_0, \varepsilon)$ and $\delta_2(t_0, \varepsilon)$ such that, for any solution $x(t; t_0, x_0)$ of perturbed system (2.7.2), the inequality

$$\|x(t; t_0, x_0)\| < \varepsilon \quad \text{for all} \quad t \in \mathcal{T}_0$$

holds for

$$\|x_0\| < \delta_1 \quad \text{and} \quad \int_{t_0}^{\infty} \sup_{\|x\| \leq \varepsilon} \|r(s, x)\| \, ds < \delta_2, \quad t_0 \in \mathcal{T}_i;$$

(ii) *uniformly integrally stable with respect to* $\mathcal{T}_i$ if condition (i) is satisfied and, for any $\varepsilon > 0$, the corresponding maximal value δ_i satisfies the condition

$$\inf[\delta_{iM}(t, \varepsilon) : t \in \mathcal{T}_i] > 0, \quad i = 1, 2.$$

The words *"with respect to $\mathcal{T}_i$"* in Definition 2.7.1 can be omitted if and only if $\mathcal{T}_i = R$.

Parallel with comparison equation

(2.7.3) $$\frac{du}{dt} = g(t, x, u), \quad u(t_0) = u_0 \geq 0,$$

we consider the perturbed extended comparison equation

(2.7.4) $$\frac{du}{dt} = g(t, x, u) + \psi(t), \quad u(t_0) = u_0 \geq 0,$$

where $\psi(t) \in C(\mathcal{T}_0, \mathcal{T}_0)$.

Definition 2.7.2. The *solution* $u = 0$ of (2.7.3) is called

(i) *integrally stable with respect to* $\mathcal{T}_i$, if for any $\varepsilon^* > 0$ and $t_0 \in \mathcal{T}_i$, there exist positive functions $\delta_1^*(t_0, \varepsilon^*)$ and $\delta_2^*(t_0, \varepsilon^*)$ such that any solution of the perturbed equation (2.7.4) satisfies the inequality $u(t; t_0, u_0) < \varepsilon^*$ for all $t \in \mathcal{T}_0$ for

$$u_0 < \delta_1^* \quad \text{and} \quad \int_{t_0}^{\infty} \psi(s)\,ds < \delta_2^*, \quad t_0 \in \mathcal{T}_i;$$

(ii) *uniformly integrally stable with respect to* $\mathcal{T}_i$, if condition (i) is satisfied and, for any $\varepsilon^* > 0$, the corresponding maximal value δ_i^* satisfies the condition

$$\inf[\delta_{iM}^*(t, \varepsilon^*) \colon t \in \mathcal{T}_i] > 0, \quad i = 1, 2.$$

The words *"with respect to $\mathcal{T}_i$"* in Definition 2.7.2 can be omitted if and only if $\mathcal{T}_i = R$.

2.7.2 Sufficient conditions for integral and asymptotic integral stability

Together with the matrix-valued function $U(t, x)$ we consider matrices and comparison functions possessing the following properties:

(H_1) $A_i(y)$, $i = 1, 2$, are nonsingular $(m \times m)$-dimensional positive definite matrices with constant elements;

(H_2) $a_k(t, \|x\|)$, $k = 1, 2, \ldots, s$, $a_k \in C(\mathcal{T}_0 \times R_+, R_+)$, $a_k(t, 0) = 0$ are monotonically increasing with respect to $t \in \mathcal{T}_0$ for any fixed w, $a_k(t, w) > 0$ for $w > 0$;

(H_3) $c_k(t, \|x\|)$, $k = 1, 2, \ldots, s$, $c_k \in C(\mathcal{T}_0 \times R_+, R_+)$, $c_k(t, 0) = 0$ are monotonically increasing with respect to $t \in \mathcal{T}_0$ for any fixed w, $c_k(t, w) > 0$ for $w > 0$.

Proposition 2.7.1. *In order that function* $v(t, x, y) = y^T U(t, x) y$ *be positive definite and decreasing on* $T_0 \times S$ *(on* $T_0 \times R^n$*),* $S \subseteq R^n$ *, it is sufficient that there exist matrices and comparison functions with properties* $(H_1) - (H_2)$ *such that*

$$a^T(t, \|x\|) A_1(y) a(t, \|x\|) \leq v(t, x, y) \leq c^T(t, \|x\|) A_2(y) c(t, \|x\|)$$

$$\text{for all} \quad (t, x) \in T_0 \times S \quad (\text{for all} \quad (t, x) \in T_0 \times R^n),$$

where the matrices $A_i(y)$, $i = 1, 2$, *are positive definite for all* $(y \neq 0) \in R^s$.

Proof. Let $\lambda_m(A_1)$ and $\lambda_M(A_2)$ be the minimal and maximal eigenvalues of the matrices A_1 and A_2, respectively, for $(y \neq 0) \in R^s$. It follows from the properties of the functions $a_k(t, w)$ and $c_k(t, w)$ that there exist functions $\alpha(t, w)$ and $\beta(t, w)$ with the same properties for which

(2.7.5) $a^T(t, w) a(t, w) \geq \alpha(t, w)$ for all $t \in T_0$, $w > 0$,

(2.7.6) $c^T(t, w) c(t, w) \leq \beta(t, w)$ for all $t \in T_0$, $w > 0$.

Taking inequalities (2.7.5) and (2.7.6) into account, for function $v(t, x, y)$ we obtain the two-sided estimate

$$\lambda_m(A_1) \alpha(t, w) \leq v(t, x, y) \leq \lambda_M(A_2) \beta(t, w)$$

for all $t \in T_0$, $w > 0$, $(y \neq 0) \in R^s$. This implies the statement of Proposition 2.7.1.

Theorem 2.7.1. *Assume that the equations of perturbed motion* (2.7.2) *are such that*

(i) *there exists a matrix-valued function* $U \in C(T_0 \times R^n, R^{s \times s})$ *that satisfies the Lipschitz condition locally in* x *and also a matrix* $A_1(y)$ *and a comparison function* $a_k(t, w)$ *with properties* (H_1) *and* (H_2) *for which the matrix* $A_1(y)$ *is positive definite;*

(ii) *there exists a generalized function* $g(t, x, v(t, x, y))$ *such that*

(2.7.7) $D^+ v(t, x, y)\big|_{(2.7.1)} \leq g(t, x, v(t, x, y))$

for all $(t, x, y) \in T_0 \times S \times R^s$ *(for all* $(t, x, y) \in T_0 \times R^n \times R^s$*);*

(iii) *the trivial solution* $u = 0$ *of the comparison equation* (2.7.3) *is integrally stable.*

Then the trivial solution $x = 0$ of system (2.7.1) is integrally stable.

Proof. Assume that $\varepsilon \in (0, H)$ and $t_0 \in \mathcal{T}_0$ are given. Let us calculate $\varepsilon^* = \lambda_m(A_1)\alpha(t_0, \varepsilon)$ and, for ε^* and $t_0 \in \mathcal{T}_0$, choose values $\delta_1 = \delta_1(t_0, \varepsilon^*) = \delta_1(t_0, \lambda_m(A_1)\alpha(t_0, \varepsilon))$ and $\delta_2 = \delta_2(t_0, \lambda_m(A_1)\alpha(t_0, \varepsilon))$ so that the solution $u(t; t_0, x_0)$ of equation (2.7.3) satisfies the inequality

$$(2.7.8) \qquad u(t; t_0, x_0) < \varepsilon^* \quad \text{for all} \quad t \geq t_0$$

for

$$(2.7.9) \qquad u_0 < \delta_1 \quad \text{and} \quad \int\limits_{t_0}^{\infty} \psi(s)\, ds < \delta_2, \quad t_0 \in \mathcal{T}_i.$$

Since the function $v(t, x, y)$ is continuous on $\mathcal{T}_0 \times S \times R^s$, one can take a value $\delta_3 = \delta_3(t_0, \lambda_m(A_1)\alpha(t_0, \varepsilon))$ such that

$$(2.7.10) \qquad v(t_0, x_0, y) < \delta_1 \quad \text{for} \quad \|x_0\| < \delta_3.$$

Let $x(t) = x(t; t_0, x_0)$ be a solution of system (2.7.2) with initial values $(t_0, x_0) \in \mathcal{T}_0 \times B_{\delta_3}$. Since the function $v(t, x, y)$ satisfies the Lipschitz condition locally with respect to x, there exists a constant $L > 0$ such that the inequality

$$(2.7.11) \qquad D^+ v(t, x, y)\big|_{(2.7.2)} \leq g(t, x, v(t, x, y)) + L\|r(t, x)\|$$

holds in addition to condition (ii) of Theorem 2.7.1 for any solution $x(t)$ which exists for all $t \geq t_0$.

Parallel with inequality (2.7.11) we consider the comparison equation

$$(2.7.12) \qquad \frac{du}{dt} = g(t, x, u) + \sigma(t), \quad u(t_0) = u_0 \geq 0,$$

where $\sigma(t) = L\|r(t, x)\|$ and $u_0 = v(t_0, x_0, y)$. By Proposition 1.5.1 for the function $v(t, x, y)$ and the maximal solution $w(t; t_0, x_0)$ of equation (2.7.12) on the interval of existence of both functions, the estimate $v(t, x(t), y) \leq w(t; t_0, u_0)$ holds. Let us calculate $\delta_4(t_0, \varepsilon) = \delta_2(t_0, \lambda_m(A_1)\alpha(t_0, \varepsilon))/L$ and assume that there exists $t_1 \in \mathcal{T}_0$ such that

$$(2.7.13) \qquad \|x(t_1; t_0, x_0)\| = \varepsilon \quad \text{and} \quad \|x(t; t_0, x_0)\| \leq \varepsilon \quad \text{for} \quad t \in [t_0, t_1]$$

if

$$\int_{t_0}^{\infty} \sup_{\|x\|<\varepsilon} \|r(t,x)\|\, dt < \delta_4.$$

For $t \in [t_0, t_1]$, we introduce a function $\sigma^*(t) = L\|r(t,x)\|$. Then

$$\int_{t_0}^{t} \sigma^*(t)\, dt = \int_{t_0}^{t_1} L\|r(t,x)\|\, dt$$

$$\leq \int_{t_0}^{t_1} \sup_{\|x\|<\varepsilon} \|r(t,x)\|\, dt < L\delta_4(t_0,\varepsilon) = \delta_2(t_0,\varepsilon).$$

The function $\sigma^*(t)$ can be extended by continuity for all $t \geq t_0$ so that

(2.7.14) $$\int_{t_0}^{t} \sigma^*(t)\, dt < \delta_2(t_0,\varepsilon).$$

Denote by $w^*(t; t_0, u_0)$ the maximal solution of the comparison equation

(2.7.15) $$\frac{dw}{dt} = g(t,x,w) + \sigma^*(t), \qquad w(t_0) = u_0.$$

It is clear that, for $t \in [t_0, t_1]$, the equality $w^*(t; t_0, u_0) = w(t; t_0, u_0)$ holds because $\sigma(t)$ and $\sigma^*(t)$ are identically equal to zero on this interval. Furthermore,

(2.7.16) $$w^*(t_1; t_0, u_0) = w(t_1; t_0, u_0).$$

According to conditions (2.7.9) and (2.7.14) the solution $u = 0$ of equation (2.7.3) is integrally stable, whence

(2.7.17) $$w^*(t; t_0, u_0) < \varepsilon^* \quad \text{for all} \quad t \geq t_0.$$

Proposition 2.7.1 implies that

$$\lambda_m(A_1)\alpha(t_0,\varepsilon) = \lambda_m(A_1)\alpha(t_0, \|x(t_1; t_0, x_0)\|)$$
$$< \lambda_m(A_1)\alpha(t_1, \|x(t_1; t_0, x_0)\|) \leq v(t, x(t), y)$$
$$\leq w(t_1; t_0, u_0) < w^*(t_1; t_0, u_0) < \varepsilon^*.$$

The obtained relations contradict the assumption that there exists $t_1 \in \mathcal{T}_0$ for which condition (2.7.14) is satisfied. Consequently, the solution $x = 0$ of system (2.7.1) is integrally stable.

Corollary 2.7.1. (Kudo [1]). *Assume that for system (2.7.1) there exist functions $V(t, x)$ and $a(t, u)$ satisfying the following properties:*

(i) $V(t, x) \in C(R_+ \times R^n, R_+)$, $V(t, x)$ is Lipschitzian in x for a constant $M > 0$;

(ii) $a(t, \|x\|) \leq V(t, x)$, where $a \in C(R_+ \times R_+, R_+)$ and monotone increasing with respect to t for each fixed u and $a(t, u) > 0$ for $u \neq 0$;

(iii) $D^+ V(t, x)\big|_{(2.7.1)} < g(t, V(t, x))$ for all $(t, x) \in R_+ \times R^n$, where $g \in C(R_+ \times R_+, R)$, $g(t, 0) \equiv 0$.

Then, the integral stability of the trivial solution $u = 0$ of

$$(2.7.18) \qquad \frac{du}{dt} = g(t, u), \qquad u(t_0) = u_0 \geq 0$$

implies the integral stability of the state $x = 0$ of system (2.7.1).

Theorem 2.7.2. *Assume that conditions (i) and (ii) of Theorem 2.7.1 are satisfied and trivial solution $u = 0$ of the extended equation (2.7.4) is uniformly integrally stable. Then the trivial solution $x = 0$ of system (2.7.1) is uniformly integrally stable.*

Proof. Let us calculate $\tilde{\varepsilon} = \lambda_m(A_1)\alpha(0, \varepsilon)$ for a given $\varepsilon \in (0, H)$. It follows from the uniform integral stability of the solution $u = 0$ of equation (2.7.3) that there exist numbers $\delta_1 = \delta_1(\tilde{\varepsilon})$ and $\delta_2 = \delta_2(\tilde{\varepsilon})$ such that any solution $u(t; t_0, u_0)$ of equation (2.7.4) satisfies the inequality $u(t; t_0, u_0) < \tilde{\varepsilon}$ for all $t \geq t_0$, provided that

$$u_0 < \delta_1(\tilde{\varepsilon}) \quad \text{and} \quad \int_{t_0}^{\infty} \psi(s)\, ds < \delta_2(\tilde{\varepsilon}), \quad t_0 \in \mathcal{T}_i.$$

Let us introduce constants $\delta_3(\tilde{\varepsilon}) = \delta_1(\tilde{\varepsilon})/L$ and $\delta_2(\tilde{\varepsilon}) = \delta_2(\tilde{\varepsilon})/L$, and assume that there exists $t_1 \geq t_0$ such that

$$(2.7.19) \qquad \|x(t_1; t_0, x_0)\| = \varepsilon, \quad \|x(t; t_0, x_0)\| \leq \varepsilon, \quad \text{for all} \quad t \in [t_0, t_1]$$

if

$$(2.7.20) \qquad \|x_0\| < \delta_3(\tilde{\varepsilon}) \quad \text{and} \quad \int_{t_0}^{t_1} \sup_{\|x\| < \varepsilon} \|r(t, x(t))\|\, dt < \delta_4(\tilde{\varepsilon}).$$

Proposition 2.7.2 and the conditions of Theorem 2.7.2 imply that

$$\lambda_m(A_1)\alpha(0,\varepsilon) < \lambda_m(A_1)\alpha(t_1,\varepsilon) \le V(t,x(t_1),y)$$
$$\le w(t_1;t_0,u_0) = w^*(t_1;t_0,u_0) < \widetilde{\varepsilon}.$$

The relation obtained contradicts the assumption that there exists $t_1 \ge t_0$ for which (2.7.19) is true. Hence, the solution $x = 0$ of system (2.7.1) is uniformly integrally stable.

Corollary 2.7.2. (Kudo [1]). *Under the assumption of Corollary 2.7.1 the uniformly integral stability of trivial solution $u = 0$ of (2.7.3) assures the uniformly integral stability of the trivial solution $x = 0$ of (2.7.1).*

2.7.3 Uniform Lipschitz stability

Taking the results of Dannan and Elaydi [1,2], and Kudo [2] into account, we formulate the following definition.

Definition 2.7.3. The *equilibrium state* $x = 0$ of system (2.7.1) is *uniformly stable in the Lipschitz sense with respect to $\mathcal{T}_i$*, if there exist constants $M \ge 1$ and $\delta > 0$ such that $\|x(t;t_0,x_0)\| \le M\|x_0\|$ for all $t_0 \in \mathcal{T}_i$ and $t \in \mathcal{T}_0$ for $\|x_0\| < \delta$.

The words *"with respect to $\mathcal{T}_i$"* in Definition 2.7.3 can be omitted if and only if $\mathcal{T}_i = R$.

Definition 2.7.4. The *solution* $u = 0$ of equation (2.7.3) is *equi-stable in the Lipschitz sense with respect to $\mathcal{T}_i$*, if for given $\varepsilon \in (0,H)$, $M \ge 1$ and $t_0 \in \mathcal{T}_i$, there exist constants $\delta_1(t_0,\varepsilon) > 0$ and $\delta_2(t_0,\varepsilon) > 0$ such that $u(t;t_0,x_0,u_0) < Mu_0$ on any interval $t_0 \le t \le t_1$ on which $\|x(t;t_0,x_0)\| \le \varepsilon$ for $\|x_0\| < \delta_1$ and $u_0 < \delta_2(t_0,\varepsilon)$.

The words *"with respect to $\mathcal{T}_i$"* in Definition 2.7.4 can be omitted if and only if $\mathcal{T}_i = R$.

Theorem 2.7.3. *Assume that the equations (2.7.1) are such that there exist*

(i) *a matrix-valued function $U \in C(\mathcal{T}_0 \times R^n, R^{s \times s})$ satisfying the Lipschitz condition locally with respect in x;*

(ii) *matrices $A_1(y)$ and $A_2(y)$, $(y \ne 0) \in R^s$, and a comparison function $a \in C(R_+ \times R_+, R^s)$ such that Proposition 2.7.1 holds and, $a(t,0) =$*

0, $a(t, w) > 0$ for $w \neq 0$, $ka(t, s) \leq a(t, ks)$ and $kc(t, s) \leq c(t, ks)$ for a certain constant k, moreover, if $a(t, q) \leq c(t, s)$, then $q \leq s$;

(iii) a generalized majorizing function $g \in C(R_+ \times R^n \times R_+, R)$ such that $D^+v(t, x, y)\big|_{(2.7.1)} \leq g(t, x, v(t, x, y))$ for all $(t, x, y) \in T_0 \times S \times R^s$.

In this case, if a trivial solution of equation (2.7.3) is equi-stable in the Lipschitz sense with respect to T_i, then the equilibrium state $x = 0$ of system (2.7.1) is uniformly stable in the Lipschitz sense with respect to T_i.

Proof. Let $\varepsilon \in (0, H)$ and $t_0 \in T_i$ be given. The Lipschitz equistability of the solution $u = 0$ of equation (2.7.3) with respect to T_i implies that there exist $\delta_1(t_0, \varepsilon) > 0$, $\delta_2(t_0, \varepsilon) > 0$ and $M \geq 1$ such that $u(t; t_0, x_0, u_0) < Mu_0$ on any interval $[t_0, t_1]$ on which $\|x(t; t_0, x_0)\| < \varepsilon$, provided that $\|x_0\| < \delta_1(t_0, \varepsilon$ and $u_0 < \delta_2(t_0, \varepsilon)$. We take $u_0 = y^{\mathrm{T}}U(t_0, x_0)y$ and $\delta^* = \delta^*(t_0, \varepsilon) > 0$ such that

$$(2.7.21) \qquad \lambda_M(A_2)\beta(t_0, \delta^*) < \delta_1(t_0, \varepsilon).$$

Let $\delta(t_0, \varepsilon) = \min(\delta^*(t_0, \varepsilon), \delta_2(t_0, \varepsilon))$ and $\|x_0\| < \delta$. Then $\|x(t; t_0, x_0)\| \leq M\|x_0\|$ for all $t_0 \in T_i$ and $t \in T_0$ for any solution $x(t; t_0, x_0)$ of system (2.7.1) with $\|x_0\| < \delta(t_0, \varepsilon)$ and $t_1 > t_0$ which satisfies the conditions $\|x(t_1; t_0, x_0)\| = \varepsilon$ and $\|x(t; t_0, x_0)\| \leq \varepsilon$, for $t_0 \leq t \leq t_1$.

Indeed, by virtue of the comparison principle and condition (iii) of Theorem 2.7.3, we have

$$(2.7.22) \qquad v(t, x(t), y) \leq u(t; t_0, x_0, u_0) \quad \text{for all} \quad t \in T_0.$$

It follows from the properties of the functions $a(t, w)$ and $c(t, w)$ and the corresponding scalar functions $\alpha(t, w)$ and $\beta(t, w)$ (see Assumptions (H_1) – (H_2)) that

$$\alpha(t_0, \lambda_m(A_1)\|x(t; t_0, x_0)\|) \leq \lambda_m(A_1)\alpha(t_0, \|x(t; t_0, x_0)\|)$$

$$\leq v(t, x(t), y) \leq u(t; t_0, x_0, u_0) \leq Mu_0 = Mv(t_0, x_0, y)$$

$$\leq M\lambda_M(A_2)\beta(t_0, \|x_0\|) \leq \beta(t_0, M\lambda_M(A_2)\|x_0\|).$$

This yields

$$(2.7.23) \qquad \|x(t; t_0, x_0)\| \leq M \frac{\lambda_M(A_2)}{\lambda_m(A_1)} \|x_0\|$$

for all $t \in T_0$. If this is not the case, then, for $t = t_1$, we have

$$(2.7.23) \qquad \varepsilon \leq M \frac{\lambda_M(A_2)}{\lambda_m(A_1)} \delta,$$

but since $M \geq 1$ and $\lambda_M(A_2)\lambda_m^{-1}(A_1) \geq 1$, we obtain $\varepsilon \leq \delta$ contrary to the choice of δ for a given $\varepsilon \in (0, H)$. Theorem 2.7.3 is thus proved.

Corollary 2.7.3. (Kudo [2]). *Suppose that for the system (2.7.1) there exist functions $V(t,x) \in C(R_+ \times R^n, R_+)$, $a \in C(R_+ \times R_+, R_+)$, $c \in C(R_+ \times R_+, R_+)$ and $g \in C(R_+ \times R_+, R)$, $g(t,0) = 0$ such that*

(i) *$V(t,x)$ is locally Lipschitzian in x and $V(t,0) = 0$;*

(ii) *$a(t, \|x\|) \leq V(t,x) \leq c(t, \|x\|)$ for all $(t,x) \in R_+ \times R^n$, where $a(t,r)$ increases monotonically with respect to t for each fixed r, $a(t,0) = 0$, $a(t,r) > 0$ for $r \neq 0$, $kc(t,s) \leq c(t,ks)$ for a positive constant k and if $a(t,r) \leq c(t,s)$, then $r \leq s$;*

(iii) *$D^+V(t,x)\big|_{(2.7.1)} \leq g(t, V(t,x))$ for all $(t,x) \in R_+ \times R^n$.*

Then the uniform Lipschitz stability of the trivial solution $u = 0$ of (2.7.3) implies the uniform Lipschitz stability of the trivial solution $x = 0$ of (2.7.1).

Corollary 2.7.4. (Dannan and Elaydi [2]). *Assume that for system (2.7.1) there exist two functions $V(t,x)$ and $g(t,u)$ satisfying the following conditions:*

(i) *$V(t,x) \in C(R_+ \times \mathcal{D}, R_+)$, $V(t,0) = 0$, $V(t,x)$ is locally Lipschitz in x and satisfies $V(t,x) \geq b(\|x\|)$, where $b(r) \in C([0,\delta], R_+)$, $b(0) = 0$ and $b(r)$ is strictly monotone increasing in r such that $b^{-1}(\alpha r) \leq rq(\alpha)$ for some function q, with $q(\alpha) \geq 1$ if $\alpha \geq 1$;*

(ii) *$g(t,u) \in C(R_+ \times R_+, R)$, and $g(t,0) = 0$ for all $t \in R_+$;*

(iii) *$D^+V(t,x)\big|_{(2.7.1)} \leq g(t, V(t,x))$ for all $(t,x) \in R_+ \times \mathcal{D}$.*

If the zero solution of (2.7.3) is uniformly Lipschitz stable, then so is the zero solution of (2.7.1).

2.8 Notes

2.1. The importance of solving the problem of Liapunov function construction has been emphasized in many surveys and monograph (see, for example, Barbashin [2], Hahn [2], Krasovskii [1], Zubov [3,4], etc.). Some progress in this problem solution is associated with the idea of the matrix-valued function and hierarchical decomposition of the equations in question (see Martynyuk [7, 10, 15, 20, 21]).

2.2. Results of this section are presented according to Krapivnyi and Martynyuk [1], Martynyuk [8], and Martynyuk and Begmuratov [1]. In our construction we use some results by Barbashin [2], Djordjević [1, 2], Michel and Miller [1], and Šiljak [1].

2.3. This section is based on works by Begmuratov and Martynyuk [1], Martynyuk, Miladzhanov and Begmuratov [1, 2], and Begmuratov, Martynyuk and Miladzhanov [1]. Hierarchical decomposition of Ikeda and Šiljak [1] is applied in context with matrix-valued Liapunov functions.

2.4. The methods of the matrix-valued function construction presented in this section were developed by Martynyuk and Slyn'ko [1]. Some results by Djordjević [2], Krasovskii [1], Krapivnyi and Martynyuk [1], Martynyuk [4], Martynyuk and Miladzhanov [1], Michel and Miller [1] were used. Note that the first order partial equation (2.4.5) proposed by us for determining the nondiagonal elements of matrix-valued function can be solved by numerical methods. In some cases the proposed algorithm leads to the construction of a sign-definite integral of a system which allows a detailed investigation of motion of the system under consideration.

2.5. This section is based on the paper by Martynyuk and Slyn'ko [2]. In the development of this method of matrix-valued function construction we used the idea of overlapping decomposition of Ikeda and Šiljak [1] and some results on algebraic equations theory (see Lankaster [1]). The method of matrix-valued function construction proposed in the section enlarges essentially the possibilities of a direct Liapunov method in nonlinear dynamics. Namely, in the framework of the method of overlapping decomposition the method of vector Liapunov function can not be always applied either to system (2.5.10) or system (2.5.2). This can be easily demonstrated by using the technique set out by Šiljak [2, p. 420 – 423]. Meanwhile, our method of matrix-valued Liapunov function construction allows stability investigation of such a system.

2.6. Motion polystability analysis is a new direction of investigation in nonlinear dynamics of systems (see Martynyuk [9, 11, 13, 14], and Vorotnikov [1]).

The results obtained in this direction involve linear stationary systems and systems with periodic coefficients (see Martynyuk and Chernetskaya [1, 2]), composite systems consisting of two, three and four subsystems (see Martynyuk [20]). In a number of papers partial polystability was studied together with exponential stability. Our presentation is based on the results by Martynyuk [14, 18], Slyn'ko [1] and Vorotnikov [1].

2.7. Integral stability was first considered by Vrkoč [1]. Some results in this direction are presented by Kudo [1, 2]. The investigation of Lipschitz stability was undertaken by Dannan and Elaydi [1, 2]. This section is based on the works by Martynyuk [14, 18] and Kudo [1, 2] and develops the general concept of Liapunov matrix-valued function application in integral and Lipschitz stability theory. Corollaries 2.7.1 – 2.7.3 correspond to the results by Kudo [1, 2], Corollary 2.7.4 is due to Dannan and Elaydi [2].

The reader can find other results in this direction in works by Chen [1, 2], El-Sheikh and Soliman [1], Fausset and Koksal [1], Kim, Kye and Lee [1], Peng [2], Jin [1], etc.

The problem of Liapunov function construction is one of the central problems of nonlinear dynamics of systems. There are a lot of papers dealing with this problem; however no survey of recent results in this direction exists. We shall cite only some papers that demonstrate the variety of approaches in solution of this problem. They are: Foster and Davies [1], Fu and Abed [1], Galperin and Skowronski [1], Kinnen and Chen [1], Levin [1], Liu and Zhang [1], Mejlakhs [1], Michel, Sarabudla, et al. [1], Noldus, Vingerhoeds, et al. [1], Olas [1], Mukhametzyanov [1], Pota and Moylan [1], Rosier [1], Schwartz and Yan [1], Skowronski [1], Zubov [5], etc.

3

QUALITATIVE ANALYSIS
OF DISCRETE-TIME SYSTEMS

3.1 Introduction

Discrete systems appear to be effective mathematical models in the investigation of many processes and phenomena of real world. Recall however that in the papers by Euler and Lagrange the so-called recurrent series and some problems of probability theory were studied, being described by discrete (finite difference) equations. The intensive study of discrete systems for the last three decades has been evoked by new problems of technical progress. The discrete equations proved to be an efficient model in describing mechanical systems with impulse control as well as systems containing digital computing devices. Recently the discrete systems have been applied in modeling the process of population dynamics, macroeconomics, chaotic dynamics of economical systems as well as in modeling of recurrent neural networks and chemical reactions, and also in the investigation of the dynamics of discrete Markov processes, finite and probable automatic machines, calculation processes, etc.

One of the main problems of nonlinear dynamics of systems of kind is the problem on stability of solutions to the corresponding systems of equations in the Liapunov's or other sense.

In Section 3.2 we formulate the problems of qualitative analysis of nonlinear discrete equations studied in this section.

Section 3.3 presents the results of qualitative analysis of discrete systems in terms of matrix-valued function.

In Sections 3.4–3.5 the method of mixed decomposition is applied and the general theorems from Section 3.3 are used based on the constructive matrix-valued functions construction.

Section 3.6 contains sufficient stability and instability conditions for autonomous discrete systems obtained in terms of the semi-definite positive matrix-valued function.

In Section 3.7 sufficient conditions of connective stability are established in terms of the hierarchical Liapunov function.

In Section 3.8 some general results of the Section 3.6 are applied to the investigation of the discrete controlled system.

3.2 Systems Described by Difference Equations

Consider a system with a finite number of degrees of freedom described by the system of difference equations in the form

$$(3.2.1) \qquad\qquad x(\tau + 1) = f(\tau,\, x(\tau)),$$

where $\tau \in \mathcal{N}_\tau^+ \triangleq \{\tau_0 + k,\ \tau_0 \geq 0,\ k = 0, 1, 2, \ldots\}$, $x \in R^n$, $f\colon \mathcal{N}_\tau^+ \times R^n \to R^n$, and $f(\tau, x)$ is continuous in x. Let solution $x(\tau; \tau_0, x_0)$ of system (3.2.1) be definite for all $\tau \in \mathcal{N}_\tau^+$ и and $x(\tau_0; \tau_0, x_0) = x_0$. Assume that $f(\tau, x) = x$ for all $\tau \in \mathcal{N}_\tau^+$ iff $x = 0$. Besides, system (3.2.1) admits zero solution $x = 0$ and it corresponds to the unique equilibrium state of system (3.2.1).

Definition 3.2.1. The *equilibrium state* $x = 0$ of system (3.2.1) is called:

(a) *stable in the sense of Liapunov* iff for any $\tau_0 \in \mathcal{N}_\tau^+$ and any $\varepsilon > 0$ there exists $\delta = \delta(\tau_0, \varepsilon) > 0$ such that $\|x(\tau; \tau_0, x_0)\| < \varepsilon$ for all $\tau \geq \tau_0$, $\tau \in \mathcal{N}_\tau^+$ whenever $\|x_0\| < \delta$;

(b) *uniformly stable* iff the conditions of Definition 3.2.1(a) are satisfied and for any $\varepsilon > 0$ the corresponding value of δ_M satisfies the condition
$$\inf\,[\delta_M(\tau, \varepsilon)\colon \tau \in \mathcal{N}_\tau^+] > 0;$$

(c) *stable in the whole* iff the conditions of Definition 3.2.1(a) and $\delta_M(\tau, \varepsilon) \to +\infty$, as $\varepsilon \to +\infty$, forall $\tau \in \mathcal{N}_\tau^+$;

(d) *uniformly stable in the whole* iff the conditions of Definition 3.2.1(b) и 3.2.1(c) are satisfied simultaneously;

(e) *unstable*, iff there exist $\tau^* \in \mathcal{N}_\tau^+$, $\tau^* > \tau_0$ and $\varepsilon \in (0, +\infty)$, such that for any $\delta \in (0, +\infty)$ an x_0, $\|x_0\| < \delta$, is found such that $\|x(\tau^*; \tau_0, x_0)\| \geq \varepsilon$.

Further we designate by $B_\Delta(\tau_0) = \{x\colon \|x\| < \Delta(\tau_0)\}$ a sphere with the center at the origin and radius $\Delta(\tau_0)$.

Definition 3.2.2. The *equilibrium state* $x = 0$ of system (3.2.1) is called:

(a) *attractive*, iff for any $\tau_0 \in \mathcal{N}_\tau^+$ there exists $\Delta(\tau_0) > 0$ and for any $\xi > 0$ there exists $\tilde{\tau}(\tau_0; x_0, \xi) \in \mathcal{N}_\tau^+$ such that condition $\|x_0\| < \Delta(\tau_0)$ implies $\|x(\tau; \tau_0, x_0)\| < \xi$ for all $\tau \geq \tau_0 + \tilde{\tau}$, $\tau \in \mathcal{N}_\tau^+$;

(b) x_0-*attractive*, iff the conditions of Definition 3.2.2(a), are satisfied for any $\tau_0 \in \mathcal{N}_\tau^+$ there exists $\Delta(\tau_0) > 0$ and for any $\xi \in (0, +\infty)$ there exists $\tau_n(\tau_0, \Delta(\tau_0), \xi) \in \mathcal{N}_\tau^+$ such that

$$\sup\left[\tau_m(\tau_0; x_0, \xi): x_0 \in B_\Delta(\tau_0)\right] = \tau_n(\tau_0, \Delta(\tau_0), \xi);$$

(c) τ_0-*uniformly attractive*, iff the conditions of Definition 3.2.2(a), are satisfied, there exists $\Delta > 0$ and for any $(x_0, \xi) \in B_\Delta \times (0, +\infty)$ there exists $\tau_n(x_0, \xi) \in \mathcal{N}_\tau^+$ such that

$$\sup\left[\tau_m(\tau_0, x_0, \xi): \tau_0 \in \mathcal{N}_\tau^+\right] = \tau_n(x_0, \xi);$$

(d) *uniformly attractive*, iff all conditions of Definition 3.2.2(b) and 3.2.2(c), are satisfied, i.e. there exists $\Delta > 0$ and for any $\xi \in (0, +\infty)$ there exists $\tau_n(\Delta, \xi) \in \mathcal{N}_\tau^+$ such that

$$\sup\left[\tau_m(\tau_0, x_0, \xi): (\tau_0, x_0) \in \mathcal{N}_\tau^+ \times B_\Delta\right] = \tau_n(\Delta, \xi).$$

The attraction properties (a)–(d) of the state $x = 0$ of system (3.2.1) take place in the whole, if the conditions of Definition 3.2.2(a) are satisfied for any $\Delta(\tau_0) \in (0, +\infty)$ and any $\tau_0 \in \mathcal{N}_\tau^+$.

The definitions of the properties of asymptotic stability of solutions to the discrete systems in terms of the definitions of stability and attraction are presented below in the chapter when necessary.

In the investigation of concrete problems it often turns out to be important not only to determine whether the state $x = 0$ of system (3.2.1) is stable or attractive, but also to estimate the stability or attraction domains of this state.

The stability (attraction) of the equilibrium state $x = 0$ of system (3.2.1) is sometimes studied by reducing system (3.2.1) to the form

(3.2.2) $$x(\tau + 1) = Ax(\tau) + g(\tau, x(\tau)),$$

where A is $n \times n$ constant matrix, the vector function $g: \mathcal{N}_\tau^+ \times R^n \to R^n$ is continuous in x and satisfies certain conditions of smallness. In this case, under some additional restrictions on the properties of matrix A, the stability of state $x = 0$ of system (3.2.2) can be studied in terms of the first approximation equations.

It is of considerable interest when the order of system (3.2.1) is quite high or when the system is a composition of more simple subsystems. In this case the finite-difference systems of the type

(3.2.3)
$$x_i(\tau + 1) = f_i(\tau, x_i(\tau)) + g_i(\tau, x_1(\tau), \ldots, x_m(\tau)),$$
$$i = 1, 2, \ldots, m,$$

are considered, where $x_i \in R^{n_i}$, $f_i \colon \mathcal{N}_\tau^+ \times R^{n_i}$, $g_i \colon \mathcal{N}_\tau^+ \times R^{n_i} \times \cdots \times R^{n_m} \to R^{n_i}$. We designate

$$n = \sum_{i=1}^m n_i, \qquad x^{\mathrm{T}} = (x_1^{\mathrm{T}}, \ldots, x_m^{\mathrm{T}})^{\mathrm{T}},$$
$$f(t, x) = (f_1^{\mathrm{T}}(\tau, x_1), \ldots, f_m^{\mathrm{T}}(\tau, x_m))^{\mathrm{T}},$$

and
$$g(\tau, x) = (g_1^{\mathrm{T}}(\tau, x_1, \ldots, x_m), \ldots, g_m^{\mathrm{T}}(\tau, x_1, \ldots, x_m))^{\mathrm{T}}.$$

Now system (3.2.3) can be represented in the vector form

(3.2.4) $$x(\tau + 1) = f(\tau, x(\tau)) + g(\tau, x(\tau)) \triangleq H(\tau, x(\tau)).$$

Formally, system (3.2.4) coincides in form with the system (3.2.1), but if $g(\tau, x(\tau)) \equiv 0$, then system (3.2.4) falls apart into the independent subsystems

(3.2.5) $$x_i(\tau + 1) = f_i(\tau, x_i(\tau)), \quad i = 1, 2, \ldots, m.$$

Each of the subsystems may possess the same degree of complexity of the solutions behavior as system (3.2.1). Because of this the investigation of system (3.2.4) requires the development of the above mentioned fact. Such methods are developed in the qualitative theory of stability of large scale systems.

The discrete systems of more complex structure represent the (i, j)-pairs of subsystems (cf. Djordjević [3])

(3.2.6)
$$x_i(\tau + 1) = f_{ij}(\tau, x_i(\tau), x_j(\tau)) + g_{ij}(\tau, x(\tau)),$$
$$x_j(\tau + 1) = f_{ji}(\tau, x_j(\tau), x_i(\tau) + g_{ji}(\tau, x(\tau)), \quad i \neq j,$$

where $f_{ij} \colon \mathcal{N}_\tau^+ \times R^{n_i} \times R^{n_j} \to R^{n_i}$, $g_{ij} \colon \mathcal{N}_\tau^+ \times R^n \to R^{n_i}$, $f_{ji} \colon \mathcal{N}_\tau^+ \times R^{n_j} \times R^{n_i} \to R^{n_j}$, $g_{ji} \colon \mathcal{N}_\tau^+ \times R^n \to R^{n_j}$.

Designate $x_{ij} = (x_i^{\mathrm{T}}, x_j^{\mathrm{T}})^{\mathrm{T}}$, $F_{ij}(\tau, x_{ij}) = (f_{ij}^{\mathrm{T}}(\tau, x_i, x_j), f_{ji}^{\mathrm{T}}(\tau, x_j, x_i))^{\mathrm{T}}$, $G_{ij}(\tau, x(\tau)) = (g_{ij}^{\mathrm{T}}(\tau, x(\tau)), g_{ji}(\tau, x(\tau)))^{\mathrm{T}}$. Besides, the pair (i, j) of subsystems (3.2.6) are written in a more compact form

$$(3.2.7) \qquad x_{ij}(\tau + 1) = F_{ij}(\tau, x_{ij}(\tau)) + G_{ij}(\tau, x(\tau))$$

If the interconnection functions $G_{ij}(\tau, x(\tau)) \equiv 0$, then the difference system (3.2.7) falls apart into (i, j)-pairs of independent subsystems

$$(3.2.8) \qquad x_{ij}(\tau + 1) = F_{ij}(\tau, x_{ij}(\tau)), \quad (i \neq j) \in [1, m],$$

where $x_{ij} \in R^{n_i \times n_j}$ and $F_{ij} \colon \mathcal{N}_\tau^+ \times R^{n_i \times n_j} \to R^{n_i \times n_j}$.

It is supposed on systems (3.2.6) – (3.2.8) that the state $x_{ij} = 0$ $(i \neq j) \in [1, m]$ is a unique equilibrium state.

The dynamical properties of subsystems (3.2.5) or the pairs (i, j) of subsystems (3.2.8) are determined for the investigation of dynamics of the whole system (3.2.3) or (3.2.7). The fact will be demonstrated while constructing various sufficient conditions of stability-like properties of solutions to the finite difference equations.

3.3 Matrix-Valued Liapunov Functions Method

3.3.1 Auxiliary results

The direct Liapunov's method for the system (3.3.1) in terms of matrix-valued function $U(\tau, x)$ presupposes the existence of the mapping $U \colon \mathcal{N}_\tau^+ \times R^n \to R^{s \times s}$ and the *first difference*

$$(3.3.1) \qquad \Delta U(\tau, x(\tau)) = U(\tau + 1, x(\tau + 1)) - U(\tau, x(\tau))$$

along solutions of system (3.3.1). Here the first difference is understood element-wise for the matrix-valued function U.

These functions are characterized by positive (negative) definiteness, radial unboundedness, decreasing and positive (negative) semi-definiteness according to Definitions 1.4.7 – 1.4.9 from Chapter 1, where $t \in \mathcal{T}_0$ is replaced by $\tau \in \mathcal{N}_\tau^+$.

By means of the vector $\eta \in R_+^s$, $\eta > 0$, and the matrix-valued function $U(\tau, x)$ we construct the function

$$(3.3.2) \qquad\qquad v(\tau, x, \eta) = \eta^T U(\tau, x)\eta$$

that is important for the investigation of system (3.3.1). It is clear that

$$(3.3.3) \qquad\qquad \Delta v(\tau, x, \eta)\big|_{(3.2.4)} = \eta^T U(\tau, x(\tau))\eta,$$

the sign $\big|_{(*)}$ means that the difference is computed by virtue of system $(*)$.

If there is a function $\omega: \mathcal{N}_\tau^+ \times R_+ \to R_+$ such that

$$(3.3.4) \qquad\qquad \Delta v(\tau, x, \eta)\big|_{(3.2.4)} \leq \omega(\tau, v(\tau, x, \eta))$$

then we shall consider the inequality

$$(3.3.5)\quad \begin{aligned} v(\tau + 1, x(\tau + 1), \eta) &\leq v(\tau, x(\tau), \eta) + \omega(\tau, v(\tau, x(\tau), \eta)) \\ &\stackrel{\text{def}}{=} g(\tau, v(\tau, x(\tau), \eta)). \end{aligned}$$

For the inequality (3.3.5) the comparison equation

$$(3.3.6) \qquad u(\tau + 1) = g(\tau, u(\tau)) = u(\tau) + \omega(\tau, u(\tau))$$

is considered.

Further we need the following assertion (see Lakshmikantham, Leela, *et al.* [1]).

Proposition 3.3.1. *Let function $g(\tau, u)$ be definite on $\mathcal{N}_\tau^+ \times R_+$ and nondecreasing in u for a fixed $\tau \in \mathcal{N}_\tau^+$. Assume that for $\tau \geq t_0$*

$$(3.3.7) \qquad y(\tau + 1) \leq g(\tau, y(\tau)), \quad u(\tau + 1) \geq g(\tau, u(\tau))$$

and there exists solution $u(\tau)$ of the comparison equation (3.3.6).

Then, condition $y(t_0) \leq u(t_0)$ implies

$$(3.3.8) \qquad\qquad y(\tau) \leq u(\tau) \quad \text{for all} \quad \tau \geq t_0.$$

Proof. Let under condition of Proposition 3.3.1 estimate (3.3.8) be violated. Then, for $y(t_0) \leq u(t_0)$ there exists the $k \in \mathcal{N}_\tau^+$ such that $y(k) \leq u(k)$ and $y(k + 1) \geq u(k + 1)$. Inequality (3.3.7) and function g monotonicity yield the estimate

$$g(k, u(k)) \leq u(k + 1) < y(k + 1) \leq g(k, y(k)) \leq g(k, u(k)).$$

The obtained contradiction proves Proposition 3.3.1.

3.3.2 Comparison principle application

We extend Theorem 2.4.1 from Martynyuk [20] for system (3.3.1).

Theorem 3.3.1. *Let an n-vector function f in system (3.3.1) be continuous in the second argument and definite on $\mathcal{N}_\tau^+ \times \mathcal{N}$ (on $\mathcal{N}_\tau^+ \times R^n$). Let there exist*

    (1) *an open connected discrete time invariant neighborhood $\mathcal{G} \subseteq \mathcal{N}$ of point $x = 0$;*

    (2) *the matrix-valued function $U \colon \mathcal{N}_\tau^+ \times \mathcal{G} \to R^{m \times m}$, $U(\tau, 0) = 0$ for all $\tau \in \mathcal{N}_\tau^+$, and a vector $y \in R^m$ such that the function*

$$(3.3.9) \qquad v(\tau, x, y) = y^{\mathrm{T}} U(\tau, x) y$$

       is positive definite, radially unbounded and continuous in the second argument;

    (3) *function $g \colon \mathcal{N}_\tau^+ \times R_+ \to R$, $g(\tau, 0) = 0$, $g(\tau, u)$ is nondecreasing in u and such that*

$$\Delta v(\tau, x, y)\big|_{(3.2.4)} \leq g(\tau, v(\tau, x, y))$$

for all $(\tau, x, y) \in \mathcal{N}_\tau^+ \times \mathcal{G} \times R^m$ (for all $(\tau, x, y) \in \mathcal{N}_\tau^+ \times R^n \times R^m$).

Then

    (a) *stability (in the whole) of solution $u(\tau) = 0$ of the equation (3.3.6) yields stability (in the whole) of state $x(\tau) = 0$ of the system (3.3.1);*

    (b) *asymptotic stability (in the whole) of solution $u(\tau) = 0$ of the equation (3.3.6) yields asymptotic stability (in the whole) of the state $x(\tau) = 0$ of the system (3.3.1).*

Proof. By Proposition 3.3.1 we have

$$(3.3.10) \qquad v(\tau, x(\tau), y) \leq u(\tau), \quad \tau \in \mathcal{N}_\tau^+$$

whenever

$$v(t_0, x(t_0), y) \leq u(t_0).$$

Since by condition (2) of Theorem 3.3.1 the function (3.3.9) is positive definite and radially unbounded, there exists the function $\pi(\|x\|) \in K(KR)$ such that

$$\pi(\|x\|) \leq v(\tau, x(\tau), y) \leq u(\tau) \quad \text{for all} \quad \tau \in \mathcal{N}_\tau^+.$$

Let solution $u = 0$ of the equation (3.3.6) be stable. Then condition $u(t_0) < \eta(\varepsilon, t_0)$ implies estimate $u(\tau) < \pi(\varepsilon)$ which, by (3.3.10) yields the inequalities

$$(3.3.11) \qquad \pi(\|x(\tau)\|) \leq v(\tau, x(\tau), \eta) < \pi(\varepsilon).$$

It follows from (3.3.11)

(3.3.12) $\|x(\tau)\| < \varepsilon.$

Using the assumption on function (3.3.9) continuity in the second argument one can find a $\delta(\varepsilon, t_0)$ such that for $\|x(t_0)\| < \delta(\varepsilon, t_0)$ the upper bound $v(\tau, x(t_0), y) \leq u_0$ is satisfied.

Now we are to show that the inequality (3.3.12) is satisfied for any $\tau \in \mathcal{N}_\tau^+$. Let this be not true. Then there exists a $k \in \mathcal{N}_\tau^+$ such that $u(k) > \varepsilon$ and $u(k-1) < \varepsilon$. Therefore, $v(\tau, x(k), y) \geq \pi(\varepsilon)$ and $\pi(\varepsilon) \leq v(\tau, x(k), \eta) \leq u(k) < \pi(\varepsilon)$. The obtained contradiction proves the assertion (a) of theorem.

In case of the assertion (b) of Theorem 3.3.1 we have from inequalities

$$\pi(\|x(\tau)\|) \leq v(\tau, x(\tau), y) \leq u(\tau)$$

that

$$\lim_{\tau \to +\infty} \pi(\|x(\tau)\|) = 0$$

and, therefore $\lim \|x(\tau)\| = 0$ for $\tau \to +\infty$.

Theorem 3.3.2 *Assume that conditions (1) – (3) of Theorem 3.3.1 are satisfied, and the function (3.3.9) is decreasing. Then*

(a) *uniform stability (in the whole) of solution $u(\tau) = 0$ of the equation (3.3.6) yields uniform stability (in the whole) of state $x(\tau) = 0$ of the system (3.3.1);*

(b) *uniform asymptotic stability (in the whole) of solution $u(\tau) = 0$ of the equation (3.3.6) yields uniform asymptotic stability (in the whole) of state $x(\tau) = 0$ of the system (3.3.1).*

Proof. In addition to the arguments used in the proof of assertion (a) of Theorem 3.3.1 the value $\delta(\varepsilon, t_0)$ can be chosen independent of $t_0 \in \mathcal{N}_\tau^+$. This can be done in view of function (3.3.9) decreasing. In fact, there exists a function $\mu(\|x\|) \in K(KR)$ such that $v(\tau, x(\tau), y) \leq \mu(\|x(\tau)\|)$ for all $(\tau, x, y) \in \mathcal{N}_\tau^+ \times \mathcal{G} \times R^m$ (for all $(\tau, x, y) \in \mathcal{N}_\tau^+ \times R^n \times R^m$). As before, we have the estimate

$$\pi(\|x(\tau)\|) \leq v(\tau, x(\tau), y) \leq u(\tau)$$

being valid for $v(t_0, x(t_0), y) \leq u(t_0) \leq \eta(\varepsilon)$. If we take $\mu(\|x(t_0)\|) < \eta(\varepsilon)$ which yields $\|x(t_0)\| < \delta(\varepsilon) \equiv \mu^{-1}(\eta(\varepsilon))$, then $\|x(\tau)\| < \varepsilon$ for all $\tau \in \mathcal{N}_\tau^+$.

3.3.3 General theorems on stability

General Theorems 1.8.1–1.8.3 for nonautonomous systems (1.2.7) can be extended for discrete time system (3.3.1). We present only few of them.

Theorem 3.3.3. *Let an n-vector function f in the system (3.3.1) be continuous in the second argument and definite on a region in $\mathcal{N}_\tau^+ \times \mathcal{N}$. If there exist*

(1) *an open connected discrete time-invariant neighborhood $\mathcal{G} \subseteq \mathcal{N}$, $\mathcal{N} \subseteq R^n$, of point $x = 0$;*

(2) *the matrix-valued function $U \colon \mathcal{N}_\tau^+ \times G \to R^{m \times m}$, $U(\tau, 0) = 0$ for all $\tau \in \mathcal{N}_\tau^+$, and vector $y \in R^m$ such that function $v(\tau, x, y) = y^T U(\tau, x) y$ is continuous in the second argument;*

(3) *functions $\psi_{i1}, \psi_{i2}, \psi_{i3} \in K$, $\tilde{\psi}_{i1} \in \tilde{C}K$, $i = 1, 2, \dots, m$;*

(4) *the $m \times m$-matrices $A_j(y)$, $j = 1, 2, 3$, $\tilde{A}_2(y)$ such that*

    (a) $\psi_1^T(\|x\|) A_1(y) \psi_1(\|x\|) \leq v(\tau, x, y)$
    $\leq \tilde{\psi}_2^T(\tau, \|x\|) \tilde{A}_2(y) \tilde{\psi}_2(\tau, \|x\|)$
    for all $(\tau, x, y) \in \mathcal{N}_\tau^+ \times \mathcal{G} \times R^m$;

    (b) $\psi_1^T(\|x\|) A_1(y) \psi_1(\|x\|) \leq v(\tau, x, y) \leq \psi_2^T(\|x\|) A_2(y) \psi_2(\|x\|)$
    forall $(\tau, x, y) \in \mathcal{N}_\tau^+ \times \mathcal{G} \times R^m$;

    (c) $\Delta v(\tau, x, y)\big|_{(3.2.4)} \leq \psi_3^T(\|x\|) A_3(y) \psi_3(\|x\|)$
    for all $(\tau, x, y) \in \mathcal{N}_\tau^+ \times \mathcal{G} \times R^m$.

Then, if the matrices $A_1(y)$, $A_2(y)$, $\tilde{A}_2(y)$ for all $(y \neq 0) \in R^m$ are positive definite and $A_3(y)$ is negative semi-definite, then

(a) *state $x(\tau) = 0$ of the system (3.3.1) is stable under condition (4a);*

(b) *state $x(\tau) = 0$ of the system (3.3.1) is uniformly stable under condition (4b).*

Proof of this theorem is similar to that of Theorem 1.8.2 with obvious changes. Therefore, we omit it.

Theorem 3.3.4. *Let an n-vector function f in the system (3.3.1) be continuous in the second argument and definite on a region in $\mathcal{N}_\tau^+ \times \mathcal{N}$. If there exist:*

(1) *an open discrete time-invariant neighborhood $\mathcal{G} \subseteq \mathcal{N}$, $\mathcal{N} \subseteq R^n$, of point $x = 0$;*

(2) *the matrix-valued function $U \colon \mathcal{N}_\tau^+ \times \mathcal{G} \to R^{m \times m}$, $U(\tau, 0) = 0$ for all $\tau \in \mathcal{N}_\tau^+$, and vector $y \in R$ such that function $v(\tau, x, y) = y^T U(\tau, x) y$ is continuous in the second argument;*

(3) *the functions* η_{1i}, η_{2i}, $\eta_{3i} \in K$, $\tilde{\eta}_{2i} \in CK$, $i = 1, 2, \ldots, m$;

(4) *the* $m \times m$-*matrices* $C_j(y)$, $j = 1, 2, 3$, *and* $\tilde{C}_2(y)$ *such that*

   (a) $\eta_1^{\mathrm{T}}(\|x\|)C_1(y)\,\eta_1(\|x\|) \leq v(\tau, x, y)$
      $\leq \tilde{\eta}_2^{\mathrm{T}}(\tau, \|x\|)\tilde{C}_2(y)\,\tilde{\eta}_2(\tau, \|x\|)$
      for all $(\tau, x, y) \in \mathcal{N}_\tau^+ \times \mathcal{G} \times R^m$;

   (b) $\eta_1^{\mathrm{T}}(\|x\|)\,C_1(y)\,\eta_1(\|x\|) \leq v(\tau, x, y) \leq \eta_2^{\mathrm{T}}(\|x\|)\,C_2(y)\,\eta_2(\|x\|)$
      for all $(\tau, x, y) \in \mathcal{N}_\tau^+ \times \mathcal{G} \times R^m$;

   (c) $\Delta v(\tau, x, y)\big|_{(3.2.4)} \leq \eta_3^{\mathrm{T}}(\|x\|)\,C_3(y)\,\eta_3(\|x\|) + m(\tau, \eta_3(\|x\|))$
      for all $(\tau, x, y) \in \mathcal{N}_\tau^+ \times \mathcal{G} \times R^m$, *where the function* $m(\tau, \cdot)$
      satisfies the condition

$$\lim \frac{|m(\tau, \eta_3(\|x\|))|}{\|\eta_3(\|x\|)\|} = 0 \quad \text{for} \quad \|\eta_3(\|x\|)\| \to 0$$

      uniformly in $\tau \in \mathcal{N}_\tau^+$.

Then, if matrices $C_1(y)$, $C_2(y)$, $\tilde{C}_2(y)$ *are positive definite and matrix* $C_3(y)$ *for all* $(y \neq 0) \in R^m$ *is negative definite, then*

   (a) *state* $x(\tau) = 0$ *of the system* (3.3.1) *is asymptotically stable under condition* (4a);

   (b) *state* $x(\tau) = 0$ *of the system* (3.2.1) *is uniformly asymptotically stable under condition* (4b).

Proof of this theorem is similar to that of Theorem 1.8.3.

Theorem 3.3.5. *Let* n-*vector function* f *in the system* (3.2.1) *be continuous in the second argument and definite on a region in* $\mathcal{N}_\tau^+ \times \mathcal{N}$. *If there exist*

(1) *an open connected discrete time invariant neighborhood* $\mathcal{G} \subseteq \mathcal{N}$, $\mathcal{N} \subseteq R^n$, *of point* $x = 0$;

(2) *the matrix-valued function* $U: \mathcal{N}_\tau^+ \times \mathcal{G} \to R^{m \times m}$ *and a vector* $y \in R^m$ *such that function* $v(\tau, x, y) = y^{\mathrm{T}}U(\tau, x)y$ *is continuous in the second argument;*

(3) *functions* ψ_{i11}, ψ_{i2}, $\psi_{i3} \in K$, $i = 1, 2, \ldots, m$, *and* $m \times m$-*matrices* $A_1(y)$, $A_2(y)$, $G(y)$ *such that*

   (a) $\psi_1^{\mathrm{T}}(\|x\|)\,A_1(y)\,\psi_1(\|x\|) \leq v(\tau, x, y) \leq \psi_2^{\mathrm{T}}(\|x\|)\,A_2(y)\,\psi_2(\|x\|)$
      for all $(\tau, x, y) \in \mathcal{N}_\tau^+ \times \mathcal{G} \times R^m$;

   (b) $\Delta v(\tau, x, y)\big|_{(3.2.4)} \geq \psi_3^{\mathrm{T}}(\|x\|)\,G(y)\,\psi_3(\|x\|)$
      for all $(\tau, x, y) \in \mathcal{N}_\tau^+ \times \mathcal{G} \times R^m$;

(4) *point* $x = 0$ *belongs to the boundary* $\mathcal{G}$, *i.e.* $(x = 0) \in \partial\mathcal{G}$;

(5) $v(\tau, x, y) = 0$ *on* $\mathcal{N}_\tau^+ \times (\partial\mathcal{G} \cap B_\Delta)$, *where* $\overline{B}_\Delta \subset \mathcal{N}$.

Then, *if the matrices* $A_1(y)$, $A_2(y)$ *and* $G(y)$ *for all* $(y \neq 0) \in R^m$ *are positive definite, then state* $x(\tau) = 0$ *of system (3.2.1) is unstable.*

Proof of this theorem is similar to that of Theorem 1.8.6.

3.4 Large Scale System Decomposition

We assume on the system (3.2.1) that it can be naturally decomposed into s interconnected subsystems

$$(3.4.1) \qquad x_i(\tau + 1) = f_i(x_i(\tau), \tau) + g_i(x(\tau), \tau), \quad i = 1, 2, \ldots, s,$$

where $x_i \in R^{n_i}$, $f_i \colon R^{n_i} \times \mathcal{N}_\tau^+ \to R^{n_i}$, $g_i \colon R^{n_1} \times R^{n_2} \times \cdots \times R^{n_s} \times \mathcal{N}_\tau^+ \to R^{n_i}$, $x(\tau) = (x_1^T(\tau), \ldots, x_s^T(\tau))^T$. Formally setting $g_i = 0$, $i = 1, \ldots, s$, in the system (3.4.1) we get a set of independent subsystems

$$(3.4.2) \qquad x_i(\tau + 1) = f_i(x_i(\tau), \tau), \quad i = 1, 2, \ldots, s.$$

The functions $g_i \colon R^n \times \mathcal{N}_\tau^+ \to R^{n_i}$ connecting independent subsystems (3.4.2) in large scale discrete system (3.4.1)

$$(3.4.3) \qquad g_i = g_i(x_1(\tau), \ldots, x_s(\tau), \tau), \quad i = 1, 2, \ldots, s,$$

are assumed vanishing for all $\tau \in \mathcal{N}_\tau^+$, if and only if $x(\tau) = 0$. Thus, the point $x = 0$ is the only equilibrium point of the large scale discrete system (3.4.1) and points $x_i = 0$, $i = 1, 2, \ldots, s$, are the only equilibrium states of independent subsystems (3.4.2).

The decomposition of the large scale discrete system (3.4.2), prescribed by (3.4.1)–(3.4.3) and accomplished so that subsystems (3.4.2) satisfy existence and uniqueness condition of the solution, is called the *first level decomposition of a large scale discrete system.*

We suppose that the system (3.4.1) allows the decomposition of a higher level. We call pair (i, j) the couple of systems made from the set of subsystems (3.4.1) for $(i \neq j)$:

$$(3.4.4) \qquad \begin{aligned} x_i(\tau + 1) &= f_{ij}(x_i(\tau), x_j(\tau), \tau) + g_{ij}(x(\tau), \tau), \\ x_j(\tau + 1) &= f_{ji}(x_j(\tau), x_i(\tau), \tau) + g_{ji}(x(\tau), \tau), \end{aligned}$$

where $f_{ij} \colon R^{n_i} \times R^{n_j} \times \mathcal{N}_\tau^+ \to R^{n_i}$ and $g_{ij} \colon R^n \times \mathcal{N}_\tau^+ \to R^{n_i}$. We introduce the designations $x_{ij} = (x_i^T, x_j^T)^T$, $\overline{f}_{ij}(x_{ij}(\tau), \tau) = (f_{ij}^T, f_{ji}^T)^T$, $\overline{g}_{ij}(x(\tau), \tau) =$

$(g_{ij}^{\mathrm{T}}, g_{ji}^{\mathrm{T}})^{\mathrm{T}}$. Then the couple (i,j) of interconnected subsystems (3.4.4) is written as

$$(3.4.5) \qquad x_{ij}(\tau+1) = \overline{f}_{ij}(x_{ij}(\tau),\,\tau) + g_{ij}(x(\tau),\,\tau), \quad (i \neq j) \in [1,s].$$

The pairs of subsystems

$$(3.4.6) \qquad x_{ij}(\tau+1) = \overline{f}_{ij}(x_{ij}(\tau),\tau), \quad (i \neq j) \in [1,s],$$

are independent couples (i,j) that are united into large scale discrete system by couple (i,j) of the functions

$$(3.4.7) \qquad \overline{g}_{ij} = \overline{g}_{ij}(x_1(\tau),\ldots,x_s(\tau),\tau), \quad (i \neq j) \in [1,s].$$

We suppose that decomposition (3.4.5)–(3.4.7) can be made so that the functions $\overline{g}_{ij}(x(\tau),\tau)$, $(i \neq j) \in [1,s]$, vanish for any $\tau \in \mathcal{N}_\tau^+$, if and only if $x(\tau) = 0$, so that the point $x_{ij} = 0$ is the equilibrium point of independent couple (i,j) (3.4.6).

The decomposition of large scale discrete system (3.4.1), prescribed by (3.4.5)–(3.4.7) and fulfilled so that the independent couples (i,j) (3.4.6) satisfy the existence and uniqueness condition of the solution, is called the *second level decomposition of a large scale discrete system.*

Remark 3.4.1. If for the system (3.4.1) there exists a couple (p,q) of interconnected subsystems of the second level decomposition for some $(p \neq q) \in [1,s]$, then, obviously, there exists a couple (q,p) as well, and, moreover, these couples coincide. Therefore, for the large scale discrete system (3.4.1) admitting the first level decomposition into s interconnected subsystems, not more than $s(s-1)/2$ different couples of subsystems can exist. To consider all possible different couples (i,j) of the second level decomposition it is sufficient to consider the couples of subsystems for all $(i < j) \in [1,s]$.

Remark 3.4.2. We assume that in the first level decomposition of the system (3.4.1) it is always possible to distinguish in explicit form the independent subsystems and the functions, defined by (3.4.2) and (3.4.3). On the other hand in the second level decomposition of the large scale discrete system (3.4.1) we admit the existence of subsystems p and q, for which the couple (p,q) can not be made in the explicit form, i.e. in the correlations of the form of (3.4.3)–(3.4.7). We shall also assume that in some large scale discrete system it is possible not only to express couples (i,j) explicitly for all $(i < j) \in [1, s]$, but the functions $\overline{g}_{ij}(x, \tau)$ of independent couples with the other independent subsystems become

$$\overline{g}_{ij} = g_{ij}(x_1, \ldots, x_{i-1}, x_{i+1}, x_{j-1}, x_{j+1}, \ldots, x_s, \tau),$$

i.e. they will not depend on vector x_{ij}, $(i < j) \in [1, s]$. Such a decomposition of the large scale discrete system (3.4.1) is called a *complete second level decomposition*.

As well as in the case of a continuous system, our aim is to construct a hierarchical matrix-valued Liapunov function and to establish various sufficient conditions for stability, uniform stability, uniform asymptotic stability, uniform asymptotic stability in the whole and instability of equilibrium state $x = 0$ of the system (3.4.1).

3.5 Stability and Instability of Large Scale Systems

3.5.1 Auxiliary estimates

Let large scale discrete system (3.2.1) allow the first and second level decompositions (3.4.1) – (3.4.3) and (3.4.5) – (3.4.7). In order to formulate sufficient stability conditions we introduce a series of assumptions.

Assumption 3.5.1. *Let for the large scale discrete system (3.4.1) there exist*

(a) *a discrete time-invariant neighborhoods $\mathcal{N}_i \subseteq R^{n_i}$, $i = 1, 2, \ldots, s$, of the equilibrium states $x_i = 0$ of independent subsystems (3.9.2) of the first level decomposition;*

(b) *a discrete time-invariant neighborhoods $\mathcal{N}_{ij} \subseteq R^{n_i} \times R^{n_j}$, $(i < j) = [1, s]$, of the equilibrium states $x_{ij} = 0$ of independent couples (i, j) of (3.4.6) of the second level decomposition;*

(c) *the matrix-valued function*

(3.5.1)
$$U(x, \tau) = \begin{pmatrix} v_{11}(x_1, \tau) & v_{12}(x_{12}, \tau) & \cdots & v_{1s}(x_{1s}, \tau) \\ v_{21}(x_{21}, \tau) & v_{22}(x_2, \tau) & \cdots & v_{2s}(x_{2s}, \tau) \\ \cdots\cdots\cdots\cdots\cdots\cdots\cdots\cdots\cdots\cdots\cdots\cdots \\ v_{s1}(x_{s1}, \tau) & v_{s2}(x_{s2}, \tau) & \cdots & v_{ss}(x_s, \tau) \end{pmatrix}$$

the elements of which satisfy the estimates

$$c_{ii}^0 \psi_i^2(\|x_i\|) \leq v_{ii}(x_i, \tau) \leq c_{ii}^* \psi_i^2(\|x_i\|),$$

$$c_{ij}^0 \psi_i(\|x_i\|)\psi_j(\|x_j\|) \leq v_{ij}(x_{ij}, \tau) \leq c_{ij}^* \psi_i(\|x_i\|)\psi_j(\|x_j\|),$$

$$(3.5.2) \qquad v_{ij}(x_{ij}, \tau) = v_{ji}(x_{ij}, \tau),$$

$$\text{for all} \quad ((x_i, \tau) \in \mathcal{N}_i \times \mathcal{N}_\tau^+, \quad (x_{ij}, \tau) \in \mathcal{N}_{ij} \times \mathcal{N}_\tau^+,$$

$$(i \neq j) \in [1, s]),$$

where c_{ij}^0 and c_{ij}^*, $(i, j) \in [1, s]$, are positive constants, $\psi_i(\|x_i\|)$ are components of a vector function

$$\psi(\|x\|) = (\psi_1(\|x_1\|), \dots, \psi_s(\|x_s\|))^{\mathrm{T}},$$

$$\psi_i \in K(KR) \ (\psi_i(\|x_i\|) \colon R_+ \to R_+ \ \text{for all} \ i \in [1, s]).$$

By means of real vector $\eta \in \overset{\circ}{R}_+^s$ and the matrix-valued function (3.5.1) we introduce the scalar function

$$(3.5.3) \qquad v(x, \tau, \eta) = \eta^{\mathrm{T}} U(x, \tau)\eta.$$

Proposition 3.5.1. *If all conditions of Assumption 3.5.1 are satisfied, then for function (3.5.3) the bilateral estimate*

$$(3.5.4) \quad \begin{aligned} \psi^{\mathrm{T}}(\|x\|)H^{\mathrm{T}}C^0 H\psi(\|x\|) &\leq v(x, \tau, \eta) \leq \psi^{\mathrm{T}}(\|x\|)H^{\mathrm{T}}C^* H\psi(\|x\|), \\ \text{for all} \quad (x, \tau) &\in \mathcal{N} \times \mathcal{N}_\tau^+, \quad \mathcal{N} = \mathcal{N}_1 \times \mathcal{N}_2 \times \cdots \times \mathcal{N}_s \subseteq R^n \end{aligned}$$

is valid, where $H^{\mathrm{T}} = H = \mathrm{diag}\,(\eta_1, \eta_2, \dots, \eta_s)$, $C^0 = [c_{ij}^0]$, $C^* = [c_{ij}^*]$ *are matrices composed of constants* c_{ij}^0 *and* c_{ij}^*, $(i, j) \in [1, s]$.

To prove Proposition 3.5.1 we use the form of the function $v(x, \tau, \eta)$ definition (3.5.3) and the conditions of Assumption 3.5.1. We have

$$v(x, \tau, \eta) = \eta^{\mathrm{T}} U(x, \tau)\eta = \sum_{i=1}^s \eta_i^2 v_{ii}(x_i, \tau) + \sum_{i=1}^s \sum_{\substack{j=1 \\ (j \neq i)}}^s \eta_i \eta_j v_{ij}(x_{ij}, \tau)$$

$$\leq \sum_{i=1}^s \eta_i^2 c_{ii}^* \psi_i^2(\|x_i\|) + \sum_{i=1}^s \sum_{\substack{j=1 \\ (j \neq i)}}^s \eta_i \eta_j c_{ij}^* \psi_i(\|x_i\|)\psi_j(\|x_j\|),$$

and hence the estimate from above in (3.5.4). The estimate from below is obtained similarly.

Remark 3.5.1. Estimates (3.5.4) remain valid as well if instead of the functions of class $K(KR)$, $(\psi_i, \psi_j) \in K(KR)$, positive definite functions $u_i = u_i(x_i)$, $u_i \colon R^{n_i} \to R_+$ and $u_j = u_j(x_j)$, $u_j \colon R^{n_j} \to R_+$, $(i,j) \in [1, s]$, are considered, or if $\psi_i(\|x_i\|) = \|x_i\|$, $\psi_j(\|x_j\|) = \|x_j\|$, $(i, j) \in [1, s]$.

By virtue of large scale discrete system (3.4.1) the first difference for function $v(x, \tau)$

$$\Delta v(x, \tau, \eta)\big|_{(3.4.1)} = v(x(\tau + 1),\, \tau + 1, \eta) - v(x(\tau), \tau, \eta),$$

is computed without incorporation of solutions of system (3.4.1). In view of (3.5.3) we have

$$\Delta v(x, \tau, \eta) = \eta^{\mathrm{T}} \Delta U(x, \tau) \eta,$$

where $\Delta U(x, t)$ is a matrix-valued function with elements $\Delta v_{ij}(x_{ij}, t)$, $(i, j) \in [1, s]$, defined by the equalities for $i = j$

$$(3.5.5) \qquad \Delta v_{ii}(x_i, \tau) = \Delta v_{ii}(x_i, \tau)\big|_{(3.4.2)} + \Delta v_{ii}(x_i, \tau)\big|_{(3.4.3)},$$

where

$$\Delta v_{ii}(x_i, \tau)\big|_{(3.4.2)} = v_{ii}(f_i(x_i, \tau),\, \tau + 1) - v_{ii}(x_i, \tau);$$
$$\Delta v_{ii}(x_i, \tau)\big|_{(3.4.3)} = v_{ii}(f_i(x_i, \tau) + g_i(x, \tau),\, \tau + 1) - v_{ii}(f_i(x_i, \tau),\, \tau + 1),$$

and for $i \neq j$

$$(3.5.6) \qquad \Delta v_{ij}(x_{ij}, \tau) = \Delta v_{ij}(x_{ij}, \tau)\big|_{(3.4.6)} + \Delta v_{ij}(x_{ij}, \tau)\big|_{(3.4.7)},$$

where

$$\Delta v_{ij}(x_{ij}, \tau)\big|_{(3.4.6)} = v_{ij}(\overline{f}_{ij}(x_{ij}, \tau),\, \tau + 1) - v_{ij}(x_{ij}, \tau);$$
$$\Delta v_{ij}(x_{ij}, \tau)\big|_{(3.4.7)} = v_{ij}(\overline{f}_{ij}(x_{ij}, \tau) + \overline{g}_{ij}(x, \tau),\, \tau + 1)$$
$$- v_{ij}(\overline{f}_{ij}(x_{ij}, \tau),\, \tau + 1).$$

Expression (3.5.5) is the first difference of the function $v_{ii}(x_i, \tau)$ by virtue of the ith interconnected subsystem (3.4.1), and (3.5.6) is the first difference of the function $v_{ij}(x_{ij}, \tau)$ by virtue of couple (i, j) of interconnected subsystems (3.4.6), $(i \neq j) \in [1, s]$, in view of Remark 3.4.1 and the matrix-valued function $U(x, \tau)$ being symmetric.

Assumption 3.5.2. *Let there exist*

(a) *real constant values* ρ_{ii}^0, ρ_{ij}^k, $(i \neq j) \in [1, s]$, $k = 1, 2, 3$;

(b) *real constant vectors*

$$\mu_i^k = (\mu_{i1}^k, \mu_{i2}^k, \ldots, \mu_{is}^k)^{\mathrm{T}},$$

$$\xi_{ij} = (\xi_{ij1}, \xi_{ij2}, \ldots, \xi_{ijs})^{\mathrm{T}},$$

$$\nu_{ij} = (\nu_{ij1}, \nu_{ij2}, \ldots, \nu_{ijs})^{\mathrm{T}}, \quad (i \neq j) \in [1, s], \quad k = 1, 2,$$

such that the estimates

$$\Delta v_{ii}(x_i, \tau)\big|_{(3.4.2)} \leq \rho_{ii}^0 \varphi_i^2(\|x_i\|) \quad \text{for all} \quad (x_i, \tau) \in \mathcal{N}_i \times \mathcal{N}_\tau^+, \quad i \in [1, s],$$

$$\Delta v_{ii}(x_i, \tau)\big|_{(3.4.3)} \leq \langle \mu_i^1, \varphi(\|x\|) \rangle \langle \mu_i^2, \varphi(\|x\|) \rangle,$$

$$\Delta v_{ij}(x_{ij}, \tau)\big|_{(3.4.6)} \leq \rho_{ij}^1 \varphi_i^2(\|x_i\|) + 2\rho_{ij}^2 \varphi_i(\|x_i\|) \varphi_j(\|x_j\|)$$
$$+ \rho_{ij}^3 \varphi_j^2(\|x_j\|),$$
$$\text{for all} \quad (x_{ij}, \tau) \in \mathcal{N}_{ij} \times \mathcal{N}_\tau^+, \quad (i \neq j) \in [1, s],$$

$$\Delta v_{ij}(x_{ij}, \tau)\big|_{(3.4.7)} \leq \langle \xi_{ij}, \varphi(\|x\|) \rangle \langle \nu_{ij}, \varphi(\|x\|) \rangle,$$
$$\text{for all} \quad (x_{ij}, \tau) \in \mathcal{N}_{ij} \times \mathcal{N}_\tau^+, \quad (i \neq j) \in [1, s]$$

are valid, where $\varphi_i(\|x_i\|)$ *are components of the vector function* $\varphi(\|x\|) = (\varphi_1(\|x_1\|), \ldots, \varphi_s(\|x_s\|))^{\mathrm{T}}$, $\varphi_i \in K$, $i \in [1, s]$ *and* $\langle \cdot, \cdot \rangle$ *is a scalar product of vectors.*

Remark 3.5.2. The functions $v_{ii}(x_i, \tau)$, $i \in [1, s]$, and $v_{ij}(x_{ij}, \tau)$, $(i \neq j) \in [1, s]$, defined in Assumption 3.5.1 and its first differences $\Delta v_{ii}(x_i, \tau)\big|_{(3.4.2)}$ and $\Delta v_{ij}(x_{ij}, \tau)\big|_{(3.4.6)}$ reflect qualitative properties of the ith independent subsystem (3.4.2) of the first level decomposition and independent couple (i, j) (3.4.6) of the second level decomposition respectively.

Thus, if for some $p \in [1, s]$ it will turn out that $\rho_{pp}^0 \leq 0$, then the pth independent subsystem of the first level decomposition is stable.

If for some $(p < q) \in [1, s]$ the matrix

$$P_{pq} = \begin{pmatrix} \rho_{pq}^1 & \rho_{pq}^2 \\ \rho_{pq}^3 & \rho_{pq}^4 \end{pmatrix}$$

is conditionally negative semi-definite in R_+^2, then the independent couple (p, q) is stable.

Definition 3.5.1. The $n \times n$-matrix A, $A = A^{\mathrm{T}}$, is called

(a) *conditionally positive semi-definite*, if $x^{\mathrm{T}}Ax \geq 0$ for any $x \in R_+^n$;

(b) *conditionally positive definite*, if $x^{\mathrm{T}}Ax > 0$ for any $x \in R_+^n \setminus \{0\}$ and $x^{\mathrm{T}}Ax = 0$ whenever $x = 0$;

(c) *conditionally negative semi-definite*, if matrix $-A$ is conditionally positive semi-definite;

(d) *conditionally negative definite*, if matrix $-A$ is conditionally positive definite.

We need the following matrices for further presentation:

(a) an $s \times s$-matrix $P = [\rho_{ij}]$ with elements

$$
\rho_{ij} = \begin{cases} \eta_i^2 \rho_{ii}^0 + \eta_i \displaystyle\sum_{\substack{k=1 \\ (k \neq i)}}^{s} \eta_k (\rho_{ki}^1 + \rho_{ik}^3), & i = j \\[2em] 2\eta_i \eta_j \rho_{ij}^2, & i \neq j; \end{cases}
$$

(b) the matrix

$$
M = \sum_{i=1}^{s} \eta_i^2 \mu_i^1 {\mu_i^2}^{\mathrm{T}};
$$

(c) the matrix

$$
K = \sum_{i=1}^{s} \sum_{\substack{j=1 \\ (j \neq i)}}^{s} \eta_i \eta_j \xi_{ij} \nu_{ij}^{\mathrm{T}};
$$

(d) the matrix $\overline{S} = P + M + K$;

(e) symmetric matrix $S = \frac{1}{2}(\overline{S} + \overline{S}^{\mathrm{T}})$.

Proposition 3.5.2. *If all conditions of Assumption 3.10.2 are satisfied, then for the first difference $\Delta v(x, \tau)|_{(3.4.1)}$ estimate*

$$
(3.5.7) \quad \Delta v(x, \tau, \eta)|_{(3.4.1)} \leq \varphi^{\mathrm{T}}(\|x\|) S \varphi(\|x\|) \quad \text{for all} \quad (x, \tau) \in \mathcal{N} \times \mathcal{N}_\tau^+
$$

takes place.

Proof. We have for the first difference $\Delta v(x, \tau, \eta)|_{(3.4.1)}$ by virtue of the system (3.4.1) according to (3.5.4)–(3.5.7) and estimates of Proposition 3.5.1

$$\Delta v(x,\tau,\eta)\big|_{(3.4.1)} = \eta^{\mathrm{T}}\Delta U(x,\tau)\eta$$

$$= \sum_{i=1}^{s} \eta_i \Delta v_{ii}(x_i,\tau) + \sum_{i=1}^{s}\sum_{\substack{j=1\\(j\neq i)}}^{s} \eta_i\eta_j \Delta v_{ij}(x_{ij},\tau)$$

$$= \sum_{i=1}^{s} \eta_i^2 \Big(\Delta v_{ii}(x_i,\tau)\big|_{(3.4.2)} + \Delta v_{ii}(x_i,\tau)\big|_{(3.4.3)}\Big)$$

$$+ \sum_{i=1}^{s}\sum_{\substack{j=1\\(j\neq i)}}^{s} \eta_i\eta_j \Big(\Delta v_{ij}(x_{ij},\tau)\big|_{(3.4.6)} + \Delta v_{ij}(x_{ij},\tau)\big|_{(3.4.7)}\Big)$$

$$\leq \sum_{i=1}^{s} \eta_i^2 \Big(\rho_{ii}^0\varphi_i^2(\|x_i\|) + \langle\mu_i^1,\ \varphi(\|x\|)\rangle\langle\mu_i^2,\ \varphi(\|x\|)\rangle\Big)$$

$$+ \sum_{i=1}^{s}\sum_{\substack{j=1\\(j\neq i)}}^{s} \eta_i\eta_j \Big(\rho_{ij}^1\varphi_j^2(\|x_i\|) + 2\rho_{ij}^2\varphi_i(\|x_i\|)\varphi_j(\|x_j\|)$$

$$+ \rho_{ij}^3\varphi_j^2(\|x_j\|) + \langle\xi_{ij},\ \varphi(\|x\|)\rangle\langle\nu_{ij},\ \varphi(\|x\|)\rangle\Big)$$

for all $(x_i,\tau) \in \mathcal{N}_i \times I$, for all $(x_{ij},\tau) \in \mathcal{N}_{ij} \times I$.

With regard to

$$\sum_{i=1}^{s} \eta_i^2\rho_{ii}^0\varphi_i^2(\|x_i\|) + \sum_{i=1}^{s}\sum_{\substack{j=1\\(j\neq i)}}^{s} \eta_i\eta_j \Big(\rho_{ij}^1\varphi_i^2(\|x_i\|) + 2\rho_{ij}^2\varphi_i(\|x_i\|)\varphi_j(\|x_j\|)$$

$$+ \rho_{ij}^3\varphi_j^2(\|x_j\|)\Big) = \sum_{i=1}^{s} \Big(\rho_{ii}^0\eta_i^2 + \eta_i \sum_{\substack{k=1\\(k\neq i)}}^{s} \eta_k(\rho_{ki}^1 + \rho_{ik}^3)\Big)\varphi_i^2(\|x_i\|)$$

$$+ \sum_{i=1}^{s}\sum_{\substack{j=1\\(j\neq i)}}^{s} 2\eta_i\eta_j\rho_{ij}^2\varphi_i(\|x_i\|)\varphi_j(\|x_j\|) = \varphi^{\mathrm{T}}(\|x\|)P\varphi(\|x\|);$$

$$\sum_{i=1}^{s} \eta_i^2\langle\mu_i^1,\ \varphi(\|x\|)\rangle\langle\mu_i^2,\ \varphi(\|x\|)\rangle = \sum_{i=1}^{s} \eta_i^2\varphi^{\mathrm{T}}(\|x\|)\mu_i^1{\mu_i^2}^{\mathrm{T}}\varphi(\|x\|)$$

$$= \varphi^{\mathrm{T}}(\|x\|)\Big(\sum_{i=1}^{s} \eta_i^2\mu_i^1{\mu_i^2}^{\mathrm{T}}\Big)\varphi(\|x\|) = \varphi^{\mathrm{T}}(\|x\|)M\varphi(\|x\|);$$

$$\sum_{\substack{i=1}}^{s} \sum_{\substack{j=1 \\ (j \neq i)}}^{s} \eta_i \eta_j \langle \xi_{ij}, \varphi(\|x\|) \rangle \langle \nu_{ij}, \varphi(\|x\|) \rangle$$

$$= \sum_{\substack{i=1}}^{s} \sum_{\substack{j=1 \\ (j \neq i)}}^{s} \eta_i \eta_j \varphi^{\mathrm{T}}(\|x\|) \xi_{ij} \nu_{ij}^{\mathrm{T}} \varphi(\|x\|) = \varphi^{\mathrm{T}}(\|x\|) K \varphi(\|x\|),$$

we receive for the first difference $\Delta v(x, \tau, \eta)$

$$\Delta v(x, \tau, \eta)\big|_{(3.4.1)} \leq \varphi^{\mathrm{T}}(\|x\|)(P + M + K)\varphi(\|x\|)$$

$$= \varphi^{\mathrm{T}}(\|x\|)\overline{S}\varphi(\|x\|) = \frac{1}{2}\varphi^{\mathrm{T}}(\|x\|)(\overline{S} + \overline{S}^{\mathrm{T}})\varphi(\|x\|) = \varphi^{\mathrm{T}}(\|x\|)S\varphi(\|x\|),$$

that proves the assertion of Proposition 3.5.2.

3.5.2 Stability and instability conditions

Estimate (3.5.4) for the function $v(x, \tau)$ and estimate (3.5.7) for the first difference $\Delta v(x, \tau, \eta)$ enable us to establish existence conditions for various dynamical properties of the equilibrium state $x = 0$ of the system (3.4.1) that can be easily verified.

Theorem 3.5.1. *Let for large scale discrete system (3.4.1)*

(1) *all conditions of Assumptions 3.5.1 and 3.5.2 be satisfied;*

(2) *matrix S be*

    (a) *conditionally negative semi-definite in R_+^s;*

    (b) *conditionally negative definite in R_+^s.*

Then the equilibrium state $x = 0$ of the large scale discrete system (3.4.1) is

(1) *stable;*

(2) *asymptotically stable.*

If conditions of Assumptions 3.5.1 and 3.5.2 are satisfied for $\mathcal{N}_i = R^{n_i}$ and $\mathcal{N}_{ij} = R^{n_i} \times R^{n_j}$ and the vector function $\varphi(\|x\|)$ components are radially unbounded functions, then the equilibrium state $x = 0$ of the large scale discrete system (3.4.1) is

    (a) *stable in the whole;*

    (b) *asymptotically stable in the whole.*

Proof. We shall prove at the beginning the first part of the assertion. We have for the scalar function $v(x, \tau, \eta) = \eta^T U(x, \tau)\eta$, $\eta \in R_+^s \setminus \{0\}$ and its first difference $\Delta v(x, \tau, \eta)$ by virtue of system (3.4.1) according to Propositions 3.5.1 and 3.5.2

$$(3.5.8) \qquad \psi^T(\|x\|)H^T C^0 H \varphi(\|x\|) \le \psi^T(\|x\|)H^T C^* H \varphi(\|x\|)$$

and

$$(3.5.9) \qquad \Delta v(x, \tau, \eta)\big|_{(3.4.1)} \le \varphi^T(\|x\|)S\varphi(\|x\|).$$

Since by Assumption 3.5.1 the elements of the matrix C^0 are positive constants and $H = \operatorname{diag}(\eta_1, \eta_2, \ldots, \eta_s)$, $\eta \in R_+^s \setminus \{0\}$, then the matrix $H^T C^0 H$ is composed of positive elements. As it is known this is sufficient for the form $\psi^T(\|x\|)H^T C^0 H \psi(\|x\|)$ to be conditionally positive definite in R_+^s. Therefore, the function $v(x, \tau, \eta)$ is positive definite for any $(x, \tau) \in \mathcal{N} \times \mathcal{N}_\tau^+$. Condition (2a) of Theorem 3.5.1 is sufficient for the form $\varphi^T(\|x\|)S\varphi(\|x\|)$ to be negative semi-definite, that yields negativeness of the first difference $\Delta v(x, \tau, \eta)$.

For function $v(x, \tau, \eta)$ and its first difference $\Delta v(x, \tau, \eta)$ by virtue of the large scale discrete system (3.4.1) all conditions of the theorem on stability for discrete system are satisfied, and, hence, the state $x = 0$ of the large scale discrete system (3.4.1) is stable.

Under condition (2b) of the theorem the first difference $\Delta v(x, \tau, \eta)$ of the function $v(x, \tau, \eta)$ by virtue of system (3.4.1) is a negative definite function and all conditions of Theorem 3.5.1 on asymptotic stability are satisfied. Therefore, the state $x = 0$ of the large scale discrete system (3.4.1) is asymptotically stable.

If additional conditions of the theorem are satisfied, i.e. $\mathcal{N}_i = R^{n_i}$, $\mathcal{N}_{ij} = R^{n_i} \times R^{n_j}$ and $\psi_i(\|x_i\|)$ are radially unbounded functions, then the function $v(x, \tau, \eta)$ is positive definite and radially unbounded and its first difference $\Delta v(x, \tau, \eta)$ is negative semi-definite (definite). This is sufficient, as it is known, (see Hahn [1]) for the stability in the whole (asymptotic stability in the whole). The theorem is proved.

To formulate sufficient conditions for large scale discrete system instability we introduce the assumption.

Assumption 3.5.3. *Let there exist real vectors* $\tilde{\mu}_i^1$, $\tilde{\mu}_i^2$, $\tilde{\xi}_{ij}$, $\tilde{\nu}_{ij}$, *and real constants* $\tilde{\rho}_{ii}^0$, $\tilde{\rho}_{ij}^k$, $k = 1, 2, 3$, $(i \ne j) \in [1, s]$, *for which estimates in Assumption 3.5.2 are satisfied with the inequality sign "$\ge$".*

We define the matrices $\tilde{P}$, $\tilde{M}$ and $\tilde{K}$ similarly to matrices P, M and K, matrices $\tilde{\overline{S}} = \tilde{P} + \tilde{M} + \tilde{K}$ and $\tilde{S} = \frac{1}{2}(\tilde{\overline{S}} + \tilde{\overline{S}}^T)$.

Theorem 3.5.2. *Let for the large scale discrete system (3.4.1)*

(1) *conditions of Assumptions 3.5.1 and 3.5.3 be satisfied;*

(2) *matrix $\widetilde{S}$ be conditionally positive definite in R_+^s.*

Then, the equilibrium state $x = 0$ of the large scale discrete system (3.4.1) is unstable.

Proof. Under condition (1) of Theorem 3.5.2 we can show in the same way as in the proof of Theorem 3.5.1 that for scalar function $v(x, \tau, \eta) = \eta^T U(x, \tau)\eta$ and its first difference $\Delta v(x, \tau, \eta)$ by virtue of system (3.4.1) the estimates

$$(3.5.10) \qquad v(x, \tau, \eta) \geq \psi^T(\|x\|) H^T C^0 H \psi(\|x\|),$$

$$(3.5.11) \qquad \Delta v(x, \tau, \eta)\big|_{(3.4.1)} \geq \varphi^T(\|x\|) \widetilde{S} \varphi(\|x\|)$$

take place. Under condition (2) of Theorem 3.5.2 all conditions of Proposition 2.8 by LaSalle [1] on instability are satisfied, and, therefore, the equilibrium state $x = 0$ of the large scale discrete system (3.4.1) is unstable.

3.6 Autonomous Large Scale Systems

Consider an autonomous discrete time large scale system allowing the first level decomposition into s interconnected subsystems

$$(3.6.1) \qquad x_i(\tau + 1) = f_i(x_i(\tau)) + g_i(x(\tau)) \overset{\text{def}}{=} h_i(x_i(\tau), x(\tau)),$$
$$i = 1, 2, \ldots, s,$$

where $\tau \in I = \{0, 1, 2, \ldots\}$, $x_i \in R^{n_i}$, $f_i \colon R^{n_i} \to R^{n_i}$, $g_i \colon R^{n_1} \times R^{n_2} \times \cdots \times R^{n_s} \to R^{n_i}$. We suppose on the right side parts of the system (3.6.1) that the vector function $h_i(x_i, x) = (h_1(x_1, x), \ldots, h_s(x_s, x))^T$ is continuous in the ball $B_H = \{x \in R^n \colon \|x\| < H, \ 0 < H \leq \infty\}$ and for any point $x_0 \in B_H$ there is only one positive semi-trajectory. Further we designate by $x^+(\tau, x_0)$ $(x^-(\tau, x_0))$ the solutions of the system (3.11.1) that are continuable to the right (left); if they are bounded, we shall designate this by "$\hat{}$", i.e. $\hat{x}^+(\tau, x_0)$ $(\hat{x}^-(\tau, x_0))$. If the solution is infinitely continuable to the right and to the left, we shall write $x(\tau, x_0)$.

Consider the matrix-valued function

$$(3.6.2) \qquad U(x) = V(x)V^{\mathrm{T}}(x),$$

where

$$(3.6.3) \qquad V(x) = \begin{pmatrix} v_{11}(x_1) & v_{12}(x_1) & \cdots & v_{1s}(x_1) \\ v_{21}(x_2) & v_{22}(x_2) & \cdots & v_{2s}(x_2) \\ \hdotsfor{4} \\ v_{s1}(x_s) & v_{s2}(x_s) & \cdots & v_{ss}(x_s) \end{pmatrix}$$

is composed of the linear forms $v_{ij}(x_i)$, defined by the equalities

$$v_{ij}(x_i) = \langle c_i^j, x_i \rangle \quad \text{for all} \quad (i,j) \in [1,s],$$

where c_i^j is a real n_i-dimensional vector, $c_i^j = (c_{i1}^j, c_{i2}^j, \ldots, c_{in_i}^j)^{\mathrm{T}}$, $(i,j) \in [1,s]$.

By means of vector $\eta \in \overset{\circ}{R}{}_+^s$ and matrix-valued function (3.6.2) we construct a scalar function $v(x,\eta)$ in the form

$$(3.6.4) \qquad v(x,\eta) = \eta^{\mathrm{T}} U(x)\eta = \eta^{\mathrm{T}} V(x) V^{\mathrm{T}}(x)\eta.$$

The first difference for this function by virtue of the system (3.11.1) is defined as

$$(3.6.5) \qquad \Delta v(x,\eta) = \sum_{i,j,k=1}^{s} \eta_i \eta_j \Delta(v_{ki}(x_k) v_{ji}(x_j)).$$

It is easy to show that function $v(x,\eta)$ is of constant sign, positive (positive semi-definite).

Similar to the continuous case (see Martynyuk and Krapivnyi [2]), in the investigation of stability via semi-definite functions it is important to consider the sets where functions $v(x,\eta)$ and $\Delta(x,\eta)$ can vanish. We introduce these sets and in view of special character of the function (3.6.4) (cf. Bulgakov [1])

$$m = \{(x,\eta) \in R^n \times \overset{\circ}{R}{}_+^s : v(x,\eta) = 0\},$$

$$M = \{(x,\eta) \in R^n \times \overset{\circ}{R}{}_+^s : \Delta v(x,\eta) = 0\},$$

$$\overset{\circ}{m} = \{(x,\eta) \in \overset{\circ}{R}{}^n \times \overset{\circ}{R}{}_+^s : v(x,\eta) = 0\},$$

$$\overset{\circ}{M} = \{(x,\eta) \in \overset{\circ}{R}{}^n \times \overset{\circ}{R}{}^s_+ \colon \ \Delta v(x,\eta) = 0\},$$

$$m^r = \{(x,\eta) \in \mathcal{B}_r \times \overset{\circ}{R}{}^s_+ \colon \ v(x,\eta) = 0\},$$

$$M^r = \{(x,\eta) \in \mathcal{B}_r \times \overset{\circ}{R}{}^s_+ \colon \ \Delta v(x,\eta) = 0\},$$

$$\overset{\circ}{m}{}^r = \{(x,\eta) \in \overset{\circ}{\overline{\mathcal{B}}}_r \times \overset{\circ}{R}{}^s_+ \colon \ v(x,\eta) = 0\},$$

where $\overset{\circ}{R}{}^n = R^n \setminus \{0\}$, $\overset{\circ}{R}{}^s_+ = R^s_+ \setminus \{0\}$, $\overset{\circ}{\overline{\mathcal{B}}}_r = \overline{\mathcal{B}}_r \setminus \{0\}$.

It has been noticed that under conditions $v(x,\eta) \geq 0$ and $\Delta v(x,\eta) \leq 0$ the inclusion $m \subset M$ takes place.

Further the designation $\overset{\circ}{M} \ni \hat{x}(\tau)$ means that the set $\overset{\circ}{M}$ does not contain bounded trajectories of the system in question and the designation $\overset{\circ}{m}{}^h \ni x^-(\tau) \to 0$, $\tau \to -\infty$, shows that the set $\overset{\circ}{m}{}^h$ does not contain negative semi-trajectories entering the origin as $\tau \to -\infty$.

Assumption 3.6.1. *There exist time-invariant neighborhood $\mathcal{B}_h \subseteq \mathcal{B}_{h_1} \times \mathcal{B}_{h_2} \times \cdots \times \mathcal{B}_{h_s}$ of the equilibrium state $x = 0$ of the system (3.6.1), where function $v(x,\eta)\colon R^n \times \overset{\circ}{R}{}^s_+ \to R$, $v(0,\eta) = 0$ and $v(x,\eta) \geq 0$ for all $x \in \mathcal{B}_h$. Here $\mathcal{B}_{h_i} \subseteq R^{n_i}$, $0 < h \leq H$.*

Assumption 3.6.2. *There exist time-invariant neighborhoods $\mathcal{N}_i \subseteq \mathcal{B}_{h_i}$ of the equilibrium states $x_i = 0$, $i \in [1,s]$, of the independent subsystems*

(3.6.6)
$$x_i(\tau + 1) = f_i(x(\tau)),$$

and real constants ρ_{pk} and μ_{pk} $((p,k) \in [1,s]$, $\rho_{pk} = \rho_{kp}$, $\mu_{pk} = \mu_{kp})$ such that the estimates

$$\eta_i \eta_j \big[f_i^{\mathrm{T}}(x_i) C_i^{\mathrm{T}} C_j f_j(x_j) - x_i^{\mathrm{T}} C_i^{\mathrm{T}} C_j x_j \big] \leq \rho_{ij} \|x_i\| \, \|x_j\|,$$

$$\sum_{i=1}^{s} \sum_{j=111}^{s} \eta_i \eta_j \big[h_i^{\mathrm{T}}(x_i, x) C_i^{\mathrm{T}} C_j h_j(x_j, x) - f_i^{\mathrm{T}}(x_i) C_i^{\mathrm{T}} C_j f_j(x_j) \big]$$

$$\leq \sum_{p=1}^{s} \sum_{k=1}^{s} \mu_{pk} \|x_p\| \, \|x_k\|,$$

for all $(x_i \neq 0) \in \mathcal{N}_i$, $(x \neq 0) \in \mathcal{N}$, $\mathcal{N} = \mathcal{N}_1 \times \mathcal{N}_2 \times \cdots \times \mathcal{N}_s$ hold valid, where η_i, η_j are components of vector $\eta \in \overset{\circ}{R}{}^s_+$, and matrices of dimensions $s \times n_k$ are composed of the coefficients of the linear forms $\langle c_k^i, x_k \rangle$, $(i, k) \in [1, s]$ so that

$$C_k = \begin{pmatrix} c_{k1}^1 & c_{k2}^1 & \cdots & c_{kn_k}^1 \\ \cdots\cdots\cdots\cdots\cdots\cdots \\ c_{k1}^s & c_{k2}^s & \cdots & c_{kn_k}^s \end{pmatrix}.$$

We introduce the following matrices:

(a) the matrix $B = \{B_{ij}\}$, $(i, j) \in [1, s]$, composed of blocks B_{ij} with dimensions $n_i \times n_j$, that are defined by the expressions

$$B_{ij} = \eta_i \eta_j C_i^{\mathrm{T}} C_j, \quad B_{ij} = B_{ji}^{\mathrm{T}}, \quad \text{for all} \quad (i, j) \in [1, s];$$

(b) the symmetric matrix S with dimensions $s \times s$ and elements

$$s_{pk} = \rho_{pk} + \mu_{pk},$$
$$s_{pk} = s_{kp}, \quad (p, k) \in [1, s].$$

We shall formulate sufficient stability conditions for the large scale discrete system (3.6.1) based on the following definitions.

Definition 3.6.1. The *equilibrium state* $x = 0$ of the system (3.6.1) is

(a) *uniformly U-stable*, if and only if for any $\varepsilon > 0$ there exists a $\delta = \delta(\varepsilon) > 0$ such that condition $x_0 \in m^\delta$ implies $x(\tau, x_0) \in \mathcal{B}_\varepsilon$ for all $\tau \in \mathcal{N}_\tau^+$;

(b) *uniformly U-attractive*, if and only if there exists a $\Delta > 0$ and for any $(x_0, \zeta) \in m^\Delta \times (0, +\infty)$ there exists a $\tau^*(x_0, \zeta) \in [0, +\infty)$ such that $x(\tau, x_0) \in \mathcal{B}_\zeta$ for all $\tau \geq \tau_0 + \tau^*(x_0, \zeta)$;

(c) *uniformly asymptotically U-stable*, if and only if conditions of definitions (a) and (b) are satisfied;

(d) *U-unstable*, if and only if there exists a $\varepsilon > 0$ such that for any $\delta > 0$ there exists a $x_0 \in \overset{\circ}{m}{}^\delta$ and there exists a $T \in \mathcal{N}_\tau^+$ such that $x(\tau, x_0) \in \overset{\circ}{\mathcal{B}}_\varepsilon$ for $0 \leq \tau \leq T$ and $x(\tau, x_0) \in \mathrm{ext}\,\overset{\circ}{\mathcal{B}}_\varepsilon$.

The following assertions take place.

Theorem 3.6.1. *Let the large scale discrete system (3.4.1) be such that*

(1) *conditions of Assumptions 3.6.1 and 3.6.2 are satisfied;*

(2) *the matrix S is conditionally negative semi-definite in R_+^s;*

(3) *one of the following conditions is satisfied*

  (a) *there exists a $0 < \sigma < h$: $\overset{\circ}{m}{}^\sigma \ni \hat{x}(\tau, x_0)$ and $\overset{\circ}{m}{}^h \ni x^-(\tau, x_0)$ $\to 0$ as $\tau \to -\infty$;*

  (b) *there exists a $0 < \tilde{\sigma} \leq h$: $\overset{\circ}{m}{}^{\tilde{\sigma}} \ni x^-(\tau, x_0)$.*

Then the equilibrium state $x = 0$ of the system (3.6.1) is stable in the sense of Liapunov.

If one more condition is added to conditions of Theorem 3.6.1:

(4) *there exists a $0 < \delta \leq h$: $(M^\delta \backslash m^\delta) \ni \hat{x}(\tau, x_0)$, then the equilibrium state $x = 0$ of the system (3.6.1) is asymptotically stable in the sense of Liapunov.*

Theorem 3.6.2. (Test of U-asymptotic stability). *Let for system (3.6.1) conditions (1) – (2) of Theorem 3.6.1 be satisfied. The equilibrium state $x = 0$ of the system (3.6.1) is U-asymptotically stable, if and only if:*

(1) *there exists a $0 < \sigma \leq h$: $\overset{\circ}{m}{}^\sigma \ni \hat{x}(\tau, x_0)$;*

(2) $\overset{\circ}{m}{}^h \ni x^-(\tau, x_0) \to 0$ *as* $\tau \to -\infty$.

Proof. We start with the proof of Theorem 3.6.1. Consider the function $v(x, \eta)$ defined by expression (3.6.4)

$$v(x, \eta) = \sum_{i,j,k=1}^{s} \eta_i \eta_j v_{ik}(x_i) v_{jk}(x_j)$$

$$= \sum_{i,j,k=1}^{s} \eta_i \eta_j \langle c_i^k, x_i \rangle \langle c_j^k, x_j \rangle = \sum_{i,j=1}^{s} x_i^{\mathrm{T}} \left(\sum_{k=1}^{s} \eta_i \eta_j c_i^k c_j^{k\mathrm{T}} \right) x_j$$

$$= \sum_{i,j=1}^{s} x_i^{\mathrm{T}} \left(\eta_i \eta_j C_i^{\mathrm{T}} C_j \right) x_j = \sum_{i,j=1}^{s} x_i^{\mathrm{T}} B_{ij} x_j = x^{\mathrm{T}} B x.$$

It is easy to notice that the block matrix B is a Gram matrix of the system of vectors $c_{ki} = (c_{ki}^1, c_{ki}^2, \ldots, c_{ki}^s)^{\mathrm{T}} \in R^s$, $k = [1, s]$, $i = [1, n_k]$. It is clear that for $n_k \neq 1$, for all $k \in [1, s]$ (trivial case) the system of vectors c_{ki} is linearly dependent. Therefore, its Gram determinant is equal to zero and the form $v(x, \eta)$ is positive semi-definite for any $x_i \in \mathcal{B}_{h_i}$ ($x_j \in \mathcal{B}_{h_j}$), $0 < h_i, h_j \leq +\infty$.

We have for the first difference of function $v(x, \eta)$ by virtue of system (3.6.1)

$$\Delta v(x, \eta) = \sum_{i,j,k=1}^{s} \eta_i \eta_j \Delta(v_{ik}(x_i) v_{jk}(x_j)) = \sum_{i,j=1}^{s} \eta_i \eta_j \Delta(x_i^{\mathrm{T}} C_i^{\mathrm{T}} C_j x_j)$$

$$= \sum_{i,j=1}^{s} \eta_i \eta_j \left(h_i^{\mathrm{T}}(x_i, x) C_i^{\mathrm{T}} C_j h_j(x_j, x) - x_i^{\mathrm{T}} C_i^{\mathrm{T}} C_j x_j \right)$$

$$= \sum_{i,j=1}^{s} \eta_i \eta_j \left(f_i(x_i) C_i^{\mathrm{T}} C_j f_j(x_j) - x_i^{\mathrm{T}} C_i^{\mathrm{T}} C_j x_j \right.$$
$$+ \left. (h_i^{\mathrm{T}}(x_i, x) C_i^{\mathrm{T}} C_j h_j(x_j, x) - f_i(x_i) C_i^{\mathrm{T}} C_j f_j(x_j)) \right).$$

By virtue of Assumption 3.6.2 we get

(3.6.7) $$\Delta v(x, \eta) \leq \sum_{i,j=1}^{s} (\rho_{ij} + \mu_{ij}) \|x_i\| \|x_j\| = u^{\mathrm{T}} S u,$$

where $u = (\|x_1\|, \|x_2\|, \ldots, \|x_s\|)^{\mathrm{T}}$ and matrix S is previously defined. Therefore, under the conditions $(1)-(2)$ of Theorem 3.6.1 for the function $v(x, \eta)$ and its first difference $\Delta v(x, \eta)$ by virtue of the system (3.6.1) estimates $v(x, \eta) \geq 0$ and $\Delta v(x, \eta) \leq 0$ take place. Under the conditions $(1)-(3)$ all conditions of Theorem 1 by Bulgakov [1] are satisfied, and the equilibrium state is stable in the sense of Liapunov. Under the conditions $(1)-(4)$ of Theorem 3.6.1 all conditions of Theorems $1-2$ (see Bulgakov [1]) are satisfied, and the equilibrium state $x = 0$ of the large scale discrete system (3.6.1) is asymptotically stable in the sense of Liapunov.

Theorem 3.6.1 is proved.

The proof of Theorem 3.6.2 is similar to the proof of Theorem 2 by Bulgakov [1] in view of conditions $(1)-(2)$ of Theorem 3.6.1.

Example 3.6.1. We consider a large scale discrete system consisting of the subsystems

(3.6.8) $$x_1(\tau + 1) = \begin{pmatrix} 0.5 & \\ 0 & 0.1 \end{pmatrix} x_1(\tau) + \begin{pmatrix} -1 \\ 0 \end{pmatrix} x_{21} x_{11}^2,$$

and

$$x_2(\tau + 1) = 0.5 \, x_2(\tau) - x_{11}^3,$$

where $x_1(\tau) = (x_{11}, x_{12})^{\mathrm{T}} \in R^2$, $x_2(\tau) = (x_{21})^{\mathrm{T}} \in R^1$. The independent subsystems of the first level decomposition of large scale discrete system (3.6.8) are

$$x_1(\tau + 1) = \begin{pmatrix} 0.5 & \\ 0 & 0.1 \end{pmatrix} x_1(\tau),$$
$$x_2(\tau + 1) = 0.5 \, x_2(\tau)$$

and, obviously, are asymptotically stable.

We introduce the matrix-valued function

$$V(x) = \begin{pmatrix} \langle c_1^1, x_1 \rangle & \langle c_1^2, x_1 \rangle \\ \langle c_2^1, x_2 \rangle & \langle c_2^2, x_2 \rangle \end{pmatrix},$$

where vectors c_i^j for all $(i, j) \in [1, 2]$ are defined as

$$c_1^1 = (1, 0)^{\mathrm{T}}, \quad c_1^2 = (0, 1)^{\mathrm{T}}, \quad c_2^1 = 1, \quad c_2^2 = 0.$$

We take vector $\eta = (1, 1)^{\mathrm{T}} \in \overset{\circ}{R}{}_+^2$ and define, according to (3.6.4) the scalar function

$$v(x, \tau) = \eta^{\mathrm{T}} V(x) V^{\mathrm{T}}(x) \eta = (x_{11} + x_{21})^2 + x_{12}^2,$$

which is, obviously, positive semi-definite. The set m for the function $v(x, \eta)$ is defined as

$$m = \{x = (x_1^{\mathrm{T}}, x_2^{\mathrm{T}}) \in R^3 : \ x_{11} = -x_{21}, \ x_{12} = 0\}.$$

The set m is a straight line $x_{11} + x_{21} = 0$ in the cross-section of phase space by the plane $x_{12} = 0$ for $\eta = (1, 1)^{\mathrm{T}}$.

Now we define the values of constants ρ_{pk} and μ_{pk} for which the Assumption 3.6.5 takes place.

For $p = k = 1$ we have

$$f_1^{\mathrm{T}}(x_1) C_1^{\mathrm{T}} C_1 f_1(x_1) - x_1^{\mathrm{T}} C_1^{\mathrm{T}} C_1 x_1$$

$$= x_1^{\mathrm{T}} \begin{pmatrix} 0.5 & 0 \\ 0 & 0.1 \end{pmatrix} \begin{pmatrix} 1 & 0 \\ 0 & 1 \end{pmatrix} \begin{pmatrix} 0.5 & 0 \\ 0 & 0.1 \end{pmatrix} x_1 - x_1^{\mathrm{T}} \begin{pmatrix} 1 & 0 \\ 0 & 1 \end{pmatrix} x_1$$

$$= x_1^{\mathrm{T}} \begin{pmatrix} -0.75 & 0 \\ 0 & -0.99 \end{pmatrix} x_1 \le -0.75 \|x_1\|^2.$$

Similarly we obtain

$$f_2^{\mathrm{T}}(x_2) C_2^{\mathrm{T}} C_2 f_2(x_2) - x_2^{\mathrm{T}} C_2^{\mathrm{T}} C_2 x_2 \le -0.75 \|x_2\|^2,$$

$$f_1^{\mathrm{T}}(x_1) C_1^{\mathrm{T}} C_2 f_2(x_2) - x_1^{\mathrm{T}} C_1^{\mathrm{T}} C_2 x_2 \le -0.75 \|x_1\| \|x_2\|,$$

$$\sum_{i,j=1}^{s} \left(h_i^{\mathrm{T}}(x_i, x) C_i^{\mathrm{T}} C_j h_j(x_j, x) - f_i^{\mathrm{T}}(x_i) C_i^{\mathrm{T}} C) j f_j(x_j) \right)$$

$$= -x_{11}^4 + x_{11}^6 - 2x_{11}^3 x_{21} + 2x_{11}^5 x_{21} - x_{21}^2 x_{11}^2 + x_{21}^2 x_{11}^4$$

$$= -(x_{11} + x_{21})^2 (x_{11}^2 - x_{11}^4) \le 0.$$

In the notation of the last estimate it is taken into account that for sufficiently small x_{11} the form $(x_{11}^2 - x_{11}^4)$ is positive definite. Therefore, the conditions of Assumption 3.6.2 are valid with constants $\rho_{11} = \rho_{22} = -0.75$ and $\mu_{pk} = 0$ for all $(p, k) \in [1, s]$ and the matrix

$$S = \begin{pmatrix} -0.75 & 0/75 \\ 0.75 & -0.75 \end{pmatrix}$$

is (conditionally) negative semi-definite.

As has been shown earlier, the set m is a straight line in phase space of the system (3.6.8). We conclude by the form of the right side part of the system that neither straight line of phase space can contain the whole trajectory of the system. Then, $m \ni x(\tau, x_0)$ for all $x_0 \in \overset{\circ}{R}{}_+^3$, i.e. it does not contain whole trajectories of system (3.6.8), but trajectory $x \equiv 0$. Hence, it is clear, that for large scale discrete system (3.6.8) condition (3) of Theorem 3.6.1 is valid.

Moreover, one can show that for the large scale discrete system (3.6.8) the correlation $M = m$ takes place. Therefore, $M \setminus m = \emptyset$ and the condition (4) of Theorem 3.6.1 is satisfied.

Thus, it has been shown that for the large scale discrete system (3.6.8) conditions (1)-(4) of Theorem 3.6.1 are satisfied and the equilibrium state $x = 0$ of the system (3.6.8) is asymptotically stable in the sense of Liapunov.

3.7 Hierarchical Analysis of Stability

3.7.1 Hierarchical decomposition and stability conditions

Consider the system

(3.7.1) $x(\tau + 1) = f(\tau, x(\tau))$,

where $\tau \in \mathcal{N}_\tau^+$, $x \in R^n$, $f \colon \mathcal{N}_\tau^+ \times R^n \to R^n$, function f is such that the solution $x(\tau; \tau_0, x_0)$ of system (3.7.1) exists and is unique for all $\tau \in \mathcal{N}_\tau^+$ when any $(\tau_0, x_0) \in \mathcal{N}_\tau^+ \times R^n$. Moreover, assume that $f(\tau, x) = x$ for all $\tau \in \mathcal{N}_\tau^+$ if and only if $x = 0$ and the state $x = 0$ is a unique state of equilibrium of system (3.7.1).

System (3.7.1) is decomposed into s interconnected systems

(3.7.2) $x_i(\tau + 1) = g_i(\tau, x_i(\tau)) + h_i(\tau, x(\tau))$, $i = 1, 2, \ldots, s$, where $x_i \in R^{n_i}$, $x = (x_1^T, x_2^T, \ldots, x_s^T)^T$, $R^n = R^{n_1} \times R^{n_2} \times \cdots \times R^{n_s}$, $g_i \colon \mathcal{N}_\tau^+ \times R^{n_i} \to R^{n_i}$, $h_i \colon \mathcal{N}_\tau^+ \times R^n \to R^{n_i}$.

The equations

$$(3.7.3) \qquad x_i(\tau + 1) = g_i(\tau, x_i(\tau)), \quad i = 1, 2, \ldots, s$$

describe the dynamics of independent subsystems of system (3.7.2). Equations (3.7.3) are derived from the equations (3.7.2) when the connections h_i are equal to zero. Assume that $g_i(\tau, 0) = 0$ for all $\tau \in \mathcal{N}_\tau^+$ and the states $x_i = 0$ are the unique equilibrium states of subsystems (3.7.3).

Further each of subsystems (3.7.3) is decomposed into m_i interconnected components

$$(3.7.4) \qquad \begin{aligned} x_{ij}(\tau + 1) &= p_{ij}(\tau, x_{ij}(\tau)) + q_{ij}(\tau, x_i(\tau)), \\ i &= 1, 2, \ldots, s, \quad j = 1, 2, \ldots, m_i, \end{aligned}$$

where $x_{ij} \in R^{n_{ij}}$, $x_i = (x_{i1}^T, x_{i2}^T, \ldots, x_{im_i}^T)^T$, $R^{n_i} = R^{n_{i1}} \times R^{n_{i2}} \times \cdots \times R^{n_{im_i}}$, $p_{ij} \colon \mathcal{N}_\tau^+ \times R^{n_{ij}} \to R^{n_{ij}}$, $q_{ij} \times R^{n_i} \to R^{n_{ij}}$.

The equations

$$(3.7.5) \qquad x_{ij}(\tau + 1) = p_{ij}(\tau, x_{ij}(\tau))$$

describe the dynamics of independent components of subsystems (3.7.3). Equations (3.7.5) are derived from the equations (3.7.4) when the connections q_{ij} are equal to zero. Assume that $p_{ij}(\tau, 0) = 0$ for all $\tau \in \mathcal{N}_\tau^+$ and the states $x_{ij} = 0$ are the unique equilibrium states of components (3.7.5).

To study stability of system (3.7.1) we use two-level construction of the Liapunov function. Assume that for each component (3.7.5) there exists the Liapunov function $v_{ij}(\tau, x_{ij})$. For subsystems (3.7.3) we construct auxiliary functions

$$(3.7.6) \qquad v_i(\tau, x_i) = \sum_{j=1}^{m_i} d_{ij} v_{ij}(\tau, x_{ij}),$$

where d_{ij} are positive constants. Similarly for the whole system (3.7.1) the function

$$(3.7.7) \qquad V(\tau, x) = \sum_{i=1}^{s} d_i v_i(\tau, x_i)$$

is constructed, where d_i are positive constants. Under certain conditions the function $V(\tau, x)$ constructed by formulas (3.7.6) – (3.7.7) is the vector hierarchical Liapunov function for the system (3.7.1).

The first difference $\Delta V(\tau, x(\tau))|_{(3.7.1)}$ of the function $V(\tau, x)$ along solutions of system (3.7.1) is specified by the formula

$$\Delta V(\tau, x(\tau))|_{(3.7.1)} = V(\tau + 1, f(\tau, x(\tau))) - V(\tau, x(\tau)).$$

To formulate sufficient stability conditions we introduce some assumptions.

Assumption 3.7.1. *There exist*

(1) *discrete time invariant neighborhoods* $\mathcal{N}_{ij} \subset R^{n_{ij}}$ *of the states* $x_{ij} = 0$ *of components (3.7.5),* $i = 1, 2, \ldots, s$, $j = 1, 2, \ldots m_i$;

(2) *functions* $\chi_{ij}, \varphi_{ij}, \psi_{ij} \in K$, $i = 1, 2, \ldots, s$, $j = 1, 2, \ldots, m_i$;

(3) *functions* $v_{ij} \colon \mathcal{N}_\tau^+ \times R^{n_{ij}} \to R_+$, *satisfying the inequalities:*

   (a) $\underline{\alpha}_{ij} \varphi_{ij}(\|x_{ij}\|) \leq v_{ij}(\tau, x_{ij}) \leq \overline{\alpha}_{ij} \chi_{ij}(\|x_{ij}\|)$,
       for all $(\tau, x_{ij}) \in \mathcal{N}_\tau^+ \times \mathcal{N}_{ij}$,

   (b) $\Delta v_{ij}(\tau, x_{ij})|_{(3.7.5)} \leqslant -\pi_{ij} \psi_{ij}(\|x_{ij}\|)$,
       for all $(\tau, x_{ij}) \in \mathcal{N}_\tau^+ \times \mathcal{N}_{ij}$,

   (c) $\Delta v_{ij}(\tau, x_{ij}(\tau))|_{(3.7.4)} - \Delta v_{ij}(\tau, x_{ij}(\tau))|_{(3.7.5)}$

   $\leqslant \sum\limits_{k=1}^{m_i} \xi_{jk}^i \psi_{ik}(\|x_{ik}\|)$, *for all* $(\tau, x_{ij}) \in \mathcal{N}_\tau^+ \times \mathcal{N}_{ij}$,

   where $\underline{\alpha}_{ij} > 0$, $\overline{\alpha}_{ij} > 0$, $\pi_{ij} > 0$, $\xi_{ij}^i \geqslant 0$ *are real constants,* $\|x\|$ *is the Euclidean norm of vector* x, $i = 1, 2, \ldots, s$, $j = 1, 2, \ldots, m_i$.

Assumption 3.7.2. *Assume that*

(1) *there exist discrete time-invariant neighborhoods* $\mathcal{N}_i \subset R^{n_i}$ *of the equilibrium states* $x_i = 0$ *of subsystems (3.7.3),* $i = 1, 2, \ldots, s$;

(2) *there exist functions* $\psi_i \colon R_+ \to R_+$, $\psi_i \in K$, $i = 1, 2, \ldots, s$;

(3) *the functions* $v_i \colon \mathcal{N}_\tau^+ \times R^{n_{ij}} \to R_+$, *constructed by formulas (3.7.6) satisfy the inequalities*

   (a) $\Delta v_i(\tau, x_i)|_{(3.7.3)} \leqslant -\pi_i \psi_i(\|x_i\|)$, *for all* $(\tau, x_i) \in \mathcal{N}_\tau^+ \times \mathcal{N}_i$,

   (b) $\Delta v_i(\tau, x_i(\tau))|_{(3.7.2)} - \Delta v_i(\tau, x_i(\tau))|_{S_i} \leqslant \sum\limits_{j=1}^{s} \xi_{ij} \psi_j(\|x_j\|)$,

       for all $(\tau, x_i) \in \mathcal{N}_\tau^+ \times \mathcal{N}_i$,
       where $\pi_i > 0$ *and* $\xi_{ij} \geqslant 0$ *are real constants,* $i = 1, 2, \ldots, s$.

We determine the matrices $W_i = (w_{jk}^i)$ with the elements

$$w_{jk}^i = \begin{cases} \pi_{ij} - \xi_{jj}^i, & \text{if } j = k, \\ -\xi_{jk}^i, & \text{if } j \neq k \end{cases}$$

and the matrix $W = (w_{jk})$ with the elements

$$w_{jk} = \begin{cases} \pi_j - \xi_{jj}, & \text{if } j = k, \\ -\xi_{jk}, & \text{if } j \neq k. \end{cases}$$

Sufficient stability test for system (3.7.1) is found in the following result.

Theorem 3.7.1. *Assume that the perturbed motion equation (3.7.1) admit the decomposition (3.7.2)–(3.7.5) and conditions of Assumptions 3.7.1 and 3.7.2 are satisfied. Then, if the matrices $W_1, W_2, \ldots, W_s$ and W are the M-matrices, the equilibrium state $x = 0$ of system (3.7.1) is asymptotically stable.*

If all conditions of Assumptions 3.7.1 and 3.7.2 are satisfied for $\mathcal{N}_{ij} = R^{n_{ij}}$, $\mathcal{N}_i = R^{n_i}$ and the functions $\varphi_i \in KR$, then asymptotic stability in the whole takes place.

Proof of this theorem is found in the paper by Kameneva [1]. The application example is presented below.

Example 3.7.1. Consider the system

$$(3.7.8) \qquad x(\tau + 1) = \begin{pmatrix} 0.99 & 0.001 & 0 \\ 0.002 & 0.5 & 1 \\ 0.2 & 0.2 & 0.56 \end{pmatrix} x(\tau),$$

where $\tau \in \mathcal{T}$, $x \in R^3$. Decompose system (3.7.8) into two interconnected subsystems

$$x_1(\tau + 1) = \begin{pmatrix} 0.99 & 0.001 \\ 0.002 & 0.5 \end{pmatrix} x_1(\tau) + \begin{pmatrix} 0 \\ 1 \end{pmatrix} x_2(\tau),$$

$$x_2(\tau + 1) = 0.56\, x_2(\tau) + \begin{pmatrix} 0.2 \\ 0.2 \end{pmatrix}^{\mathrm{T}} x_1(\tau),$$

where $x_1 \in R^2$, $x_2 \in R$. We arrive at two independent subsystems

$$(3.7.9) \qquad x_1(\tau + 1) = \begin{pmatrix} 0.99 & 0.001 \\ 0.002 & 0.5 \end{pmatrix} x_1(\tau),$$

$$(3.7.10) \qquad x_2(\tau + 1) = 0.56\, x_2(\tau).$$

We decompose the subsystem (3.7.9) into two interconnected components

$$x_{11}(\tau + 1) = 0.99\, x_{11}(\tau) + 0.001\, x_{12}(\tau),$$
$$x_{12}(\tau + 1) = 0.5\, x_{12}(\tau) + 0.002\, x_{11}(\tau)$$

and distinguish two independent components

$$x_{11}(\tau + 1) = 0.99\, x_{11}(\tau),$$
$$x_{12}(\tau + 1) = 0.5\, x_{12}(\tau),$$

where x_{11}, $x_{12} \in R$. Choose functions

$$v_{11} = |x_{11}|, \quad v_{12} = |x_{12}|, \quad \psi_{11} = |x_{11}|, \quad \psi_{12} = |x_{12}|.$$

We obtain the numbers

$$\pi_{11} = 0.01, \qquad \pi_{12} = 0.5,$$
$$\xi_{11}^1 = 0, \quad \xi_{12}^1 = 0.001, \quad \xi_{21}^1 = 0.002, \quad \xi_{22}^1 = 0$$

and the matrix

$$W_1 = \begin{pmatrix} 0.01 & -0.001 \\ -0.002 & 0.5 \end{pmatrix},$$

which is the M-matrix, because $\Delta_1 = 0.01 > 0$ and $\Delta_2 = 0.004998 > 0$.
We take $d_{11} = 45$ and $d_{12} = 1$. Then

$$a_1^{\mathrm{T}} W_1 = (45; 1) \begin{pmatrix} 0.01 & -0.001 \\ -0.002 & 0.5 \end{pmatrix} = (0.448; 0.455),$$

$$v_1(x_1) = 45\,|x_{11}| + |x_{12}|$$

and

$$\Delta v_1(x_1)\big|_{(3.7.9)} \leqslant -0.488|x_{11}| - 0.455|x_{12}|.$$

We choose functions

$$v_2(x_2) = |x_2|, \quad \psi_1 = |x_{11}| + |x_{12}|, \quad \psi_2 = |x_2|.$$

Then

$$\pi_1 = 0.455, \qquad \pi_2 = 0.44,$$
$$\xi_{11} = 0, \quad \xi_{12} = 1, \quad \xi_{21} = 0.2, \quad \xi_{22} = 0.$$

The matrix

$$W = \begin{pmatrix} 0.455 & -1 \\ -0.2 & 0.44 \end{pmatrix}$$

is the M-matrix, because $\Delta_1 = 0.455 > 0$, $\Delta_2 = 0.0002 > 0$. We take $d_1 = 128$ and $d_2 = 291$. Then

$$a^T W = (128; 291) \begin{pmatrix} 0.445 & -1 \\ -0.2 & 0.44 \end{pmatrix} = (0.04; 0.04)$$

and the function

$$V(x) = 128(45|x_{11}| + |x_{12}|) + 291|x_2|$$

is the hierarchical vector Liapunov function establishing asymptotic stability of system (3.7.8).

Let us study the system (3.7.8) by means of one-level construction of function $V(x)$.

Decompose system (3.7.8) into three interconnected subsystems

$$x_1(\tau + 1) = 0.99\, x_1(\tau) + 0.001\, x_2(\tau),$$

$$x_2(\tau + 1) = 0.5\, x_2(\tau) + 0.002\, x_1(\tau) + x_3(\tau),$$

$$x_3(\tau + 1) = 0.56\, x_3(\tau) + 0.2\, x_1(\tau) + 0.2\, x_2(\tau)$$

and distinguish three independent subsystems

$$x_1(\tau + 1) = 0.99\, x_1(\tau),$$

$$x_2(\tau + 1) = 0.5\, x_2(\tau),$$

$$x_3(\tau + 1) = 0.56\, x_3(\tau).$$

We choose the functions

$$v_i = |x_i|, \quad \psi_i = |x_i|, \quad i = 1, 2, 3.$$

and get the matrix

$$\widetilde{W} = \begin{pmatrix} 0.01 & -0.001 & 0 \\ -0.002 & 0.5 & -1 \\ -1 & -1 & 0.44 \end{pmatrix},$$

which is not the M-matrix, because

$$\Delta_1 = 0.01 > 0, \quad \Delta_2 = 0.0498 > 0, \quad \Delta_3 = -0.00000088 < 0.$$

Using matrix $\widetilde{W}$ one cannot reach a conclusion on stability of system (3.7.8); however matrices W_1 and W allow the conclusion that system (3.7.8) is asymptotically stable.

3.7.2 Novel tests for connective stability

Assume that for system (3.7.1) the decomposition (3.7.2)–(3.7.5) takes place. It is known (see Šiljak [1]) that the connection functions between the independent subsystems of system (3.7.2) can be represented

$$(3.7.11) \qquad h_i(\tau, x) = h_i(\tau, \bar{e}_{i1}x_1, \bar{e}_{i2}x_2, \ldots, \bar{e}_{is}x_s), \quad i = 1, 2, \ldots, s,$$

where $\overline{E} = (\bar{e}_{ij})$ is the fundamental matrix of connections of system (3.7.2) with the elements

$$\bar{e}_{ij} = \begin{cases} 1, & \text{if } x_j \text{ is contained in } h_i, \\ 0, & \text{if } x_j \text{ is not contained in } h_i. \end{cases}$$

Let the functions of the discrete argument $e_{ij} \colon \mathcal{N}_\tau^+ \to [0,1]$ for all $\tau \in \mathcal{N}_\tau^+$ satisfy the inequalities

$$e_{ij}(\tau) \leqslant \bar{e}_{ij}.$$

The constants $\bar{e}_{ij}$ determine the degree of connection between the independent subsystems (3.7.3), and the matrix $E(\tau) = (e_{ij}(\tau))$ describes the structural perturbations of system (3.7.1).

If $E(\tau) \equiv 0$, then the system (3.7.1) is decomposed into s independent subsystems (3.7.3) each of which is a composition of the interconnected components (3.7.4). The connection functions between the independent components (3.7.5) can be written as

$$q_{ij}(\tau, x_i) = q_{ij}(\tau, \bar{\ell}^i_{j1}x_{i1}, \bar{\ell}^i_{j2}x_{i2}, \ldots, \bar{\ell}^i_{jm_i}x_{im_i}),$$
$$i = 1, 2, \ldots, s, \quad j = 1, 2, \ldots, m_i,$$

where

$$\bar{\ell}^i_{jk} = \begin{cases} 1, & \text{if } x_{ik} \text{ is contained in } q_{ij}, \\ 0, & \text{if } x_{ik} \text{ is not contained in } q_{ij}. \end{cases}$$

Let $\ell_{jk}^i : \mathcal{N}_\tau^+ \to [0,1]$ and for all $\tau \in \mathcal{N}_\tau^+$

(3.7.12) $\ell_{jk}^i(\tau) \leqslant \bar{\ell}_{jk}^i$,   for all $i = 1, 2, \ldots, s$, $j = 1, 2, \ldots, m_i$.

The matrices $\bar{L}_i = (\bar{\ell}_{jk}^i)$ are fundamental matrices of connections for subsystems (3.7.3) and describe the initial connections between the independent components (3.7.5), and the matrices $L_i(\tau) = (\ell_{jk}^i(\tau))$ describe the structural perturbations of subsystems (3.7.5).

Similarly to the continuous case the notion of the hierarchical connective stability of discrete system (3.7.1) is as follows (cf. Ikeda and Šiljak [2]).

Definition 3.7.1. *Discrete system* (3.7.1) *is called hierarchically connective stable, if*

(1) *for* $E(\tau) \equiv 0$ *the equilibrium state* $x_i = 0$ *of subsystems* (3.7.3) *are asymptotically stable in the whole for any structural matrices* $L_i(\tau)$, $i = 1, 2, \ldots, s$;

(2) *for* $L_i(\tau) \equiv \bar{L}_i$ *the equilibrium state* $x = 0$ *of system* (3.7.1) *is asymptotically stable in the whole for any structural matrix* $E(\tau)$.

In order that to formulate sufficient conditions for the hierarchical connective stability of system (3.7.1) we introduce some assumptions.

Assumption 3.7.3. *Assume that*

(1) *conditions* (1)–(4)(b) *of Assumption 3.7.1 are satisfied for* $\mathcal{N}_{ij} = R^{n_{ij}}$ *and functions* φ_{ij} *are of Hahn class* KR, $i = 1, 2, \ldots, s$, $j = 1, 2, \ldots, m_i$;

(2) *the first differences of functions* v_{ij} *satisfy the inequalities*

$$\Delta v_{ij}(\tau, x_{ij}(\tau))\big|_{(3.7.4)} - \Delta v_{ij}(\tau, x_{ij}(\tau))\big|_{(3.7.5)} \leqslant \sum_{k=1}^{m_i} \ell_{jk}^i(\tau)\xi_{jk}^i\psi_{ik}(\|x_{ik}\|)$$

for all $(\tau, x_{ij}) \in \mathcal{N}_\tau^+ \times R^{n_{ij}}$, *where* $\xi_{jk}^i \geqslant 0$ *are real constants,* $i = 1, 2, \ldots, s$, $j = 1, 2, \ldots, m_i$.

Assumption 3.7.4. *Assume that*

(1) *conditions* (1)–(3)(a) *of Assumption 3.7.2 are satisfied for* $\mathcal{N}_i = R^{n_i}$, $i = 1, 2, \ldots, s$;

(2) *the first differences of functions* v_i *satisfy the inequalities*

$$\Delta v_i(\tau, x_i(\tau))\big|_{(3.7.2)} - \Delta v_i(\tau, x_i(\tau))\big|_{(3.7.3)} \leqslant \sum_{j=1}^{s} e_{ij}(\tau)\xi_{ij}\psi_j(\|x_j\|)$$

for all $(\tau, x_i) \in \mathcal{N}_\tau^+ \times R^{n_i}$, *where* $\xi_{ij} \geqslant 0$ *are real constants,* $i = 1, 2, \ldots, s$.

In this case the elements of matrices $W_i(\tau) = (w^i_{jk}(\tau))$ and $W(\tau) = (w_{ij}(\tau))$ depend on discrete time, i.e.

$$
w^i_{jk}(\tau) = \begin{cases} \pi_{ij} - \ell^i_{jj}(\tau)\,\xi^i_{jj}, & \text{if } j = k, \\ -\ell^i_{jk}(\tau)\,\xi^i_{jk}, & \text{if } j \neq k, \end{cases}
$$

$$
w_{ij}(\tau) = \begin{cases} \pi_j - e_{jj}(\tau)\,\xi_{jj}(\tau), & \text{if } j = i, \\ -e_{ij}(\tau)\,\xi_{ij}, & \text{if } j \neq i. \end{cases}
$$

Now we designate by $\overline{W}_1, \overline{W}_2, \ldots, \overline{W}_s$ and $\overline{W}$ the matrices corresponding to the fundamental matrices of connections $\overline{L}_1, \overline{L}_2, \ldots, \overline{L}$ and $\overline{E}$.

We shall formulate one more test for connective stability of system (3.7.1).

Theorem 3.7.2. *Assume that the perturbed motion equations (3.7.1) admit decomposition (3.7.2) – (3.7.5) and all conditions of Assumptions 3.7.3 and 3.7.4 are satisfied. Then, if the matrices $\overline{W}_1, \overline{W}_2, \ldots, \overline{W}_s$ and $\overline{W}$ are the M-matrices, then the equilibrium state $x = 0$ of system (3.7.1) is hierarchically connective stable.*

Proof. By virtue of condition (1) of Assumption 3.7.3

$$
v_i(\tau, x_i) = \sum_{j=1}^{m_i} d_{ij} v_{ij}(\tau, x_{ij}) \geqslant \sum_{j=1}^{m_i} d_{ij}\alpha_{ij}\varphi_{ij}(\|x_{ij}\|).
$$

Passing to the level of the whole system we get

$$
V(\tau, x) = \sum_{i=1}^{s} d_i v_i(\tau, x_i) \geqslant \sum_{i=1}^{s} d_i \varphi_i(\|x_i\|)
$$

for all $(\tau, x) \in \mathcal{N}^+_\tau \times \mathcal{N}$, $\mathcal{N} = \mathcal{N}_1 \times \mathcal{N}_2 \times \cdots \times \mathcal{N}_s$.

Since $\varphi_i \in KR$-class for all $i = 1, 2, \ldots, s$, one can find a function $\varphi \in KR$ such that

$$
\sum_{i=1}^{s} \varphi_i(\|x_i\|) \geqslant \varphi(\|x\|)
$$

for all $x \in \mathcal{N}$ and therefore

$$
V(\tau, x) \geqslant \sum_{i=1}^{s} d_i\varphi_i(\|x_1\|) \geqslant d^*\varphi(\|x\|),
$$

where $d^* = \min_i \{d_i\}$. This proves sign definiteness of the function $V(\tau, x)$.

Using conditions of Assumption 3.7.3 we get for the first difference $\Delta v_{ij}(\tau, x_{ij}(\tau))\big|_{(3.7.4)}$ the estimate

$$\Delta v_{ij}(\tau, x_{ij}(\tau))\big|_{(3.7.4)} = \Delta v_{ij}(\tau, x_{ij}(\tau))\big|_{(3.7.5)} + \Delta v_{ij}(\tau, x_{ij}(\tau))\big|_{(3.7.4)}$$

$$- \Delta v_{ij}(\tau, x_{ij}(\tau))\big|_{(3.7.5)} \leqslant -\pi_{ij}\psi_{ij}(\|x_{ij}\|) + \sum_{k=1}^{m_i} \ell_{jk}^i(\tau)\xi_{jk}^i \psi_{ik}(\|x_{ik}\|)$$

$$= -(\pi_{ij} - \ell_{jj}^i(\tau)\xi_{jj}^i)\psi_{ij}(\|x_{ij}\|) + \sum_{k=1, k \neq j}^{m_i} \ell_{jk}^i(\tau)\xi_{jk}^i \psi_{ik}(\|x_{ik}\|)$$

for all $(\tau, x_{ij}) \in \mathcal{N}_\tau^+ \times \mathcal{N}_{ij}$. Then for the first difference $\Delta v_i(\tau, x(\tau))\big|_{(3.7.3)}$ the inequality

$$\Delta v_i(\tau, x_i(\tau))\big|_{(3.7.3)} = \sum_{j=1}^{m_i} d_{ij}\Delta v_{ij}(\tau, x_{ij}(\tau))\big|_{(3.7.4)}$$

$$\leqslant \sum_{j=1}^{m_i} d_{ij}\big(-(\pi_{ij} - \ell_{jj}^i(\tau)\xi_{jj}^i)\psi_{ij}(\|x_{ij}\|)$$

$$+ \sum_{k=1, k \neq j}^{m_i} \ell_{jk}^i(\tau)\xi_{jk}^i \psi_{ik}(\|x_{ik}\|)\big) = -a_i^{\mathrm{T}} W_i(\tau) z_i,$$

is true for all $(\tau, x_i) \in \mathcal{N}_\tau^+ \times \mathcal{N}_i$, where $a_i = (d_{i1}, d_{i2}, \ldots, d_{im_i})^{\mathrm{T}}$, $z_i = (\psi_{i1}, \psi_{i2}, \ldots, \psi_{im_i})^{\mathrm{T}}$. Inequalities (3.7.11) imply that $W_i(\tau) \leqslant \overline{W}_i$. Then

$$(3.7.13) \qquad \Delta v_i(\tau, x_i(\tau))\big|_{(3.7.3)} \leqslant -a_i^{\mathrm{T}} \overline{W}_i z_i, \quad i = 1, 2, \ldots, s.$$

It is known (see Šiljak [1]) that if the matrix $\overline{W}_i$ is the M-matrix, then there exists a vector a_i with positive components such that the vector $a_i^{\mathrm{T}} \overline{W}_i$ has positive components. Hence, the first differences of functions $v_i(\tau, x_i)$ along solutions of subsystems (3.7.3) are negative definite and consequently, the state $x_i = 0$ of subsystems (3.7.3) are asymptotically stable in the whole for $E(\tau) \equiv 0$ and all structural matrices $L_i(\tau)$, $\tau \in \mathcal{N}_\tau^+$. Condition (1) of Definition 3.7.1 is satisfied.

Assume that the connections between the components of (3.7.5) are fixed, i.e. $L_i(\tau) \equiv \overline{L}_i$, $i = 1, 2, \ldots, s$. Similarly to the above, for the first difference of function $V(\tau, x)$ by virtue of the whole system, using the inequalities (3.7.13), we get

$$\Delta v(\tau, x(\tau))\big|_{(3.7.1)} \leqslant -a^{\mathrm{T}} \overline{W} z$$

for all $(\tau, x) \in \mathcal{N}_\tau^+ \times R^n$ and any structural matrix $E(\tau)$. Since by condition of the theorem the matrix $\overline{W}$ is the M-matrix, the equilibrium state $x = 0$ of system (3.7.1) is asymptotically stable in the whole for $L(\tau) \equiv \overline{L}$ and matrix of connections $E(\tau)$. The theorem is proved.

Example 3.7.2. To illustrate the application of Theorem 3.7.1 we consider a numerical example

$$(3.7.14) \qquad x(\tau + 1) = \begin{pmatrix} 0.2 & 0.2 & 0.5 \\ 0.4 & 0.5 & 0.5 \\ 0.05 & 0.05 & 0.9 \end{pmatrix} x(\tau),$$

where $\tau \in \mathcal{N}_\tau^+$, $x \in R^3$. Distinguish two independent subsystems

$$(3.7.15) \qquad x_1(\tau + 1) = \begin{pmatrix} 0.2 & 0.2 \\ 0.4 & 0.5 \end{pmatrix} x(\tau),$$

$$(3.7.16) \qquad x_2(\tau + 1) = 0.9\, x_2(\tau).$$

The decomposition made corresponds to the fundamental matrix of connections

$$\overline{E} = \begin{pmatrix} 0 & 1 \\ 1 & 0 \end{pmatrix}$$

and the matrix of structural perturbations

$$E(\tau) = \begin{pmatrix} 0 & e_{12}(\tau) \\ e_{21}(\tau) & 0 \end{pmatrix},$$

where $e_{12}, e_{21} \colon \mathcal{N}_\tau^+ \to [0, 1]$.

Decompose subsystem (3.7.15) and single out two independent components

$$x_{11}(\tau + 1) = 0.2\, x_{11}(\tau),$$
$$x_{12}(\tau + 1) = 0.5\, x_{12}(\tau).$$

Such decomposition corresponds to the fundamental matrix of connections

$$\overline{L}_1 = \begin{pmatrix} 0 & 1 \\ 1 & 0 \end{pmatrix}$$

and the matrix of structural perturbations

$$L_1(\tau) = \begin{pmatrix} 0 & \ell_{12}^1(\tau) \\ \ell_{21}^1(\tau) & 0 \end{pmatrix},$$

where $\ell_{12}^1, \ell_{21}^1 \colon \mathcal{N}_\tau^+ \to [0,1]$. For subsystem (3.7.15) we take the functions

$$v_{1j} = |x_{1j}|, \quad \psi_{1j} = |x_{1j}|, \quad j = 1, 2$$

and get the matrix

$$W_1(\tau) = \begin{pmatrix} 0.8 & -0.2\,\ell_{12}^1(\tau) \\ -0.4\,\ell_{21}^1(\tau) & 0.5 \end{pmatrix},$$

corresponding to the matrix of connections L_1. The matrix

$$\overline{W}_1(\tau) = \begin{pmatrix} 0.8 & -0.2 \\ -0.4 & 0.5 \end{pmatrix},$$

corresponding to the fundamental matrix of connections $\overline{L}_1$ is the M-matrix. We choose $d_1 = d_2 = 1$. For the function

$$v_1(x_1) = |x_{11}| + |x_{12}|$$

the inequality

$$\Delta v_1(x_1)\big|_{(3.7.15)} \leqslant -0.4\,|x_{11}| - 0.3\,|x_{12}|$$

take place.

We take the functions

$$v_2 = |x_2|, \quad \psi_1 = |x_{11}| + |x_{12}|, \quad \psi(|x_2|) = |x_2|$$

and fix the matrix $L_1(\tau) \equiv \overline{L}_1$. We get the matrix

$$W(\tau) = \begin{pmatrix} 0.3 & -0.5\,e_{12}(\tau) \\ -0.5\,e_{21}(\tau) & 0.1 \end{pmatrix},$$

which corresponds to the structural matrix $E(\tau)$. The matrix

$$\overline{W} = \begin{pmatrix} 0.3 & -0.5 \\ -0.05 & 0.1 \end{pmatrix},$$

corresponding to the fundamental matrix of connections $\overline{E}$ is the M-matrix. For system (3.7.14) all conditions of Theorem 3.7.2 are satisfied, therefore the equilibrium state $x = 0$ of system (3.7.14) is hierarchically connected stable.

The equilibrium state $x = 0$ of system (3.7.14) is not connectedly stable in the ussual sense. Actually, let us decompose system (3.7.14) into three independent subsystems

$$x_1(\tau + 1) = 0.2\,x_1(\tau),$$
$$x_2(\tau + 1) = 0.5\,x_2(\tau),$$
$$x_3(\tau + 1) = 0.9\,x_3(\tau).$$

Then

$$\overline{E} = \begin{pmatrix} 0 & 1 & 1 \\ 1 & 0 & 1 \\ 1 & 1 & 0 \end{pmatrix}$$

and

$$E(\tau) = \begin{pmatrix} 0 & e_{12}(\tau) & e_{13}(\tau) \\ e_{21}(\tau) & 0 & e_{23}(\tau) \\ e_{31}(\tau) & e_{32}(\tau) & 0 \end{pmatrix},$$

where $e_{ij}\colon \mathcal{N}_\tau^+ \to [0, 1]$, $i, j = 1, 2, 3$, $i \neq j$. We take the structural matrix

$$E = \begin{pmatrix} 0 & 0 & 1 \\ 1 & 0 & 1 \\ 1 & 1 & 0 \end{pmatrix}.$$

The matrix E corresponds to the system (3.7.14).

We compose for system (3.7.14) the characteristic polynomial

$$f(\lambda) = \begin{vmatrix} 0.2 - \lambda & 0 & 0.5 \\ 0.4 & 0.5 - \lambda & 0.5 \\ 0.05 & 0.05 & 0.9 - \lambda \end{vmatrix} = -\lambda^3 + 1.6\,\lambda^2 - 0.68\,\lambda + 0.0825.$$

Since $f(1) = 0.0025$ and $f(2) = -2.8775$, then by the theorem on intermediate value there exists a number $\lambda_0 \in (1; 2)$ such that $f(\lambda_0) = 0$. This means that system (3.7.14) has an eigenvalue larger than one and, therefore, its equilibrium state is unstable.

3.8 Controlled Systems

Mathematical models of numerical control systems assigned to computer application are one of the application areas for the theory of discrete time systems stability. Incidentally, a real physical process remains continuous, and the discrete model shows the process at the quantification moments only, synchronized with the computer timer. Quantification of the continuous system can be made in several ways, in particular, for linear control systems by the method of zero order approximation, that results in the linear discrete control system.

Consider a discrete large scale control system given by the model in state space

$$(3.8.1) \qquad \begin{aligned} x_i(\tau+1) &= A_i x_i(\tau) + \sum_{\substack{k=1 \\ (k \neq i)}}^{s} C_{ik} x_k(\tau) + B_i u(\tau), \\ y(\tau) &= C x(\tau), \quad i \in [1, s], \end{aligned}$$

where $x(\tau) = (x_1^{\mathrm{T}}, \ldots, x_s^{\mathrm{T}})^{\mathrm{T}}$; A_i, C_{ik}, B_i, C are constant matrices with dimensions $n_i \times n_i$, $n_i \times n_k$, $n_i \times q$, $p \times n$ respectively, $u(\tau) \in R^q$ is a vector of outer control, $y(\tau) \in R^p$ is an output vector. For the large scale discrete system (3.8.1) the decomposition into subsystems and connection functions is obvious.

We shall illustrate using large scale discrete system (3.8.1) the decomposition into couples of subsystems and application of Theorem 3.5.1.

Introduce the block matrices

$$A_{ij} = \begin{pmatrix} A_i & C_{ij} \\ C_{ji} & A_j \end{pmatrix}, \quad C_{ij}^k = \begin{pmatrix} C_{ik} \\ C_{jk} \end{pmatrix}, \quad B_{ij} = \begin{pmatrix} B_i \\ B_j \end{pmatrix}.$$

Then the large scale discrete system (3.8.1) can be written as

$$(3.8.2) \qquad x_{ij}(\tau+1) = A_{ij} x_{ij}(\tau) + \sum_{\substack{k=1 \\ (k \neq i, \ k \neq j)}}^{s} C_{ij}^k x_k(\tau) + B_{ij} u(\tau).$$

We shall consider the problem on large scale discrete system (3.8.1) stability without outer control, i.e. when $u(\tau) = 0$. The decomposition of the form of (3.4.5) can be easily made for system (3.8.2) if we set $h_{ij}(x, \tau) = C_{ij} x_j$, $l_{ij} = y_{ij} = 0$ for all $(i \neq j) \in [1, s]$.

The independent subsystems are

$$(3.8.3) \qquad\qquad x_i(\tau + 1) = A_i x_i(\tau)$$

and functions connecting

$$g_i(x_1, \ldots, x_s) = \sum_{\substack{k=1 \\ (k \neq i)}}^{s} C_{ik} x_k(\tau).$$

We establish conditions under which the Assumptions 3.5.1 and 3.5.2 are satisfied for the system (3.8.1). To this end we chose positive definite quadratic forms

$$(3.8.5) \qquad \begin{aligned} w_{ii}(x_i) = x_i^{\mathrm{T}} H_{ii} x_i \quad \text{and} \quad w_{ij}(x_{ij}) = x_{ij}^{\mathrm{T}} H_{ij} x_{ij}, \\ (i \neq j) \in [1, s], \end{aligned}$$

as the components of matrix-valued function $U(x)$. Then

$$\Delta w_{ii}(x_i)\big|_{(3.8.3)} = x_i^{\mathrm{T}} [A_{ii}^{\mathrm{T}} H_{ii} A_{ii} - H_{ii}] x_i \leq \lambda_M \left(A_{ii}^{\mathrm{T}} H_{ii} A_{ii} - H_{ii}\right) \|x_i\|^2,$$

$$\Delta w_{ii}(x_i)\big|_{(3.8.4)} = \sum_{\substack{k=1 \\ (k \neq i)}}^{s} 2^{\mathrm{T}} x_i A_{ii}^{\mathrm{T}} C_{ik} x_k + \sum_{\substack{k,p=1 \\ (k,p \neq i)}}^{s} x_k^{\mathrm{T}} C_{ik}^{\mathrm{T}} C_{ip} x_p,$$

where $\lambda_M(\cdot)$ is a maximal eigenvalue of the matrix $(\cdot)$. The first two estimates in Assumption 3.5.2 are satisfied provided that $\varphi_i(\|x_i\|) = \|x_i\|$, $\rho_{ii}^0 = \lambda_M(A_{ii}^{\mathrm{T}} H_{ii} A_{ii} - H_{ii})$ and the vector μ_i^1 and μ_i^2 components are determined from the correlations

$$\rho_{kp} = \begin{cases} \lambda_M(C_{ik}^{\mathrm{T}} C_{ik}) & \text{for all} \quad k = p, \\ \|C_{ik}^{\mathrm{T}} C_{ik}\| & \text{for all} \quad k \neq p, \quad k, p \neq i, \end{cases}$$

$$\mu_{ii}^1 \cdot \mu_{ip}^2 = \mu_{ip}^1 \cdot \mu_{ii}^2 = \|A_{ii} C_{ip}\| \quad \text{for all} \quad (p \neq i) \in [1, s].$$

With the same arguments the last two estimates hold true, if we set

$$\rho_{ii}^2 = \frac{1}{2} \lambda_M(A_{ij}^{\mathrm{T}} H_{ij} A_{ij} - H_{ij}), \quad \rho_{ij}^1 = \rho_{ij}^3 = 0$$

$$\text{for all} \quad (i \neq j) \in [1, s],$$

define vector ξ_{ij} and ν_{ij} components

$$\xi_{ij}^i \nu_{ij}^k = \xi_{ij}^k \nu_{ij}^i = \|A_{ii}^T C_{ik} + C_{ji}^T C_{ji}\| \quad \text{for all} \quad (k,\, j,\, i) \in [1,s],$$

$$\xi_{ij}^j \nu_{ij}^k = \xi_{ij}^k \nu_{ij}^j = \|A_{jj}^T C_{jk} + C_{ij}^T C_{ik}\| \quad k \neq i, \quad k \neq j, \quad i \neq j,$$

and let them for all the rest combinations of indices equal to zero.

For the matrix-valued function $U(x) = [w_{ij}(\cdot)]$, $i, j = 1, 2, \ldots, s$ with positive definite quadratic forms (3.8.5) the validity of Assumption 3.5.1 is obvious. Thus, condition (1) of Theorem 3.5.1 is satisfied for the system (3.8.1) and the problem on zero solution stability can be solved in terms of conditional definiteness of matrix S computed for the system (3.8.1).

Remark 3.8.1. Let us discuss some advantages of the approach based on the matrix-valued function application. Making use of the additional information provided by the analysis of the couples of subsystems interaction, it is possible to weaken sufficient stability conditions that, as a rule, proved to be "too sufficient" in the method of the vector Liapunov functions. The independent subsystems may possess no stability properties, provided they form stable couples. Moreover, the perspective of the approach in a computer application, is that for the large scale discrete system of large dimensions the problem falls into several partial problems, each of which is often of a considerably smaller dimension, and can be solved independently of each other up to a certain extent.

3.9 Notes

3.1. General problems of qualitative analysis of real systems modeled by difference equations are discussed in a number of known monographs and textbooks (see, for example, Abdullin, Anapolskii, *et al.* [1], Bromberg [1], Furasov [1], Gupta and Hausdorff [1], Hahn [1], Kuo [1], Lakshmikantham, Leela, *et al.* [1], LaSalle [1, 2], Michel, Wang, *et al.* [1], Porter [1], Šiljak [1, 2], Tzypkin [1], etc.). In addition to the above mentioned works the readers can find the examples of real phenomena modelling by means of difference equations in the papers by Araki, Ando, *et al.* [1], Basson and Fogarty [1], Dash and Cressman [1], Hsieh [1], Rondoni [1], Sedaghat [1], Simonovits [1], Tchuente and Tindo [1], etc.

3.2. The description of discrete time systems in this section is presented according to the results by Martynyuk and Krapivnyi [1], and Martynyuk [12] (see also Michel and Miller [1], etc.)

3.3. Theorems 3.3.1 – 3.3.3 are new.

3.4 – 3.6. The original results of these sections incorporate the paper by Krapivnyi and Martynyuk [2], and Martynyuk and Krapivnyi [1].

3.7. This section is based on the results by Lukyanova and Martynyuk [1] (see also Kameneva [1]). The notion of connective stability used in this section is due to Šiljak [1] (see also Ikeda and Šiljak [1]).

3.8. This section is based on the results by Martynyuk and Krapivnyi [1].

4

NONLINEAR DYNAMICS OF IMPULSIVE SYSTEMS

4.1 Introduction

The investigation of systems with a finite number of degrees of freedom and impulsive perturbations is a new direction in nonlinear dynamics. These systems model satisfactorily many processes and phenomena in which change of the process properties occurs suddenly at some fixed instants of time. The model of clocks is a simple classical example of system of the kind.

This chapter deals with the development of a new approach in nonlinear dynamics of impulse systems which is based on the method of matrix-valued Liapunov functions.

Section 4.2 provides general information on large scale impulsive systems. In terms of one-level decomposition a method of analysis is developed for the dynamical properties of stability type.

Short Section 4.3 introduced the reader to the class of the so-called hierarchically impulsive systems. Actually, these are the impulsive systems of high dimension which admit homogeneous decomposition in the sense of Ikeda and Šiljak [1].

Section 4.4 sets out the construction technique for hierarchical Liapunov functions in terms of homogeneous hierarchical decomposition of impulsive system.

In Sections 4.5 – 4.7 the Liapunov functions constructed in Section 4.4 are applied and conditions of uniqueness and extendability of solutions are established as well as the boundedness conditions and conditions of various types of stability of solutions to an impulsive system.

4.2 Large Scale Impulsive Systems in General

4.2.1 Notations and definitions

The *impulsive system* of differential equations of general type

(4.2.1)
$$\frac{dx}{dt} = f(t, x), \quad t \neq \tau_k(x),$$

$$\Delta x = I_k(x), \quad t = \tau_k(x), \quad k = 1, 2, \ldots,$$

has the meaning of a large scale impulsive system, if it can be decomposed into m *interconnected impulsive subsystems*

(4.2.2)
$$\frac{dx_j}{dt} = f_j(t, x_j) + f_j^*(t, x), \quad t \neq \tau_k(x), \quad j = 1, 2, \ldots, m,$$

$$\Delta x_j = I_{kj}(x_j) + I_{kj}^*(x), \quad t = \tau_k(x), \quad k = 1, 2, \ldots.$$

We assume on system (4.2.1) that

(1) $x \in R^n$, $f(t, x) = 0$ iff $x = 0$;
(2) $0 < \tau_k(x) < \tau_{k+1}(x)$, $\tau_k(x) \to +\infty$ as $k \to \infty$;
(3) $I_k \colon R^n \to R^n$ and $I_k = 0$ iff $x = 0$;
(4) functions $f(t, x)$ and $I_k(x)$ are definite and continuous in the domain

$$\mathcal{T}_0 \times \mathcal{S}(\rho) = [t_0, \infty) \times \{x \colon \|x\| \leq \rho \leq \rho_0\}, \quad t_0 \geq 0;$$

(5) functions $\tau_k(x)$, $k = 1, 2, \ldots$, and number ρ satisfy conditions excluding beating of solutions of system (4.2.1) against the hypersurfaces $S_i \colon t = \tau_k(x)$, $k = 1, 2, \ldots$, $t \geq 0$.

We assume on system (4.2.2) that

(1) $x_j = (0, \ldots, 0, x_j^T, 0, \ldots, 0)^T \in R^n$, $x_j \in R^{n_j}$,
    $f = (f_1^T, \ldots, f_m^T)^T$, $f_j^*(t, x) = f_j(t, x) - f_j(t, x_j)$;
(2) $I_{kj} = (I_{k1}^T, I_{k2}^T, \ldots, I_{km}^T)^T$, $I_{kj}^*(x) = I_{kj}(x) - I_{kj}(x_j)$, $n = n_1 + \cdots + n_m$.

The state of the jth *noninteracting impulsive subsystem* is described by the equations

(4.2.3)
$$\frac{dx_j}{dt} = f_j(t, x_j), \quad t \neq \tau_k(x_j);$$

$$\Delta x_j = I_{kj}(x_j), \quad t = \tau_k(x_j).$$

The problem on stability for large scale impulsive system (4.2.1) is formulated as follows:

To establish conditions under which stability of equilibrium state $x = 0$ of system (4.2.2) is derived from the properties of stability of impulse subsystems (4.2.3) and properties of connection functions $f_j^*(t, x)$ and $I_{kj}^*(x)$.

Let $x_0(t) = x(t; t_0, y_0)$ $(y_0 \neq x_0)$ be a given solution of the system (4.2.1). Since the times of impulsive effects on solution $x_0(t)$ may not coincide with those on any neighboring solution $x(t)$ of system (4.2.1), the smallness requirement for the difference $\|x(t) - x_0(t)\|$ for all $t \geq t_0$ seems not natural. Therefore the stability definitions presented in Chapter 2 for the system of ordinary differential equations should be adapted to system (4.2.1).

We designate by Ξ a set of functions continuous from the left with discontinuities of the first kind, defined on R_+ with the values in R^n. Let the set of the discontinuity point of each of these functions be no more than countable and do not contain finite limit points in R^1. Let $\zeta \geq 0$ be a fixed number.

Definition 4.2.1. A *function* $y(t) \in \Xi$ is in ζ-*neighborhood of function* $x(t) \in \Xi$, if

(1) discontinuity point of function $y(t)$ are in ζ-neighborhoods of discontinuity point of function $x(t)$;

(2) for all $t \in R_+$, that do not belong to ζ-neighborhoods of discontinuity points of function $x(t)$, the inequality $\|x(t) - y(t)\| < \zeta$ is satisfied.

The totality of ζ-neighborhoods, $\zeta \in (0, \infty)$, of all elements of the set Ξ forms the basis of topology, which is referred to as B-*topology*.

Let $x(t)$ be a solution of system (4.2.1), and $t = \tau_k$, $k \in Z$, be an ordered sequence of discontinuity points of this solution.

Definition 4.2.2. *Solution* $x(t)$ *of system* (4.3.1) *satisfies*

(1) α-*condition*, if there exists a number $\vartheta \in R_+$, $\vartheta > 0$, such that for all $k \in Z$: $\tau_{k+1} - \tau_k \geq \vartheta$;

(2) β-*condition*, if there exists a $k \geq 0$ such that every unit segment of the real axis R_+ contains no more than k points of sequence τ_k.

Let the solution $x(t)$ satisfy one of the conditions (α or β) and be definite on $[a, \infty)$, $a \in R$. Besides, the solution $x(t)$ is referred to as *unboundedly continuable to the right*.

Let the solution $x_0(t) = x(t; t_0, y_0)$ of system (4.2.1) exist for all $t \geq t_0$ and be unperturbed. We assume that $x_0(t)$ reaches the surface S_k: $t = \tau_k(x)$ at times t_k, $t_{k+1} > t_k$ and $t_k \to \infty$ as $k \to \infty$.

Definition 4.2.3. *Solution* $x_0(t)$ of system (4.2.1) is called

(1) *stable*, if for any tolerance $\varepsilon > 0$, $\Delta > 0$, $t_0 \in R_+$ a $\delta = \delta(t_0, \varepsilon, \Delta) > 0$ exists such that condition $\|x_0 - y_0\| < \delta$ implies $\|x(t) - x_0(t)\| < \varepsilon$ for all $t \geq t_0$ and $|t - t_k| > \Delta$, where $x(t)$ is an arbitrary solution of system (4.3.1) existing on interval $[t_0, \infty)$;

(2) *uniformly stable*, if δ in condition (1) of Definition 4.2.3 does not depend on t_0;

(3) *attractive*, if for any tolerance $\varepsilon > 0$, $\Delta > 0$, $t_0 \in R_+$ there exist $\delta_0 = \delta_0(t_0) > 0$ and $T = T(t_0, \varepsilon, \Delta) > 0$ such that whenever $\|x_0 - y_0\| < \delta_0$, then $\|x(t) - x_0(t)\| < \varepsilon$ for $t \geq t_0 + T$ and $|t - t_k| > \Delta$;

(4) *uniformly attractive*, if δ_0 and T in condition (3) of Definition 4.2.3 do not depend on t_0;

(5) *asymptotically stable*, if conditions (1) and (3) of Definition 4.2.3 hold;

(6) *uniformly asymptotically stable*, if conditions (2) and (3) of Definition 4.2.3 hold.

Remark 4.2.1. If $f(t, 0) = 0$ and $I_k(0) = 0$, $k \in \mathcal{Z}$, then system (4.2.1) admits zero solution. Moreover, if $\tau_k(x) \equiv t_k$, $k \in \mathcal{Z}$, are such that $\tau_k(x)$ do not depend on x, then any solution of system (4.2.1) undergoes the impulsive effect at one and the same time. This situation shows that the notion of stability for system (4.2.1) is an ordinary one.

Remark 4.2.2. Actually the condition (1) of Definition 4.2.3 means that for the solution $x_0(t)$ of system (4.2.1) to be stable in the sense of Liapunov, it is necessary that for $\|x(t_0) - x_0(t_0)\| < \delta$ any solution $x(t)$ of the system remain in the neighborhood of solution $x_0(t)$ for all $t \in [t_0, \infty)$, and point t_0 is not to be the discontinuity point of solutions $x(t)$ and $x_0(t)$.

4.2.2 Auxiliary results

Further we shall need some systematized conditions on functions similar to Liapunov functions for system (4.2.2).

Assumption 4.2.1. *There exist*

(1) *open connected time-invariant neighborhoods*

$$\mathcal{N}_{jx} = \{x_j \colon \|x_j\| < h_{j0}\} \subseteq R^{n_j}$$

of states $x_j = 0$, $j = 1, 2, \ldots, m$, $h_{j0} = \text{const} > 0$;

(2) *functions* φ_{j1}, ψ_{j1} *of class* K;

(3) *constants* a_{jl}, b_{jl}, $j, l = 1, 2, \ldots, m$, *and a matrix-valued function* $U(t, x) = [u_{jl}(t, \cdot)]$ *with elements*

(4.2.4)
$$v_{jj} = v_{jj}(t, x_j); \quad v_{jl} = v_{lj} = v_{jl}(t, x_j, x_l), \quad j \neq l,$$
$$v_{jj}(t, 0) = 0, \quad v_{jl}(t, 0, 0) = 0, \quad j, l = 1, 2, \ldots, m$$

in the domain $\mathcal{T}_0 \times S(\rho_0)$, *where* $\rho_0 = \min_j h_{j0}$, $j = 1, 2, \ldots, m$, *and satisfying estimates*

(4.2.5)
$$a_{jj}\varphi_{j1}^2(\|x_j\|) \leq v_{jj}(t, x_j) \leq b_{jj}\psi_{j1}^2(\|x_j\|)$$
$$\text{for all} \quad (t, x_j) \in \mathcal{T}_0 \times \mathcal{N}_{jx}, \quad j = 1, 2, \ldots, m;$$
$$a_{jl}\varphi_{j1}(\|x_j\|)\varphi_{l1}(\|x_l\|) \leq v_{jl}(t, x_j, x_l) \leq b_{jl}\psi_{j1}(\|x_j\|)\psi_{l1}(\|x_l\|)$$
$$\text{for all} \quad (t, x_{,j}, x_l) \in \mathcal{T}_0 \times \mathcal{N}_{jx} \times \mathcal{N}_{kx}, \quad \text{and} \quad j \neq l.$$

Here $v_{jj} \in C^{1,1}(\mathcal{T}_0 \times R^{n_j}, R_+)$ correspond to subsystems (4.2.3) and $v_{jl} \in C^{1,1}(\mathcal{T}_0 \times R^{n_j} \times R^{n_l}, R)$ take into account connections $f_j^*(t, x)$ and $I_{lj}^*(x)$ between them.

We consider the scalar function

(4.2.6)
$$v(t, x, \eta) = \eta^T U(t, x)\eta, \quad \eta \in R_+^m, \quad \eta > 0,$$

and its total derivative

(4.2.7)
$$Dv(t, x, \eta) = \eta^T DU(t, x)\eta,$$
$$DU(t, x) = Dv_{jl}(t, \cdot), \quad j, l = 1, 2, \ldots, m,$$

due to system of equations (4.2.2).

Lemma 4.2.1. *If all conditions of Assumption 4.2.1 are satisfied, then for function (4.2.6) the estimate*

(4.2.8)
$$u_1^T H^T A H u_1 \leq v(t, x, \eta) \leq u_2^T H^T B H u_2$$
$$\text{for all} \quad (t, x) \in \mathcal{T}_0 \times \mathcal{N}_x,$$

is valid, where

$$u_1 = (\varphi_{11}(\|x_1\|), \varphi_{21}(\|x_2\|), \ldots, \varphi_{m1}(\|x_m\|))^{\mathrm{T}},$$
$$u_2 = (\psi_{11}(\|x_1\|), \psi_{21}(\|x_2\|), \ldots, \psi_{m1}(\|x_m\|))^{\mathrm{T}},$$
$$H = \mathrm{diag}\,\{\eta_1, \eta_2, \ldots, \eta_m\}, \quad A = [a_{jl}], \quad B = [b_{jl}],$$
$$a_{jl} = a_{lj}, \quad b_{jl} = b_{lj}, \quad j, l = 1, 2, \ldots, m,$$

$\mathcal{N}_x \subseteq \mathcal{N}_{1x} \times \mathcal{N}_{2x} \times \cdots \times \mathcal{N}_{mx}$ *is an open connected neighborhood of state* $x = 0$, *such that* $\mathcal{N}_x = \{x\colon \|x\| < \rho_0\}$, $\rho_0 = \min_{j} h_{j0}$.

The Proof of estimate (4.2.8) is similar to that from section 2.2 for system without impulsive effects.

Assumption 4.2.2. *There exist*

(1) *open connected time-invariant neighborhoods* $\mathcal{N}_{jx} \subseteq R^{n_j}$ *of states* $x_j = 0$, $j = 1, 2, \ldots, m$, *and open connected neighborhood* $\mathcal{N}_x \subseteq \mathcal{N}_{1x} \times \cdots \times \mathcal{N}_{mx}$ *of state* $x = 0$;

(2) *functions* v_{jl}, $j, l = 1, 2, \ldots, m$, *mentioned in Assumption 4.2.1 and functions* φ_j, $j = 1, 2, \ldots, m$, φ_m, φ_M *such that in domain* $\mathcal{T}_0 \times S(\rho_0)$ *the conditions* $\varphi_j(0) = \varphi_m(0) = \varphi_M(0) = 0$ *hold, and*

$$(4.2.9) \qquad 0 < \varphi_m(v(t, x, \eta)) \leq \sum_{j=1}^{m} \varphi_j^2(v_{jj}(t, x_j)) \leq \varphi_M(v(t, x, \eta));$$

(3) *constants* $\rho_j^{(1)}$, $\rho_j^{(2)}$, ρ_{jl}, $j \neq l$, $j, l = 1, 2, \ldots, m$, *and the following conditions are satisfied*

(a) $\eta_j^2 \{D_t v_{jj} + (D_{x_j} v_{jj})^{\mathrm{T}} f_j(t, x_j)\} \leq \rho_j^{(1)} \varphi_j^2(v_{jj}(t, x_j))$
for all $(t, x_j) \in \mathcal{T}_0 \times \mathcal{N}_{jx0}$, $j = 1, 2, \ldots, m$;

(b) $\sum\limits_{j=1}^{m} \eta_j^2 (D_{x_j} v_{jj})^{\mathrm{T}} f_j^*(t, x) + 2 \sum\limits_{j=1}^{m} \sum\limits_{\substack{l=2 \\ l>j}}^{m} \eta_j \eta_l \{D_t v_{jl} + (D_{x_j} v_{jl})^{\mathrm{T}} f(t, x)$

$+ (D_{x_l} v_{il})^{\mathrm{T}} f(t, x)\} \leq \sum\limits_{j=1}^{m} \rho_j^{(2)} \varphi_j^2(v_{jj}(t, x_j))$

$+ 2 \sum\limits_{j=1}^{m} \sum\limits_{\substack{l=2 \\ l>j}}^{m} \rho_{jl} \varphi_j(v_{jj}(t, x_j)) \varphi_l(v_{ll}(t, x_l))$

for all $(t, x_j, x_l) \in \mathcal{T}_0 \times \mathcal{N}_{jx0} \times \mathcal{N}_{lx0}$, $j \neq l$.

Here $\mathcal{N}_{jx0} = \{x_j\colon x_j \in \mathcal{N}_{jx}, \ x_j \neq 0\}$, $t \neq \tau_k(x)$, $k = 1, 2, \ldots$.

Lemma 4.2.2. *If all conditions of Assumption 4.2.2 are satisfied, then for expression (4.2.7) the inequality*

(4.2.10)
$$Dv(t,x,\eta)\big|_{(4.2.2)} \leq u^{\mathrm{T}}Gu$$
$$\text{for all} \quad (t,x) \in \mathcal{T}_0 \times \mathcal{N}_{x0}, \quad t \neq \tau_k(x), \quad k = 1, 2, \ldots,$$

is valid, where

$$u^{\mathrm{T}} = (\varphi_1(v_{11}(t,x_1)), \ \varphi_2(v_{22}(t,x_2)), \ \ldots, \ \varphi_m(v_{mm}(t,x_m))),$$
$$G = [\sigma_{jl}], \quad j, l = 1, 2, \ldots, m, \quad \sigma_{jl} = \sigma_{lj},$$
$$\sigma_{jj} = \rho_j^{(1)} + \rho_j^{(2)}, \quad \sigma_{jl} = \rho_{jl}, \quad j \neq l, \quad j, l = 1, 2, \ldots, m.$$

For the Proof of estimate (4.2.10) see Martynyuk and Miladzhanov [2]. Let $\lambda_M(G)$ be maximal eigenvalue of matrix G.

Corollary 4.2.1. *If all conditions of Assumption 4.2.2 are satisfied and*

(1) $\lambda_M(G) < 0$;
(2) $\lambda_M(G) > 0$,

then the following estimates hold true

(4.2.11)
$$Dv(t,x,\eta)\big|_{(4.2.2)} \leq \lambda_M(G)\varphi_m(v(t,x,\eta))$$
$$\text{for all} \quad (t,x) \in \mathcal{T}_0 \times \mathcal{N}_{x0};$$

(4.2.12)
$$Dv(t,x,\eta)\big|_{(4.2.2)} \leq \lambda_M(G)\varphi_M(v(t,x,\eta))$$
$$\text{for all} \quad (t,x) \in \mathcal{T}_0 \times \mathcal{N}_{x0},$$

correspondingly.

Assumption 4.2.3. *There exist*

(1) functions v_{jl}, $j, l = 1, 2, \ldots, m$, mentioned in Assumption 4.2.1, and functions ψ_j, $j = 1, 2, \ldots, m$, ψ_m, ψ_M: $\psi_j(0) = \psi_m(0) = \psi_M(0) = 0$, such that in domain $\mathcal{T}_0 \times \mathcal{S}(\rho_0)$ the conditions

(4.2.13)
$$0 < \psi_m(v(\tau_k(x),x,\eta))$$
$$\leq \sum_{j=1}^{m} \psi_j^2(v_{jj}(\tau_k(x_j),x_j)) \leq \psi_M(v(\tau_k(x),x,\eta)), \quad k = 1, 2, \ldots$$

are satisfied;

(2) *constants* $\alpha_j^{(1)}$, $\alpha_j^{(2)}$, α_{jl} $(j \neq l)$, $j, l = 1, 2, \ldots, m$, *and the follow-ing inequalities are satisfied*

(a) $\eta_j^2 \{ v_{jj}(\tau_k(x_j), x_j + I_{kj}(x_j)) - v_{jj}(\tau_k(x_j), x_j) \}$

$\leq \alpha_j^{(1)} \psi_j^{(2)}(v_{jj}(\tau_k(x_j), x_j))$ *for all* $x_j \in \mathcal{N}_{jx}$, $j = 1, 2, \ldots, m$;

(b) $\displaystyle\sum_{j=1}^{m} \eta_j^2 \{ v_{jj}(\tau_k(x), x_j + I_{kj}(x)) v_{jj}(\tau_k(x), x_j + I_{kj}(x_j))$

$+ v_{jj}(\tau_k(x_j), x_j) - v_{jj}(\tau_k(x), x_j) \}$

$\displaystyle + 2 \sum_{j=1}^{2} \sum_{\substack{l=2 \\ l>j}}^{m} \eta_j \eta_l \{ v_{jl}(\tau_k(x), x_j + I_{kj}(x), x_l + J_{kl}(x))$

$\displaystyle - v_{jl}(\tau_k(x), x_j, x_l) \} \leq \sum_{j=1}^{m} \alpha_j^{(2)} \psi_j^2(v_{jj}(\tau_k(x_j), x_j))$

$\displaystyle + 2 \sum_{j=1}^{m} \sum_{\substack{l=2 \\ l>j}}^{m} \alpha_{jl} \psi_j(v_{jj}(\tau_j(x_j), x_j)) \psi_k(v_{ll}(\tau_k(x_l), x_l))$

for all $(x_j, x_l) \in \mathcal{N}_{jx} \times \mathcal{N}_{lx}$, $k = 1, 2, \ldots$.

Lemma 4.2.3. *If all conditions of Assumption 4.2.3 are satisfied, the estimate*

(4.2.14) $v(\tau_k(x), x + I_k(x), \eta) - v(\tau_k(x), x, \eta) \leq u_k^{\mathrm{T}} C u_k$

is valid, where

$$u_k = (\psi_1(v_{11}(\tau_k(x_1), x_1)), \psi_2(v_{22}(\tau_k(x_2), x_2)), \ldots, \psi_m(v_{mm}(\tau_k(x_m), x_m)))^{\mathrm{T}},$$
$$C = [c_{jl}], \quad j, l = 1, 2, \ldots, m, \quad c_{jl} = c_{lj}, \quad k = 1, 2, \ldots,$$
$$c_{jj} = \alpha_j^{(1)} + \alpha_j^{(2)}, \quad c_{jl} = \alpha_{jl}, \quad j \neq l, \quad j, l = 1, 2, \ldots, m.$$

Proof. Under all conditions of Assumption 4.2.3 we have

$$v(\tau_k(x), x + I_k(x), \eta) - v(\tau_k(x), x, \eta)$$
$$= \eta^{\mathrm{T}}[U(\tau_k(x), x + I_k(x)) - U(\tau_k(x), x)]\eta$$
$$= \sum_{j=1}^{m} \eta_j^2 \{ v_{jj}(\tau_k(x), x_j + I_{kj}(x)) - v_{jj}(\tau_k(x), x_j) \}$$

$$+ 2 \sum_{j=1}^{m} \sum_{\substack{l=2 \\ l>j}}^{m} \eta_j \eta_l \{ v_{jl}(\tau_k(x), x_j + I_{kj}(x), x_l + I_{kl}(x)) - v_{jl}(\tau_k(x), x_j, x_k) \}$$

$$= \sum_{j=1}^{m} \eta_j^2 \{ v_{jj}(\tau_k(x_j), x_j + I_{kj}(x_j)) - v_{jj}(\tau_k(x_j), x_j) \}$$

$$+ \sum_{j=1}^{m} \eta_j^2 \{ v_{jj}(\tau_k(x), x_j + I_{kj}(x_j)) - v_{jj}(\tau_k(x_j), x_j + I_{kj}(x_j))$$

$$+ v_{jj}(\tau_k(x_j), x_j) - v_{jj}(\tau_k(x), x_j) \}$$

$$+ 2 \sum_{j=1}^{m} \sum_{\substack{l=2 \\ l>j}}^{m} \eta_j \eta_l \{ v_{jl}(\tau_k(x), x_j + I_{kj}(x), x_l + I_{kl}(x)) - v_{jl}(\tau_k(x), x_j, x_l) \}$$

$$\leq \sum_{j=1}^{m} \alpha_j^{(1)} \psi_j^2 (v_{jj}(\tau_k(x_j), x_j)) + \sum_{j=1}^{m} \alpha_j^{(2)} \psi_j^2 (v_{jj}(\tau_k(x_j), x_j))$$

$$+ 2 \sum_{j=1}^{m} \sum_{\substack{l=2 \\ l>j}}^{m} \alpha_{jl} \psi_j (v_{jj}(\tau_k(x_j), x_j)) \, \psi_l(v_{ll}(\tau_k(x_l), x_l))$$

$$= \sum_{j=1}^{m} c_{kj} \psi_j (v_{jj}(\tau_k(x_j), x_j)) \, \psi_l(v_{ll}(\tau_k(x_k), x_l)) = u_k^{\mathrm{T}} C u_k, \quad k = 1, 2, \dots.$$

Corollary 4.2.2. *If all conditions of Assumption 4.2.3 are satisfied and*

(1) $\lambda_M(C) < 0$,

(2) $\lambda_M(C) > 0$,

then the following estimates hold true

(4.2.15)
$$v(\tau_k(x), x + I_k(x), \eta) - v(\tau_k(x), x, \eta)$$
$$\leq \lambda_M(C) \psi_m(v(\tau_k(x), x, \eta)) \quad \text{for all} \quad x \in \mathcal{N}_{x0};$$

(4.2.16)
$$v(\tau_k(x), x + I_k(x), \eta) - v(\tau_k(x), x, \eta)$$
$$\leq \lambda_M(C) \psi_M(v(\tau_k(x), x, \eta)) \quad \text{for all} \quad x \in \mathcal{N}_{x0},$$

correspondingly.

Here $\lambda_M(C)$ *is maximal eigenvalue of matrix* C.

Assumption 4.2.4. *There exist*

(1) *functions* v_{jl}, $j, l = 1, 2, \dots, m$, *mentioned in Assumption 4.3.1 and functions* ψ_j, $j = 1, 2, \dots, m$, ψ_m, ψ_M, *mentioned in Assumption 4.2.3;*

(2) *constants* $\beta_j^{(1)}$, $\beta_j^{(2)}$, β_{jl}, $j \neq l$, $j, l = 1, 2, \dots, m$, *and for all* $k = 1, 2, \dots$ *the following conditions are satisfied*

(a) $\eta_j^2 v_{jj}(\tau_k(x_j), x_j + I_{kj}(x_j)) \leq \beta_j^{(1)} \psi_j^2(v_{jj}(\tau_k(x_j), x_j))$
   for all $x_j \in \mathcal{N}_{jx}$, $j = 1, 2, \ldots, m$;

(b) $\sum\limits_{j=1}^{m} \eta_j^2 \{v_{jj}(\tau_k(x), x_j + I_{kj}(x)) - v_{jj}(\tau_k(x_j), x_j + I_{kj}(x_j))\}$

$\qquad + 2 \sum\limits_{j=1}^{m} \sum\limits_{\substack{l=2 \\ l>j}}^{m} \eta_j \eta_l v_{jl}(\tau_k(x), x_j + I_{kj}(x), x_l + I_{kl}(x))$

$\qquad \leq \sum\limits_{j=1}^{m} \beta_j^{(2)} \psi_j^2(v_{jj}(\tau_k(x_j), x_j)) + 2 \sum\limits_{j=1}^{m} \sum\limits_{\substack{l=2 \\ l>j}}^{m} \beta_{jl} \psi_j(v_{jj}(\tau_k(x_j), x_j))$

$\qquad \times \psi_l(v_{ll}(v_{ll}(\tau_k(x_l), x_l)))$
   for all $(x_j, x_l) \in \mathcal{N}_{jx} \times \mathcal{N}_{lx}$.

Lemma 4.2.4. *If all conditions of Assumption 4.2.4 are satisfied, then the estimate*

(4.2.17)
$$v(\tau_k(x), x + I_k(x), \eta) \leq u_k^{\mathrm{T}} C^* u_k$$
$$\text{for all} \quad x \in \mathcal{N}_x, \quad k = 1, 2, \ldots,$$

takes place, where

$$C^* = [c_{jl}^*], \quad j, l = 1, 2, \ldots, m, \quad c_{jl} = c_{lj},$$
$$c_{jj}^* = \beta_j^{(1)} + \beta_j^{(2)}, \quad c_{jl}^* = \beta_{jl}, \quad j \neq l, \quad j, l = 1, 2, \ldots, m.$$

The Proof of Lemma 4.2.4 is similar to that of Lemma 4.2.3.

Corollary 4.2.3. *If all conditions of Assumption 2.4.4 are satisfied and*

(1) $\lambda_M(C^*) < 0$;
(2) $\lambda_M(C^*) > 0$,

then for all $k = 1, 2, \ldots$

(4.2.18)
$$v(\tau_k(x), x + I_k(x), \eta) \leq \lambda_M(C^*) \psi_m(v(\tau_k(x), x, \eta))$$
$$\text{for all} \quad x \in \mathcal{N}_{x0};$$

(4.2.19)
$$v(\tau_k(x), x + I_k(x), \eta) \leq \lambda_M(C^*) \psi_M(v(\tau_k(x), x, \eta))$$
$$\text{for all} \quad x \in \mathcal{N}_{x0},$$

correspondingly.

Here $\lambda_M(C^*)$ *is the maximal eigenvalue of matrix* C^*.

Assumption 4.2.5. *Conditions (1) and (2) of Assumption 4.2.2 are satisfied and in the inequalities of condition (3) of Assumption 4.2.2 the inequality sign "$\leq$" is reversed.*

Lemma 4.2.5. *If all conditions of Assumption 4.2.5 are satisfied, then for (4.2.7) the estimate*

$$(4.2.20) \qquad Dv(t,x,\eta)\big|_{(4.2.2)} \geq u^{\mathrm{T}}Gu \quad \text{for all} \quad (t,x) \in \mathcal{T}_0 \times \mathcal{N}_{x0}$$

is valid, where u and G are defined as in Lemma 4.2.2.

Corollary 4.2.4. *If the conditions of Assumption 4.2.5 are satisfied and*

(1) $\lambda_m(G) < 0$;

(2) $\lambda_m(G) > 0$,

then

$$(4.2.21) \qquad \begin{aligned} Dv(t,x,\eta)\big|_{(4.2.2)} &\geq \lambda_m(G)\varphi_M(v(t,x,\eta)) \\ &\text{for all} \quad (t,x) \in \mathcal{T}_0 \times \mathcal{N}_{x0}; \end{aligned}$$

$$(4.2.22) \qquad \begin{aligned} Dv(t,x,\eta)\big|_{(4.2.2)} &\geq \lambda_m(G)\varphi_m(v(t,x,\eta)) \\ &\text{for all} \quad (t,x) \in \mathcal{T}_0 \times \mathcal{N}_{x0}, \end{aligned}$$

correspondingly.

Here $\lambda_m(G)$ is a minimal eigenvalue of matrix G.

Assumption 4.2.6. *Condition (1) of Assumption 4.2.3 is satisfied and in inequalities of condition (2) of Assumption 4.2.3 the inequality sign "$\leq$" is reversed.*

Lemma 4.2.6. *Under conditions of Assumption 4.2.6, the estimate*

$$(4.2.23) \qquad \begin{aligned} v(\tau_k(x),\, x + I_k(x),\eta) - v(\tau_k(x),x,\eta) &\geq u_k^{\mathrm{T}}Cu_k \\ &\text{for all} \quad k = 1,2,\ldots, \end{aligned}$$

takes place, where u_k and C are defined as in Lemma 4.2.3.

Corollary 4.2.5. *If in inequality (4.2.23) $\lambda_m(C) > 0$, then for all $k = 1,2,\ldots$ estimate*

$$(4.2.24) \qquad \begin{aligned} v(\tau_k(x),\, x + I_k(x),\eta) - v(\tau_k(x),x,\eta) &\geq \lambda_m(C)\psi_m(v(\tau_k(x),x)) \\ &\text{for all} \quad x \in \mathcal{N}_{x0}, \end{aligned}$$

takes place.

Assumption 4.2.7. *Condition (1) of Assumption 4.2.4 is satisfied and in inequalities of condition (2) of the same assumption the inequality sign "≤" is reversed.*

Lemma 4.2.7. *Under conditions of Assumption 4.2.7 the estimate*

$$(4.2.25) \qquad \begin{aligned} v(\tau_k(x), \, x + I_k(x), \eta) &\geq u_k^{\mathrm{T}} C^* u_k \\ \text{for all} \quad x \in \mathcal{N}_x, \quad \text{and} \quad k &= 1, 2, \ldots, \end{aligned}$$

takes place, where u_k and C^* are defined as in Lemma 4.2.4.

Corollary 4.2.6. *If in inequality (4.2.25) $\lambda_m(C^*) > 0$, then for all $k = 1, 2, \ldots$ the inequality*

$$(4.2.26) \qquad \begin{aligned} v(\tau_k(x), \, x + I_k(x), \eta) &\geq \lambda_m(C^*)\psi_m(v(\tau_k(x), x, \eta)) \\ \text{for all} \quad x &\in \mathcal{N}_{x0}, \end{aligned}$$

is valid.

Assumption 4.2.8. *Let*

$$\Pi_j = \{(t, x_j) \in \mathcal{T}_0 \times \mathcal{N}_{jx} : \; v_{jj}(t, x_j) > 0\}$$

be domains of positiveness of the functions $v_{jj}(t, x_j)$, $j = 1, 2, \ldots, m$, for any $t \geq t_0$ having nonzero an open cross-section by plane $t = \mathrm{const}$ adherent to the origin, and in this domain functions v_{jl}, $j, l = 1, 2, \ldots, m$, are bounded.

Lemma 4.2.8. *Under conditions of Assumptions 4.2.1 and 4.2.8 and if matrix A in estimate (4.2.8) is positive definite, i.e. $\lambda_m(H^{\mathrm{T}}AH) > 0$, then*

(1) *domain $\Pi = \{(t, x) \in \mathcal{T}_0 \times \mathcal{S}(\rho): \; v(t, x, \eta) > 0\}$ of function $v(t, x, \eta)$ positiveness for any $t \in \mathcal{T}_0$ has nonzero open cross-section by plane $t = \mathrm{const}$ adherent to the origin;*

(2) *in domain Π function $v(t, x, \eta)$ is bounded.*

Proof. Under conditions of Assumption 4.2.8 and $\lambda_m(H^{\mathrm{T}}AH) > 0$ the domain of function $v(t, x, \eta)$ positiveness is

$$\Pi = \{(t, x) \in \mathcal{T}_0 \times \mathcal{S}(\rho): \; x \neq 0\},$$

that has an open cross-section by the plane $t = \mathrm{const}$ for any $t \in \mathcal{T}_0$. Moreover, positiveness of function $v_{jj}(t, x)$ is a necessary condition for positiveness of function $v(t, x, \eta)$ and therefore, $\Pi \subseteq \bigcap\limits_{j=1}^{m} \Pi_j$. Domains Π_j

for any $j = 1, 2, \ldots, m$ have, by condition of Assumption 4.2.8, nonzero open cross-section by plane $t = \text{const}$ adherent to the origin. This proves assertion (1) of Lemma 4.2.8.

Boundedness of functions v_{jl}, $j, l = 1, 2, \ldots, m$, implies that matrix $U(t, x)$ is bounded, but then function $v(t, x, \eta)$ constructed according to (4.2.6) is also bounded.

4.2.3 Sufficient stability conditions

Assumptions 4.2.1 – 4.2.8, Lemmas 4.2.1 – 4.2.8 and Corollaries 4.2.1 – 4.2.6 allow us to formulate various stability and asymptotic stability conditions for zero solution of system (4.2.2).

Theorem 4.2.1. *If differential perturbed motion equations of large scale impulsive system (4.2.2) are such that in domain $\mathcal{T}_0 \times \mathcal{S}(\rho)$ all conditions of Assumptions 4.2.1, 4.2.2, and 4.2.3 are satisfied, and*

(1) *matrix A is positive definite $(\lambda_m(H^{\mathrm{T}}AH) > 0)$;*

(2) *matrix G is negative semi-definite $(\lambda_M(G) \leq 0)$;*

(3) *matrix C is negative semi-definite or equals to zero $(\lambda_M(C) \leq 0)$,*

then the zero solution of large scale impulsive system (4.2.2) is stable.

If instead of condition (3) the following condition is satisfied

(4) *matrix C is negative definite $(\lambda_M(C) < 0)$,*

then zero solution of large scale impulsive system (4.2.2) is asymptotically stable.

Proof. Under Assumption 4.2.1, Lemma 4.2.1 and condition (1) of Theorem 4.2.1 function $v(t, x, \eta)$, constructed by (4.2.6), is positive definite. Conditions of Assumption 4.2.2, Lemma 4.2.2 and condition (2) of Theorem 4.2.1 imply

$$(4.2.27) \qquad Dv(t, x, \eta)\big|_{(4.2.2)} \leq 0, \quad t \neq \tau_k(x), \quad (t, x) \in \mathcal{T}_0 \times \mathcal{S}(\rho).$$

From conditions of Assumption 4.2.3, Lemma 4.2.3 and condition (3) of Theorem 4.2.1 it follows that on hypersurfaces $\mathcal{S}_k$: $t = \tau_k(x)$, $x \in \mathcal{S}(\rho)$, the inequalities

$$(4.2.28) \qquad v(\tau_k(x), x + I_k(x), \eta) \leq v(\tau_k(x), x, \eta), \quad k = 1, 2, \ldots,$$

are satisfied.

Conditions (4.2.27) and (4.2.28) are sufficient for zero solution of large scale impulsive system (4.2.2) to be stable.

If instead of conditions (3) of Theorem 4.2.1 condition (4) of Theorem 4.2.1 is satisfied, then by Lemma 4.2.3 and Corollary 4.2.2 we find the estimate

(4.2.29)
$$v(\tau_k(x), \, x + I_k(x), \eta) - v(\tau_k(x), x, \eta)$$
$$\leq \lambda_M(C)\psi_m(v(\tau_k(x), x, \eta), \quad k = 1, 2, \ldots.$$

By virtue of conditions (4.2.27) and (4.2.29) the zero solution of large scale impulsive system (4.2.2) is asymptotically stable.

The theorem is proved.

Theorem 4.2.2. *Let differential perturbed motion equations of large scale impulsive system (4.2.2) be such that in domain $\mathcal{T}_0 \times S(\rho)$ the conditions of Assumptions 4.2.1, 4.2.2 and 4.2.4 are satisfied, and*

(1) *matrix A is positive definite $(\lambda_m(H^TAH) > 0)$;*
(2) *matrix G is negative definite $(\lambda_M(G) < 0)$;*
(3) $\lambda_M(C^*) > 0$;
(4) *functions $\tau_k(x)$ and constant ϑ satisfy the inequality*

$$\sup_k \left(\min_{x \in S(\rho)} \tau_{k+1}(x) - \max_{x \in S(\rho)} \tau_k(x) \right) = \vartheta > 0, \quad \text{where} \quad \rho < \rho_0;$$

(5) *functions $\varphi_m(y)$ and $\psi_M(y)$ and constant $a_0 > 0$ are such that for all $a \in (0, a_0]$ the estimate*

(4.2.30)
$$-\frac{1}{\lambda_M(G)} \int_a^{\lambda_M(C^*)\psi_M(a)} \frac{dy}{\varphi_m(y)} \leq \vartheta$$

is valid.

Then the zero solution of large scale impulsive system (4.2.2) is stable.

If instead of inequality (4.2.30) for some $\gamma > 0$ the inequality

(4.2.31)
$$-\frac{1}{\lambda_M(G)} \int_a^{\lambda_M(C^*)\psi_M(a)} \frac{dy}{\varphi_m(y)} \leq \vartheta - \gamma$$

is satisfied, the zero solution of large scale impulsive system (4.2.2) is asymptotically stable.

Proof of Theorem 4.2.2 is similar to that of Theorem 4.2.1.

Theorem 4.2.3. *Let differential perturbed motion equations of large scale impulsive system (4.2.2) be such that in domain $\mathcal{T}_0 \times \mathcal{S}(\rho)$ the conditions of Assumptions 4.2.1, 4.2.2, and 4.2.4 are satisfied, and*

(1) *matrix A is positive definite, i.e. $\lambda_m(H^TAH) > 0$;*

(2) $\lambda_M(G) > 0$;

(3) $\lambda_M(C^*) > 0$;

(4) *functions $\tau_k(x)$ and constant $\vartheta_1 > 0$ are such that for all $k = 1, 2, \ldots$ the inequality*

$$\max_{x \in \mathcal{S}(\rho)} \tau_k(x) - \min_{x \in \mathcal{S}(\rho)} \tau_{k-1}(x) \leq \vartheta_1, \quad \text{where} \quad \rho < \rho_0$$

is satisfied;

(5) *functions $\varphi_M(y)$ and $\psi_M(y)$ and constant a_0 are such that for all $a \in (0, a_0]$ the estimate*

$$(4.2.32) \qquad \frac{1}{\lambda_M(G)} \int_{\lambda_M(C^*)\psi_M(a)}^{a} \frac{dy}{\varphi_M(y)} \geq \vartheta_1$$

is satisfied.

Then zero solution of large scale impulsive system (4.2.2) is stable.

If instead of (4.2.32) the inequality

$$(4.2.33) \qquad \frac{1}{\lambda_M(G)} \int_{\lambda_M(C^*)\psi_M(a)}^{a} \frac{dy}{\varphi_M(y)} \geq \vartheta_1 + \gamma$$

is satisfied, for which $\gamma > 0$, then the zero solution of large scale impulsive system (4.2.2) is asymptotically stable.

Proof of this theorem is similar to that of Theorem 4.2.1.

4.2.4 Instability conditions

We establish sufficient instability conditions for the zero solution of large scale impulsive system (4.2.2) in terms of Assumptions 4.2.5 – 4.2.8 and Lemmas 4.2.6 – 4.2.8.

Theorem 4.2.4. *If differential perturbed motion equations of large scale impulsive system (4.2.2) are such that conditions of Assumptions 4.2.1, 4.2.5, 4.2.6, and 4.2.8 are satisfied and in the domain* Π

(1) *matrix* A *is positive definite (i.e.* $\lambda_m(H^TAH) > 0$);

(2) *matrix* G *is positive semi-definite or equals to zero (i.e.* $\lambda_m(G) \geq 0$);

(3) *matrix* C *is positive definite (i.e.* $\lambda_m(C) > 0$),

then zero solution of large scale impulsive system (4.2.2) is unstable.

Proof. Under conditions of Assumptions 4.2.1 and 4.2.8, and Lemmas 4.2.1, 4.2.8 and condition (1) of Theorem 4.2.4, the function $v(t,x)$ is positive definite and possesses properties (A) and (B). By conditions of Assumption 4.2.5, Lemma 4.2.5 and condition (2) of Theorem 4.2.4

$$(4.2.35) \qquad Dv(t,x,\eta)\big|_{(4.2.2)} \geq 0, \quad t \neq \tau_k(x) \quad \text{for all} \quad x \in \Pi.$$

From Assumption 4.2.6, Lemma 4.2.6, Corollary 4.2.5 and condition (3) of Theorem 4.2.4 it follows

$$\begin{aligned}
&v(\tau_k(x),\ x + I_k(x),\eta) - v(\tau_k(x),x,\eta) \\
(4.2.36) \qquad &\geq \lambda_m(C)\psi_m(v(\tau_k(x),x,\eta)), \\
&t = \tau_k(x), \quad \text{for all} \quad x \in \Pi, \quad k = 1,2,\dots.
\end{aligned}$$

Conditions (4.2.35) and (4.2.36) are sufficient for the zero solution of large scale impulsive system (4.2.2) to be unstable.

Theorem 4.2.5. *Let differential perturbed motion equations of large scale impulsive system (4.2.2) be such that conditions of Assumptions 4.2.1, 4.2.5, 4.2.7, and 4.2.8 are satisfied and in the domain* Π

(1) *matrix* A *is positive definite (i.e.* $\lambda_m(H^TAH) > 0$);

(2) $\lambda_m(G) < 0$;

(3) *matrix* C^* *is positive definite (i.e.* $\lambda_m(C^*) > 0$);

(4) *functions* $\tau_k(x)$ *and constant* $\vartheta_1 > 0$ *are such that for all* $k = 1,2,\dots$ *the inequality*

$$\max_{x \in S(\rho)} \tau_k(x) - \min_{x \in S(\rho)} \tau_{k-1}(x) \leq \vartheta_1$$

holds, where $\rho < \rho_0$;

(5) *functions* $\varphi_m(y)$ *and* $\psi_m(y)$ *and constant* $\gamma > 0$ *for all* $a \in (0, a_0]$ *satisfy inequality*

$$-\frac{1}{\lambda_m(G)} \int\limits_{a}^{\lambda_m(C^*)\psi_m(a)} \frac{dy}{\varphi_M(y)} \geq \vartheta_1 + \gamma.$$

Then the zero solution of large scale impulsive system (4.2.2) is unstable.

Proof of Theorem 4.2.5 is similar to that of Theorem 4.2.4.

Theorem 4.2.6. *Let differential perturbed motion equations of large scale impulsive system (4.2.2) be such that conditions of Assumptions 4.2.1, 4.2.5, 4.2.7, and 4.2.8 are satisfied and in domain Π*

(1) *matrix A is positive definite (i.e. $\lambda_m(H^T AH) > 0$);*

(2) *matrix G is positive definite (i.e. $\lambda_m(G) > 0$);*

(3) *matrix C^* is positive definite (i.e. $\lambda_m(C^*) > 0$);*

(4) *functions $\tau_k(x)$ and constant ϑ are such that*

$$\sup_k \left(\min_{x \in S(\rho)} \tau_{k+1}(x) - \max_{x \in S(\rho)} \tau_k(x) \right) = \vartheta > 0, \quad \rho < \rho_0;$$

(5) *functions $\varphi_m(y)$ and $\psi_m(y)$ and constant $\gamma > 0$ are such that for all $a \in (0, a_0]$ the inequality*

$$\frac{1}{\lambda_m(G)} \int_{\lambda_m(C^*)\psi_m(a)}^{a} \frac{dy}{\varphi_m(y)} \leq \vartheta - \gamma$$

is satisfied.

Then the zero solution of large scale impulsive (4.2.2) is unstable.

Proof of Theorem 4.2.6 is similar to that of Theorem 4.2.4.

Example 4.2.1. We analyze stability of the lower position of a pendulum subjected to impulse effect, the dynamics of which is described by the system of equations (see Samoilenko and Perestyuk [1])

$$\frac{dx}{dt} = y, \quad \frac{dy}{dt} = -\sin x, \quad t \neq \tau_k(x, y);$$

(4.2.37)

$$\Delta x = -x + \arccos(-0.5\, y^2 + \cos x), \quad t = \tau_k(x, y);$$

$$\Delta y = -y, \quad t = \tau_k(x, y).$$

For system (4.2.37) we construct matrix function $U(x, y)$ with elements

$$v_{11}(x) = 1 - \cos x, \quad v_{22}(y) = 0.5\, y^2,$$

$$v_{12}(x, y) = v_{21}(x, y) = 0,$$

satisfying the estimates

$$v_{11}(x) \geq 0.5\, \varepsilon |x|^2, \quad v_{22}(y) \geq 0.5\, |y|^2,$$

where $\varepsilon > 0$ is sufficiently small, i.e. $\varepsilon \to 0$.

Matrix

$$A = \begin{pmatrix} 0.5\,\varepsilon & 0 \\ 0 & 0.5 \end{pmatrix}$$

from (4.2.8) is positive definite.

If $\eta^{\mathrm{T}} = (1, 1)$, then given matrix function $U(x, y)$, we have

$$\rho_j^{(l)} = \rho_{12} = 0, \quad \alpha_j^{(l)} = \alpha_{12} = 0, \quad j, l = 1, 2,$$

and matrices G and C from conditions (4.2.10) and (4.2.14) are equal to zero.

Thus, no matter what properties the surfaces $\mathcal{S}_k: t_k = \tau_k(x, y)$ have, conditions of Theorem 4.2.1 are satisfied and therefore, zero solution of system (4.2.37) is stable.

Example 4.2.2. Consider the system of equations

$$\frac{dx}{dt} = (4.5 + 0.9\,\sin^2 x)y^3,$$

(4.2.38) $\qquad \dfrac{dy}{dt} = -(5 + \sin^2 x)y^3, \qquad t \neq \tau_k(x, y);$

$$\Delta x = -x + \sigma y,$$

$$\Delta y = \sigma x - y, \qquad t = \tau_k(x, y).$$

For system (4.2.38) we construct matrix function $U(x, y)$ with elements

$$v_{11}(x) = x^2, \quad v_{22}(y) = y^2,$$
$$v_{12}(x, y) = v_{21}(x, y) = 0.9\,xy,$$

for which estimates

$$v_{11}(x) \geq |x|^2, \quad v_{22}(y) \geq |y|^2,$$
$$v_{12}(x, y) \geq -0.9|x|\,|y|$$

are valid.

Matrix

$$A = \begin{pmatrix} 1 & -0.9 \\ -0.9 & 1 \end{pmatrix}$$

is positive definite.

If $\eta^T = (1,1)$, then matrices G and C from (4.2.10) and (4.2.14) are

$$G = \begin{pmatrix} 0 & 0 \\ 0 & -0.28 \end{pmatrix}, \qquad C = \begin{pmatrix} \sigma^2 - 1 & \frac{9}{10}|\sigma^2 - 1| \\ \frac{9}{10}|\sigma^2 - 1| & \sigma^2 - 1 \end{pmatrix}.$$

We note that $\lambda_M(G) = 0$, i.e. matrix G is negative semi-definite, for $\sigma = \pm 1$ matrix C is equal to zero, and for $|\sigma| < 1$ we have $\lambda_m(C) < 0$, i.e. matrix C is negative definite.

Thus all conditions of Theorem 4.2.1 are satisfied and zero solution of system of equations (4.2.38) is stable for $\sigma = \pm 1$, and asymptotically stable for $|\sigma| < 1$.

4.3 Hierarchical Impulsive Systems

We consider the impulsive system described in Section 4.2.2. Namely

(4.3.1)
$$\frac{dx}{dt} = f(t,x), \quad t \neq \tau_k(x),$$

$$\Delta x = x(t+0) - x(t) = I_k(x),$$

$$x(t_0^+) = x_0, \quad k \in \mathcal{Z}.$$

Here $x \in (x_1, \ldots, x_n)^T \in R^n$, $f \in C(R_+ \times \Omega(\rho), R^n)$, $I_k \in C(\Omega(\rho), R^n)$, $\Omega(\rho) = \{x \in R^n : \|x\| < \rho\}$, $\rho > 0$, $\Omega(\rho) \subseteq R^n$.

We take some general assumptions on system (4.3.1).

A_1. Functions $\tau_k \colon \Omega \to R_+$ are continuous in x.

A_2. The following correlations hold

$$0 < \tau_1(x) < \tau_2(x) < \cdots < \tau_k(x) < \ldots \quad \text{for} \quad x \in \Omega,$$
$$\inf\{\tau_k(x) - \tau_{k-1}(x) \colon k \geq 2, \ x \in R^n\} > 0$$

and $\lim\limits_{k \to \infty} \tau_k(x) = \infty$ uniformly in $x \in \Omega$.

A_3. The integral curve of any solution of system (4.3.1) intersect any hypersurface $S_k \colon t = \tau_k(x)$, $k \in \mathcal{Z}$, not more than once.

A_4. There exists a constant $\mu \in (0, \rho)$ such that as soon as $x \in \Omega(\mu)$, then $x + I_k(x) \in \Omega(\rho)$ for all $k \in \mathcal{Z}$.

A_5. For any division of vector $x \in R^n$ into subvectors $x_s \in R^{n_s}$, $n_1 + \cdots + n_m = n$, the values $\tau_k(x_s)$, $k \in \mathcal{Z}$, satisfy conditions $A_1 - A_4$.

Among conditions $A_1 - A_5$ the condition A_3 is the one that can be verified for system (4.3.1) in terms of the below result.

Theorem 4.3.1. *Suppose that*

(1) $f \in C(R_+ \times \Omega, R^n)$, $\tau_k \in C^1(\Omega, (0, \infty))$, $\tau_k(x) < \tau_{k+1}(x)$, $k \in Z$,
$\lim\limits_{k \to \infty} \tau_k(x) = \infty$ *uniformly in* $x \in \Omega$ *and* $I_k \in C(\Omega, R^n)$;

(2) *for all* $(t, x) \in R_+ \times \Omega$

    (a) $\dfrac{\partial \tau_k(x)}{\partial x} f(t, x) \le 0$;

    (b) $x + I_k(x) \in \Omega$ *for* $x \in \Omega$ *and*

$$\frac{\partial \tau_k}{\partial x}(x + sI_k(x)) I_k(x) \le 0, \quad 0 \le s \le 1, \quad k \in Z.$$

Then any solution of system (4.3.1) intersects hypersurface S_k: $t = \tau_k(x)$ *not more than once.*

Proof. Assume on the contrary. Let for system (4.3.1) there exist a solution $x(t, t_0, x_0)$ and some surface S_j such that $x(t)$ intersects S_j twice and more. Let the first time of surface S_j intersection take place at time $t = t_q$ for some q and the next time at $t = t^*$. At the same time we have $t_q = \tau_j(x(t_q))$ and $t^* = \tau_j(x(t^*))$, $t_0 < t_q < t^*$.

Further the two situations are to be considered.

Case 1. Time $t^* = t_{q+m}$ and solution $x(t)$ intersects the surfaces S_k $m - 1$ times for $t \in (t_q, t_{q+m})$, $t_{q+1}, \dots, t_{q+m-1}$.

Case 2. The solution $x(t)$ intersects the surfaces S_k infinite number of times for $t \in (t_q, t^*)$.

Let us discuss the Case 1. Condition (2a) implies that the function $\tau_k(x(t))$ does not increase for $t \in (t_\eta, t_{\eta+1})$ for any $k > 1$ and $k \le \eta \le k + m - 1$. Hence it follows in view of condition (2b) that

$$\tau_k(x + I_k(x)) \le \tau_k(x)$$

for any $x \in \Omega$ and $k = 1$. Assume that $x(t)$ hits on the surfaces S_{n_i} at times $t = t_i$, $i = k + 1, k + 2, \dots, k + m - 1$. Then we let $x_k = x(t_k)$,

    (a) $t_k = \tau_j(x_k) \ge \tau_j(x_k + I_k(x_k)) \ge \tau_j(x_{k+1})$ for $k+1 \le i \le k+m-1$;

    (b) $t_i = \tau_{n_i}(x_i) \ge \tau_{n_i}(x_{ik} + I_{n_i}(x_i)) \ge \tau_{n_i}(x_{i+1})$.

From (a) it follows that $j < n_{k+1}$, that leads to the contradiction

$$t_{k+1} = \tau_{n_{k+1}}(x_{k+1}) < \tau_j(x_{k+1}) < t_k.$$

Proceeding in the same way and taking (b) into account we get $j < n_{k+1} < \dots < n_{k+m-1} < j$, is a contradiction.

In Case 2 a sequence of impulsive perturbation moments $t_k, t_{k+1}, \ldots$ can be defined so that $t_k < t_{k+1} < t_{k+j} < t^*$ for $1 \le i \le j$. Here two possibilities occur.

Case 1'. The solution $x(t)$ intersects the surfaces S_i that differ one from another an infinite number of times in $\{t_{k+j} : j \ge 1\}$.

Case 2'. For some $1 < j_1 < j_2$ and j the solution $x(t)$ intersects the same surface S_j at times t_{j_1} and t_{j_2}.

In Case 1', since $\lim\limits_{k \to \infty} \tau_k(x) = \infty$ uniformly in the domain Ω, there must exist j such that $\tau_j(x) > t^*$ on Ω and some i such that $t_{k+i} = \tau_j(x(t_{k+i})) > t^*$. This contradicts the above made assumption.

In Case 2' we return back to the situation considered in the Case 1. This completes the proof of the theorem.

Assume that system (4.3.1) consists of m independent subsystems

(4.3.2)
$$\frac{dx_i}{dt} = g_i(t, x_i), \quad t \ne \tau_k(x_i), \quad i = 1, 2, \ldots, m,$$
$$\Delta x_i = \Phi_i^k(x_i), \quad t = \tau_k(x_i), \quad k \in \mathcal{Z},$$
$$x_i(t_0^+) = x_{i0},$$

where $x_i \in R^{n_i}$, $g_i \in C(R_+ \times R^{n_i}, R^{n_i})$, $g_i(t, 0) = 0$ for all $t \in R_+$ and the link functions

(4.3.3)
$$h_i(t, x_1, \ldots, x_m) \colon h \in C\left(R_+ \times R^{n_1} \times \cdots \times R^{n_m}, R^{n_i}\right),$$
$$t \ne \tau_k(x_i),$$
$$J_i^k(x) = I_i^k(x) - \Phi_i^k(x_i), \quad t = \tau_k(x_i).$$

Thus, system (4.3.1) can be transformed as follows

(4.3.4)
$$\frac{dx_i}{dt} = g_i(t, x_i) + h_i(t, x_1, \ldots, x_m), \quad t \ne \tau_k(x_i),$$
$$\Delta x_i = \Phi_i^k(x_i) + J_i^k(x), \qquad t = \tau_k(x_i).$$

Besides, we assume that subsystems (4.3.2) are disconnected and

(4.3.5)
$$X = X_1 \oplus X_2 \oplus \cdots \oplus X_m,$$

where X and X_i are state spaces of the systems (4.3.1) and (4.3.2) respectively.

Further we assume that each subsystem (4.3.2) allows the decomposition into M_i components defined by

$$
\text{(4.3.6)} \quad
\begin{aligned}
\frac{dx_{ij}}{dt} &= p_{ij}(t, x_{ij}), \quad t \neq \tau_k(x_{ij}), \quad j = 1, 2, \ldots, M_i, \\
\Delta x_{ij} &= \Phi_{ij}^k(x_{ij}), \quad t = \tau_k(x_{ij}), \quad k \in \mathcal{Z}, \quad x_{ij}(t_0^+) = x_{ij0},
\end{aligned}
$$

that after interacting form the subsystems

$$
\text{(4.3.7)} \quad
\begin{aligned}
\frac{dx_{ij}}{dt} &= p_{ij}(t, x_{ij}) + q_{ij}(t, x_i), \quad t \neq \tau_k(x_{ij}), \quad j = 1, 2, \ldots, M_i, \\
\Delta x_{ij} &= I_{ij}^k(x_{ij}) + J_{ij}^k(x_i), \quad t = \tau_k(x_{ij}), \quad k \in \mathcal{Z}, \\
x_{ij}(t_0^+) &= x_{ij0},
\end{aligned}
$$

where $x_{ij} \in R^{n_{ij}}$, $p_{ij} \in C(R_+ \times R^{n_{ij}}, R^{n_{ij}})$, $q_{ij} \in C(R_+ \times R^{n_i}, R^{n_{ij}})$ and $x_{ij} = 0$ is the only equilibrium state of subsystems (4.3.6). Assume that subsystems (4.3.6) are disconnected, i.e.

$$
\text{(4.3.8)} \qquad X_i = X_{i1} \oplus X_{i2} \oplus \cdots \oplus X_{iM_i}, \quad i = 1, 2, \ldots, m,
$$

where X_i and X_{ij} are state spaces of the subsystems (4.3.2) and (4.3.6) respectively.

Impulsive systems modeled by the equations (4.3.1) and allowing the first level decomposition (4.3.2) – (4.4.4) and the second level one (4.3.6) – (4.3.8) have *multilevel hierarchical structure*.

4.4 Analytical Construction of Liapunov Function

4.4.1 Structure of hierarchical matrix-valued Liapunov function

According to two levels of decomposition (4.3.2) – (4.3.4) and (4.3.6) – (4.3.8) of system (4.3.1) we suppose two-level construction of submatrix of matrix-valued function

$$
\text{(4.4.1)} \qquad U(t, x) = [U_{ij}(t, \cdot)], \quad U_{ij} = U_{ji},
$$

where $U_{ii}\colon R \times R^{n_i} \to R_+$, $i = 1, 2, \ldots, m$, $U_{ij}\colon R_+ \times R^{n_i} \times R^{n_j} \to R$, $i \neq j$, $j = 1, 2, \ldots, m$, for the system (4.3.1).

Further we need the class U_0 of piece-wise continuous matrix-valued functions. Let $\tau_0(x) \equiv 0$ for all $x \in R^n$. We introduce the sets (cf. Bainov and Kulev [1])

$$G_k = \{(t,x) \in R_+ \times \Omega(\rho): \tau_{k-1}(x) < t < \tau_k(x)\},$$

$$G = \bigcup_{k=1}^{\infty} G_k, \quad k \in \mathcal{Z},$$

$$\sigma_k = \{(t,x) \in R_+ \times \Omega: t = \tau_k(x)\}, \quad k \in \mathcal{Z}.$$

Definition 4.4.1. The *matrix-valued function* $U\colon R_+ \times \Omega(\rho) \to R^{m \times m}$ belongs to the *class* W_0, if $U(t,x)$ is continuous on every set $\{G_k\}$ and for $(t_0,x_0) \in \sigma_k \cap \mathcal{D} \subset R_+ \times \Omega(\rho)$, $k \in \mathcal{Z}$, there exist the limits

$$\lim_{\substack{(t,x)\to(t_0,x_0)\\(t,x)\in G_k}} U(t,x) = U(t_0^-,x),$$

$$\lim_{\substack{(t,x)\to(t_0,x_0)\\(t,x)\in G_{k+1}}} U(t,x) = U(t_0^+,x)$$

and

$$U(t_0^-,x) = U(t_0,x_0) \in R^{m \times m}.$$

The matrix-valued function (4.4.1) has the following structure. Functions $U_{ii}(t,\cdot)$ are constructed for subsystems (4.3.2), and the functions $U_{ij}(t,\cdot)$ for all $(i \neq j)$ take into account the links $h_i(t,x_1,\ldots,x_m)$ between subsystems (4.3.2).

The functions $U_{ii}(t,\cdot)$ have the explicit form

$$(4.4.2) \qquad U_{ii}(t,\cdot) = \xi_i^T B_i(t,\cdot)\xi_i, \quad i = 1,2,\ldots,m,$$

where $\xi_i \in R_+^{n_i}$, $\xi_i > 0$, and the submatrix-functions $B_i(t,\cdot) = [u_{pq}^{(i)}(t,\cdot)]$, $p, q = 1,2,\ldots,M_i$, have the elements

$$u_{pp}^{(i)}\colon R_+ \times R^{n_{ip}} \to R, \quad i = 1,2,\ldots,m,$$

$$u_{pq}^{(i)}\colon R_+ \times R^{n_{ip}} \times R^{n_{iq}} \to R, \quad u_{pq}^{(i)} = u_{qp}^{(i)}.$$

Functions $u_{pp}^{(i)}(t,\cdot)$ are constructed for subsystems (4.3.6) and $u_{pq}^{(i)}$ for all $(p \neq q)$ take into account the influence of link functions $q_{ij}(t,x_i)$ between subsystems (4.3.6).

Similarly to Definition 4.4.1 for the elements $u_{pq}^{(i)}$ for $i = 1, 2, \ldots, m$ and $p, q = 1, 2, \ldots, M_i$ we consider classes w_0 of piece-wise continuous functions.

In order to establish conditions for the function

$$(4.4.3) \qquad v(t, x, \eta) = \eta^{\mathrm{T}} U(t, x) \eta, \quad \eta \in R_+^m, \quad \eta > 0,$$

being of a fixed sign, we need some assumptions.

Assumption 4.4.1. *There exist*

(1) *open connected time-invariant neighborhoods* $\mathcal{N}_{ip} \subseteq R^{n_{ip}}$, $i = 1, 2, \ldots, m$, $p = 1, 2, \ldots, M_i$, *of states* $x_{ip} = 0$;

2) *the functions* $\varphi_{ip}, \overline{\varphi}_{ip} \in K(KR)$;

(3) *the constants* $\underline{\alpha}_{pp}^{(i)} > 0$, $\overline{\alpha}_{pp}^{(i)} > 0$, $\underline{\alpha}_{pq}^{(i)} = \underline{\alpha}_{qp}^{(i)}$, $\overline{\alpha}_{pq}^{(i)} = \overline{\alpha}_{qp}^{(i)}$, $q = 1, 2, \ldots, M_i$;

(4) *the functions* $u_{pp}^{(i)} \in w_0$ *and* $u_{pq}^{(i)} \in w_0$ *for* $p \neq q$, $p, q = 1, 2, \ldots, M_i$, *satisfying the estimates*

    (a) $\underline{\alpha}_{pp}^{(i)} \varphi_{ip}^2(\|x_{ip}\|) \leq u_{pp}^{(i)}(t, x_{ip}) \leq \overline{\alpha}_{pp}^{(i)} \varphi_{ip}^2(\|x_{ip}\|)$
      for all $t \neq \tau_k(\cdot)$, $k \in Z$ *and* $x_{ip} \in \mathcal{N}_{ip}$;

    (b) $\underline{\alpha}_{pq}^{(i)} \varphi_{ip}(\|x_{ip}\|) \varphi_{iq}(\|x_{iq}\|) \leq u_{pq}^{(i)}(t, x_{ip}, x_{iq})$
      $\leq \overline{\alpha}_{pq}^{(i)} \varphi_{ip}(\|x_{ip}\|) \varphi_{iq}(\|x_{iq}\|)$
      for all $t \neq \tau_k(\cdot)$, $k \in Z$, *and* $(x_{ip}, x_{iq}) \in \mathcal{N}_{ip} \times \mathcal{N}_{iq}$.

Proposition 4.4.1. *If all conditions of Assumption 4.4.1 are satisfied, then for functions* $U_{ii}(t, x_i)$ *the estimates*

$$(4.4.4) \qquad w_i^{\mathrm{T}} \Phi_i^{\mathrm{T}} A_{ii} \Phi_i w_i \leq U_{ii}(t, x_i) \leq \overline{w}_i^{\mathrm{T}} \Phi_i^{\mathrm{T}} B_{ii} \Phi_i \overline{w}_i$$

holds for all $t \neq \tau_k(\cdot)$, $k \in Z$, *and for all* $x_i \in \mathcal{N}_i$, $\mathcal{N}_i = \mathcal{N}_{i1} \times \mathcal{N}_{i2} \times \cdots \times \mathcal{N}_{iM_i}$, $i = 1, 2, \ldots, m$.

Here the following designations are used

$$w_i^{\mathrm{T}} = (\varphi_{i1}(\|x_{i1}\|), \varphi_{i2}(\|x_{i2}\|), \ldots, \varphi_{iM_i}(\|x_{iM_i}\|)),$$
$$\overline{w}_i^{\mathrm{T}} = (\overline{\varphi}_{i1}(\|x_{i1}\|), \overline{\varphi}_{i2}(\|x_{i2}\|), \ldots, \overline{\varphi}_{iM_i}(\|x_{iM_i}\|)),$$
$$\Phi_i^{\mathrm{T}} = \Phi_i = \mathrm{diag}\,(\xi_{i1}, \xi_{i2}, \ldots, \xi_{iM_i}),$$
$$A_{ii} = [\underline{\alpha}_{pq}^{(i)}], \quad B_{ii} = [\overline{\alpha}_{pq}^{(i)}], \quad i = 1, 2, \ldots, M_i.$$

Proposition 4.4.1 is proved by a direct substitution by the estimates (4)(a) and (4)(b) into the quadratic forms $\xi_i^{\mathrm{T}} B_i(t, \cdot) \xi_i$, $i = 1, 2, \ldots, m$, under all the rest of the conditions of Assumption 4.4.1.

Assumption 4.4.2. *There exist*

(1) *open connected time-invariant neighborhoods* $\mathcal{N}_i \subseteq R^{n_i}$, $i = 1, 2,$
$\ldots, m$, *of equilibrium states* $x_i = 0$;

(2) *the functions* w_i, $\overline{w}_i \in K(KR)$;

(3) *the constants* β_{ij}, $\overline{\beta}_{ij}$, $\beta_{ij} = \beta_{ji}$, $\overline{\beta}_{ij} = \overline{\beta}_{ji}$ *for all* $(i \neq j) =$
$1, 2, \ldots, m$ *such that*

(4.4.5) $$\beta_{ij}\|w_i\|\,\|w_j\| \leq U_{ij}(t, x_i, x_j) \leq \overline{\beta}_{ij}\|\overline{w}_i\|\,\|\overline{w}_j\|$$

for all $t \neq \tau_k(\cdot)$, $k \in Z$ *and* $(x_i, x_j) \in \mathcal{N}_i \times \mathcal{N}_j$, $(i \neq j) =$
$1, 2, \ldots, m$.

Proposition 4.4.2. *If the conditions of Assumption 4.4.2 are satisfied as well as estimates (a) and (b) of condition (4) from Assumption 4.4.1, then for function (4.4.3) the bilateral inequality*

(4.4.6) $$w^T H^T A H w \leq v(t, x, \eta) \leq \overline{w}^T H^T B H \overline{w}$$

holds true for all $t \neq \tau_k(\cdot)$, $k \in Z$, *and* $x \in \mathcal{N} = \mathcal{N}_1 \times \mathcal{N}_2 \times \cdots \times \mathcal{N}_m$.
Here

(4.4.7)
$$w^T = (w_1, w_2, \ldots, w_m), \quad \overline{w}^T = (\overline{w}_1, \overline{w}_2, \ldots, \overline{w}_m),$$
$$H^T = H = \operatorname{diag}[\eta_1, \eta_2, \ldots, \eta_m],$$
$$A = [\beta_{ij}], \quad B = [\overline{\beta}_{ij}], \quad i \neq j = 1, 2, \ldots, m,$$
$$\beta_{ii} = \lambda_m(\Phi_i^T A_{ii} \Phi_i), \quad \overline{\beta}_{ii} = \lambda_M(\Phi_i^T B_{ii} \Phi_i).$$

Proof. The direct substitution by the estimates (4.4.4) and (4.4.5) into (4.4.3) for function $v(t, x, \eta)$ yields the estimate (4.4.6).

4.4.2 Structure of the total derivative of hierarchical matrix-valued functions

Further, in order to establish the structure of the total derivative of function (4.4.3) along the solution of system (4.3.1) allowing two-level decomposition, we introduce some definitions and designations.

Definition 4.4.2. The *matrix-valued function* $U \colon R_+ \times \Omega(\rho) \to R^{m \times m}$ belongs to the *class* W_1, if the matrix function $U \in W_0$ and is continuously differentiable on the set $\bigcup\limits_{k=1}^{\infty} G_k \cap \mathcal{D}$, $\mathcal{D} \subset R_+ \times \Omega(\rho)$.

Definition 4.4.3. The *matrix-valued function* $U\colon R_+ \times \Omega(\rho) \to R^{m \times m}$ belongs to the *class* W_2, if the matrix function $U \in W_0$ and is locally Lipschitzian in the second argument.

Let $x(t)$ be any solution of (4.3.1) defined for $t \in [t_0, t_0 + a) \subset J \subset R_+$, $a = \text{const} > 0$, and such that $x(t) \in \Omega_1$, $\Omega_1 \subset \Omega(\rho)$ for all $t \in J$.

Theorem 4.4.2. *Let the matrix-valued function* $U \in W_2$, *then*

$$(4.4.8)\quad D^+U(t,x) = \lim_{\vartheta \to 0^+} \sup \left\{ \left[U(t + \vartheta,\, x + \vartheta f(t,x)) - U(t, x(t)) \right] \vartheta^{-1} \right\}$$

for all $(t, x(t)) \in \bigcup\limits_{k=1}^{\infty} G_k \cap \mathcal{D}$.

Actual computation of $D^+U(t,x)$ is made element-wise.

Assumption 4.4.3. *There exist*

(1) *open connected time-invariant neighborhoods* $\mathcal{N}_{ip}$, $\mathcal{N}_{ip} \subseteq R^{n_{ip}}$, *of states* $x_{ip} = 0$, $i = 1, 2, \ldots, m$, $p = 1, 2, \ldots, M_i$;
(2) *the functions* $u_{pq}^{(i)} \in w_0$, $q = 1, 2, \ldots, M_i$;
(3) *the functions* $\beta_{ip} \in K\,(KR)$;
(4) *the real numbers* $\rho_p^{(i)}$, $\mu_p^{(i)}$, $\mu_{pq}^{(i)}$ *such that*

(a) $\displaystyle\sum_{p=1}^{M_i} \xi_{ip}^2 \left\{ D_t^+ u_{pp}^{(i)} + \left(D_{x_{ip}}^+ u_{pp}^{(i)} \right)^{\mathrm{T}} p_{ip}(t, x_{ip}) \right\} \le \sum_{p=1}^{M_i} \rho_p^{(i)} \beta_{ip}^2(\|x_{ip}\|)$

for all $t \ne \tau_k(\cdot)$, $k \in \mathcal{Z}$ and $x_{ip} \in \mathcal{N}_{ip}$;

(b) $\displaystyle\sum_{p=1}^{M_i} \xi_{ip}^2 \left\{ \left(D_{x_{ip}}^+ u_{pp}^{(i)} \right)^{\mathrm{T}} q_{ip}(t, x) \right\}$

$\displaystyle + 2 \sum_{p=1}^{M_i - 1} \sum_{q=p+1}^{M_i} \xi_{ip}\xi_{iq} \left\{ D_t^+ u_{pq}^{(i)} + \left(D_{x_{ip}}^+ u_{pq}^{(i)} \right)^{\mathrm{T}} \right.$

$\left. \times (p_{ip}(t, x_{ip}) + q_{ip}(t, x_i)) + \left(D_{x_{iq}}^+ u_{pq}^{(i)} \right)^{\mathrm{T}} (p_{iq}(t, x_{iq}) + q_{iq}(t, x_i)) \right\}$

$\displaystyle \le \sum_{p=1}^{M_i} \mu_p^{(i)} \beta_{ip}^2(\|x_{ip}\|) + 2 \sum_{p=1}^{M_i - 1} \sum_{q=p+1}^{M_i} \mu_{pq}^{(i)} \beta_{ip}(\|x_{ip}\|)\, \beta_{iq}(\|x_{iq}\|)$

for all $t \ne \tau_k(\cdot)$, $k \in \mathcal{Z}$ and all $(x_{ip}, x_{iq}) \in \mathcal{N}_{ip} \times \mathcal{N}_{iq}$.

Proposition 4.4.3. *If all conditions of Assumption 4.4.3 are satisfied then the upper right-hand derivative of the functions* $U_{ij}(t, x)$ *along solutions of (4.3.7) satisfies the estimate*

$$(4.4.9)\qquad D^+U_{ii}(t, x_i) \le \lambda_M(S_{ii})\|\beta_i\|^2$$

for all $t \neq \tau_k(x_i)$, $k \in \mathcal{Z}$ and all $x_i \in \mathcal{N}_i$.

Here $\beta_i^{\mathrm{T}} = (\beta_{i1}(\|x_{i1}\|),\ \beta_{i2}(\|x_{i2}\|),\ \ldots,\ \beta_{iM_i}(\|x_{iM_i}\|))$, $\lambda_M(S_{ii})$ is the maximal eigenvalue of matrix S_{ii} with the elements

(4.4.10)
$$\sigma_{pp}^{(i)} = \rho_p^{(i)} + \mu_p^{(i)}, \quad p = 1, 2, \ldots, M_i,$$
$$\sigma_{pq}^{(i)} = \sigma_{qp}^{(i)} = \mu_{pq}^{(i)}, \quad (p \neq q) \in [1, M_i].$$

Proof. In view of (4.4.7) and (4.4.8) we have for expressions $D^+ u_{pq}^{(i)}(t, \cdot)$ the estimates (see Grujić, *et al.* [1])

(4.4.11) $$D^+ u_{pq}^{(i)}(t, x_{ip}) \leq D_t^+ u_{pq}(t, x_{ip}) + (D_{x_{ip}}^+ u_{pq}^{(i)}(t, x_{ip}))^{\mathrm{T}} \frac{dx_{pq}}{dt}$$

for all $i = 1, 2, \ldots, m$ and $(p, q) \in [1, M_i]$. In view of (4.4.11) and the conditions $(1) - (4)$ from Assumption 4.4.3 we arrive at estimates (4.4.9).

Assumption 4.4.4. *There exist*

(1) *some constants* $\rho_{ip}^0 > 0$, $\rho_{ip}^0 < \rho$, *such that* $x_{ip} \in \Omega(\rho_{ip}^0)$ *ensures the inclusion* $x_{ip} + I_{ip}^k(x_{ip}) \in \Omega(\rho_{ip})$ *for all* $k \in \mathcal{Z}$;

(2) *the functions* $u_{pq}^{(i)} \in w_0$ *for all* $i = 1, 2, \ldots, m$ *and* $(p, q) \in [1, M_i]$;

(3) *the functions* $\psi_{ip} \in K(KR)$;

(4) *the real numbers* $a_{pp}^{(i)}$, $b_{pp}^{(i)}$, $b_{pq}^{(i)}$ *such that*

(a) $\xi_{ip}^2 \left\{ u_{pp}^{(i)}(\tau_k(x_{ip}),\ x_{ip} + I_{ip}^k(x_{ip})) - u_{pp}^{(i)}(\tau_k(x_{ip}), x_{ip}) \right\} \leq a_{pp}^{(i)} \psi_{ip}^2(\|x_{ip}\|)$

for all $x_{ip} \in \mathcal{N}_{ip} \subseteq \Omega(\rho_{ip})$;

(b) $\displaystyle\sum_{p=1}^{M_i} \xi_{ip}^2 \Big\{ u_{pp}^{(i)}(\tau_k(x_i),\ x_i + I_i^k(x_i)) - u_{pp}^{(i)}(\tau_k(x_{ip}),\ x_{ip} + I_{ip}^k(x_{ip}))$

$\qquad + u_{pp}^{(i)}(\tau_k(x_{ip}), x_{ip}) - u_{pp}^{(i)}(\tau_k(x_i), x_i) \Big\} + 2 \displaystyle\sum_{p=1}^{M_i-1} \sum_{q=p+1}^{M_i} \xi_{ip}\xi_{iq}$

$\qquad \times \Big\{ u_{pq}^{(i)}(\tau_k(x_i),\ x_{ip} + J_{ip}^k(x_i),\ x_{iq} + J_{iq}^k(x_i)) - u_{pq}^{(i)}(\tau_k(x_i), x_{ip}, x_{iq}) \Big\}$

$\leq \displaystyle\sum_{p=1}^{M_i} b_{pp}^{(i)} \psi_{ip}^2(\|x_{ip}\|) + 2 \displaystyle\sum_{p=1}^{M_i-1} \sum_{q=p+1}^{M_i} b_{pq}^{(i)} \psi_{ip}(\|x_{ip}\|)\psi_{iq}(\|x_{iq}\|)$

for all $(x_{ip}, x_{iq}) \in \mathcal{N}_{ip} \times \mathcal{N}_{iq}$.

Proposition 4.4.4. *If all conditions of Assumption 4.4.4 are satisfied, then when* $t = \tau_k(x_i)$, $k \in \mathcal{Z}$, $i = 1, 2, \ldots, m$, *the estimates*

$$(4.4.12) \qquad U_{ii}(\tau_k(x_i),\, x_i + J_i^k(x_i)) - U_{ii}(\tau_k(x_i), x_i) \leq \lambda_M(C_{ii})\|\psi_i\|^2$$

are true for functions $U_{ii}(t, x_i)$.
    Here

$$\psi_i^{\mathrm{T}} = (\psi_{i1}(\|x_{i1}\|),\, \psi_{i2}(\|x_{i2}\|),\, \ldots,\, \psi_{iM_i}(\|x_{iM_i}\|)),$$
$$C_{ii} = [c_{pq}^{(i)}], \quad i = 1, 2, \ldots, m, \quad (p, q) \in [1, M_i],$$
$$c_{pq}^{(i)} = c_{qp}^{(i)}, \quad c_{pp}^{(i)} = \alpha_{pp}^{(i)} + b_{pp}^{(i)},$$
$$c_{pq}^{(i)} = c_{qp}^{(i)} = b_{pq}^{(i)}, \quad i = 1, 2, \ldots, m,$$

$\lambda_M(C_{ii})$ *is the maximal eigenvalue of matrix* C_{ii}.

The Proof of Proposition 4.4.4 is obvious in view of the condition (4a) and (4b) of Assumption 4.4.4.

Assumption 4.4.5. *There exist*

(1) *the functions* $u_{pq}^{(i)} \in w_0$ *for all* $i = 1, 2, \ldots, m$, $(p, q) \in [1, M_i]$;
(2) *the functions* $\psi_{ip} \in K(KR)$;
(3) *the real numbers* $d_{pp}^{(i)}$, $l_{pq}^{(i)}$, *and for* $t = \tau_k(x_{ip})$, $k \in \mathcal{Z}$, $x_{ip} \in \mathcal{N}_{ip}$ *the following conditions hold*

(a) $\xi_{ip}^2 u_{pp}^{(i)}(\tau_k(x_{ip}),\, x_{ip} + J_{ip}^k(x_{ip})) \leq d_{pp}^{(i)}\psi_{ip}^2(\|x_{ip}\|)$;

(b) $\displaystyle\sum_{p=1}^{M_i} \xi_{ip}^2 \Big\{ u_{pp}^{(i)}(\tau_k(x_{ip}),\, x_{ip} + J_{ip}^k(x_i)) - u_{pp}^{(i)}(\tau_k(x_{ip}),\, x_{ip} + J_{ip}^k(x_{ip})) \Big\}$

$$+ 2 \sum_{p=1}^{M_i-1} \sum_{q=p+1}^{M_i} \xi_{ip}\xi_{iq} u_{pq}^{(i)}(\tau_k(x_i),\, x_{ip} + J_{ip}^k(x_{ip}),\, x_{iq} + J_{iq}^k(x_{iq}))$$

$$\leq \sum_{p=1}^{M_i} l_{pp}^{(i)}\psi_{ip}^2(\|x_{ip}\|) + 2 \sum_{p=1}^{M_i-1} \sum_{q=p+1}^{M_i} l_{pq}^{(i)}\psi_{ip}(\|x_{ip}\|)\psi_{iq}(\|x_{iq}\|).$$

Proposition 4.4.5. *If all conditions of the Assumption 4.4.5 are satisfied, then for functions* $U_{ii}(t, \cdot)$ *when* $t = \tau_k(x_i)$, $k \in \mathcal{Z}$ *and* $i = 1, 2, \ldots, m$ *the estimate*

$$U_{ii}(\tau_k(x_i),\, x_i + J_i^k(x_i)) \leq \lambda_M(C_{ii}^*)\|\psi_i\|^2$$

is valid for all $x_i \in \mathcal{N}_i$, *where* $\lambda_M(\cdot)$ *is the maximal eigenvalue of the matrix* C_{ii}^* *with elements*

$$c_{pp}^{*(i)} = d_{pp}^{(i)} + l_{pp}^{(i)}, \quad i = 1, 2, \ldots, m,$$
$$c_{pq}^{*(i)} = c_{qp}^{*(i)} = l_{pq}^{*(i)}, \quad p, q = 1, 2, \ldots, M_i.$$

The Proof of Proposition 4.4.5 is similar to that of Proposition 4.4.4.

Assumption 4.4.6. *There exist*

(1) *an open connected neighborhood* $\mathcal{N} \subseteq R^n$ *of* $x = 0$;
(2) *the functions* β_{ip}, $i = 1, 2, \ldots, m$, $p = 1, 2, \ldots, M_i$, *mentioned in Assumption 4.4.3;*
(3) *the functions* $U_{ij}(t, \cdot)$ *satisfying the conditions of Assumption 4.4.2;*
(4) *the real numbers* ϑ_{ik}, $i, k = 1, 2, \ldots, m$, *such that*

$$\sum_{i=1}^{m} \eta_i^2 \left(D^+U_{ii}(t, x_i)\right)^{\mathrm{T}} h_i(t, x) + 2 \sum_{i=1}^{m-1} \sum_{k=i+1}^{m} \eta_i \eta_k \Big\{ D_t^+ U_{ik}(t, \cdot)$$
$$+ \left(D_{x_i}^+ U_{ik}(t, x_i, x_k)\right)^{\mathrm{T}} (g_i(t, x_i) + h_i(t, x)) + \left(D_{x_k}^+ U_{ik}(t, x_i, x_k)\right)^{\mathrm{T}}$$
$$\times (g_k(t, x_k) + h_k(t, x)) \Big\} \leq \sum_{i=1}^{m} \vartheta_{ii} \|\beta_i\|^2 + 2 \sum_{i=1}^{m-1} \sum_{k=i+1}^{m} \vartheta_{ik} \|\beta_i\| \|\beta_k\|$$

for all $t \neq \tau_k(x_i)$, $k \in \mathcal{Z}$ *and* $(x_i, x_k) \in \mathcal{N}_i \times \mathcal{N}_k$.

Proposition 4.4.6. *If all conditions of Assumptions 4.4.3 and 4.4.6 are satisfied, then for the function* $D^+v(t, x, \eta)$ *along solutions of (4.3.1) the estimate*

(4.4.13) $$D^+v(t, x, \eta) \leq \beta^{\mathrm{T}} S \beta$$

is valid for all $t \neq \tau_k(x_i)$, *where* $\beta^{\mathrm{T}} = (\beta_1, \beta_2, \ldots, \beta_m)$ *and the matrix* S *has the elements*

$$s_{ii} = \eta_i^2 \lambda_M(S_{ii}) + \vartheta_{ii},$$
$$s_{ik} = s_{ki} = \vartheta_{ik}, \quad i \neq k = 1, 2, \ldots, m.$$

Proof. The estimate (4.4.13) is obtained from conditions of Assumption 4.4.6 in an obvious way, in view of estimate (4.4.11).

Proposition 4.4.7. *If the inequality (4.4.13) is true, then there exist continuous functions* $H_1, H_2 \colon R_+ \to R_+$, $H_1(0) = H_2(0) = 0$, $H_1(r) > 0$, $H_2(r) > 0$ *for* $r > 0$, *such that*

$$0 < H_1(v(t,x,\eta)) \le \sum_{i=1}^{m} \beta_i^2(\|x_i\|) \le H_2(v(t,x,\eta)).$$

If, moreover,

    (a) $\lambda_M(S) < 0$;

    (b) $\lambda_M(S) > 0$,

then for all $t \ne \tau_k(x)$, $x \in \mathcal{N}$, $k \in \mathcal{Z}$, *the following estimates hold true*

    (a) $D^+v(t,x,\eta) \le \lambda_M(S)H_1(v(t,x,\eta))$,

    (b) $D^+v(t,x,\eta) \le \lambda_M(S)H_2(v(t,x,\eta))$,

respectively.

Here $\lambda_M(\cdot)$ *is the maximal eigenvalue of the matrix* S.

Proof. Estimates (a) and (b) follow from the fact that

$$\beta^{\mathrm{T}}S\beta \le \lambda_M(S)\beta^{\mathrm{T}}\beta$$

under the conditions of Proposition 4.4.7.

Assumption 4.4.7. *There exist*

    (1) *open connected time-invariant neighborhoods* $\mathcal{N} \subseteq R^{n_i}$ *of* $x_i = 0$, $i = 1, 2, \ldots, m$, *and neighborhood* $\mathcal{N} = \mathcal{N}_1 \times \mathcal{N}_2 \times \cdots \times \mathcal{N}_m$ *of* $x = 0$;

    (2) *the functions* U_{im}, $m = 1, 2, \ldots, m$, *satisfying the conditions of Assumption 4.4.2;*

    (3) *the functions* $\psi_{ip} \in K$;

    (4) *the real numbers* ν_{ip}, $p = 1, 2, \ldots, m$, *and for all* $k \in \mathcal{Z}$ *the following conditions are satisfied*

$$(4.4.14) \qquad \sum_{i=1}^{m} \eta_i^2 \big\{ U_{ii}(\tau_k(x), x_i + J_i^k(x)) - U_{ii}(\tau_k(x), x_i) \big\}$$

$$+ 2 \sum_{i=1}^{m-1} \sum_{p=i+1}^{m} \eta_i\eta_p \big\{ U_{ip}(\tau_k(x), x_i + J_i^k(x), x_p + J_p(x)) - U_{ip}(\tau_k(x), x_i, x_p) \big\}$$

$$\le \sum_{i=1}^{m} \nu_{ii}\psi_i^2(\|x_i\|) + 2 \sum_{i=1}^{m-1} \sum_{p=i+1}^{m} \nu_{ip}\psi_i(\|x_i\|)\psi_p(\|x_p\|)$$

for all $(x_i, x_p) \in \mathcal{N}_i \times \mathcal{N}_p$.

Proposition 4.4.8. *If inequalities (4.4.12) and (4.4.14) are satisfied, then the estimate*

(4.4.15) $\qquad v(\tau_k(x),\, x + J^k(x),\, \eta) - v(\tau_k(x),\, x, \eta) \le \psi^{\mathrm{T}} C \psi$

holds true for all $t = \tau_k(x)$, $k \in \mathcal{Z}$, $x \in \mathcal{N}_x$, *where*

$$\psi^{\mathrm{T}} = (\psi_1(\tau_k(x), \|x_1\|), \ldots, \psi_m(\tau_k(x), \|x_m\|))$$

and the matrix C *has the elements*

$$c_{ii} = \eta_i^2 \lambda_M(C_{ii}) + \nu_{ii},$$
$$c_{ip} = c_{pi} = \nu_{ip}, \quad i \ne p \in [1, m].$$

Proposition 4.4.9. *If the inequality (4.4.15) is satisfied, then there exist continuous functions* Q_1 *and* Q_2: $Q_1(0) = Q_2(0) = 0$ *and* $Q_1(s) > 0$, $Q_2(s) > 0$ *for* $s > 0$ *such that*

$$0 < Q_1(v(\tau_k(x), x, \eta)) \le \sum_{i=1}^{m} \psi_i^2(\tau_k, \|x_i\|) \le Q_2(v(\tau_k(x), x, \eta)).$$

If, moreover,

  (a) $\lambda_M(C) < 0$,
  (b) $\lambda_M(C) > 0$,

then for all $k \in \mathcal{Z}$, *the following estimates hold*

  (a) $v(\tau_k(x), x + J^k(x), \eta) - v(\tau_k(x), x, \eta) \le \lambda_M(C) Q_1(v(\tau_k(x), x, \eta))$,
  (b) $v(\tau_k(x), x + J^k(x), \eta) - v(\tau_k(x), x, \eta) \le \lambda_M(C) Q_2(v(\tau_k(x), x, \eta))$,

respectively.

Here $\lambda_M(C)$ *is the maximal eigenvalue of the matrix* C.

Proof. Consider the quadratic form $\psi^{\mathrm{T}} C \psi$. It is known that

$$\psi^{\mathrm{T}} C \psi \le \lambda_M(C) \psi^{\mathrm{T}} \psi = \lambda_M(C) \sum_{i=1}^{m} \psi_i^2(\tau_k(x), \|x\|).$$

Hence

$$v(\tau_k(x), x + J^k(x), \eta) - v(\tau_k(x), x, \eta)$$
$$\le \begin{cases} \lambda_M(C) Q_1(v(t_k(x), x, \eta)) & \text{if } \lambda_M(C) < 0; \\ \lambda_M(C) Q_2(v(t_k(x), x, \eta)) & \text{if } \lambda_m(C) > 0. \end{cases}$$

This proves Proposition 4.4.9.

Assumption 4.4.8. *There exist*

(1) *open connected time-invariant neighborhoods $\mathcal{N}_i \subseteq R^{n_i}$ of $x_i = 0$,*
    $i = 1, 2, \ldots, m$;
(2) *the functions U_{ip}, $i, p = [1, m]$, mentioned in Assumption 4.4.2;*
(3) *the functions $\psi_{ip}(\tau_k(x), \|x_i\|)$, $i, p = [1, m]$, $k \in \mathcal{Z}$, continuous in
    the second argument, $\psi_{ip}(\tau_k(x), 0) = 0$ and $\psi_{ip}(\tau_k(x), r) > 0$ for
    $r > 0$;*
(4) *the real numbers μ_{ip} and for all $k \in \mathcal{Z}$ estimates*

$$\sum_{i=1}^{m} \eta_i^2 \{ U_{ii}(\tau_k(x), x_i + J_i^k(x)) \}$$

$$+ 2 \sum_{i=1}^{m-1} \sum_{p=i+1}^{m} \eta_i \eta_p U_{ip}(\tau_k(x), x_i + J_i^k(x), x_p + J_p(x))$$

$$\leq \sum_{i=1}^{m} \mu_{ii} \psi_i^2(\tau_k(x), \|x_i\|) + 2 \sum_{i=1}^{m-1} \sum_{p=i+1}^{m} \mu_{ip} \psi_i(\tau_k(x), \|x_i\|) \, \psi_p(\tau_k(x), \|x_p\|)$$

are satisfied for all $(x_i, x_p) \in \mathcal{N}_i \times \mathcal{N}_p$.

Proposition 4.4.10. *If all conditions of Assumption 4.4.8 are satisfied,
then estimate*

$$(4.4.16) \qquad v(\tau_k(x), x + J^k(x), \eta) \leq \psi^{\mathrm{T}} C^* \psi$$

*is true, where $\psi^{\mathrm{T}} = (\psi_1(\tau_k(x), \|x\|), \ldots, \psi_m(\tau_k(x), \|x_m\|))$ and the matrix
C^* has the elements*

$$c_{ii}^* = \eta_i^2 \lambda_M(C_{ii}^*) + \mu_i,$$
$$c_{ip}^* = c_{pi}^* = \mu_{ip}, \quad i \neq p \in [1, m].$$

Proposition 4.4.11. *If estimate (4.4.16) is satisfied, then there exist
functions $Q_1, Q_2 \colon R_+ \to R_+$, $Q_1(0) = Q_2(0) = 0$, $Q_1(r) > 0$, $Q_2(r) > 0$
for $r > 0$ such that*

(a) *$v(\tau_k(x), x + J^k(x), \eta) - v(\tau_k(x), x, \eta) \leq \lambda_M(C^*) Q_1(v(\tau_k(x), x, \eta))$,*
(b) *$v(\tau_k(x), x + J^k(x), \eta) - v(\tau_k(x), x, \eta) \leq \lambda_M(C^*) Q_2(v(\tau_k(x), x, \eta))$,*

for

(a) *$\lambda_M(C^*) < 0$,*
(b) *$\lambda_M(C^*) > 0$,*

respectively.

It is easy to see that estimates (4.4.13) together with inequalities (4.4.16) allow us to establish some conditions for function $v(t, x, \eta)$ decreasing along the solutions of (4.3.1) and by the same token to apply this function in the behavior investigation of the solutions to system (4.3.1).

4.5 Uniqueness and Continuability of Solutions

The matrix-valued function constructed in Section 4.4 allows constructive uniqueness and continuability conditions to be established for solutions of system (4.3.1). We recall the following.

Definition 4.5.1. *Function* $x \colon (t_0, t_0 + a) \to R^n$, $t_0 \geq 0$, $a > 0$, *is called a solution of (4.3.1), if*

(1) $x(t_0^+) = x_0$ *and* $(t, x(t)) \in D$ *for all* $t \in [t_0, t_0 + a)$;

(2) $x(t)$ *is continuously differentiable function, satisfies the condition* $dx/dt = f(t, x(t))$ *for all* $t \in [t_0, t_0 + a)$ *and* $t \neq \tau_k(x(t))$, $k \in Z$;

(3) *if* $t \in [t_0, t_0 + a)$ *and* $t = \tau_k(x(t))$, *then* $x(t^+) = x(t) + I_k(x(t))$ *for all* $k \in Z$ *and at* $s \neq \tau_k(x(s))$ *for all* $k \in Z$, $t < s < \delta$, $\delta > 0$, *the solution* $x(t)$ *is left continuous.*

We note that, instead of ordinary initial condition $x(t_0) = x_0$ for system (4.3.1) the limiting condition $x(t_0^+) = x_0$ is given. This is more natural for system (4.3.1), since (t_0, x_0) can be such that $t_0 = \tau_k(x_0)$ for some k. If $t_0 \neq \tau_k(x_0)$ for any $k \in Z$ we interpret the condition $x(t_0^+) = x_0$ in the usual sense, i.e. $x(t_0) = x_0$.

Theorem 4.5.1. *Assume that*

(1) *conditions (A) are satisfied;*

(2) *vector function* $f(t, x)$ *is definite and continuous on* $[0, T] \times \Omega(\rho)$ *and* $f(t, 0) = 0$ *for all* $t \in [0, T]$;

(3) *conditions of Asumptions 4.4.1 and 4.4.2 are satisfied and in estimate (4.4.6) the matrix A is positive definite;*

(4) *Assumptions 4.4.3 and 4.4.6 hold true and in estimate (4.4.13) the matrix S is negative semi-definite;*

(5) *conditions of Assumptions 4.4.4, 4.4.5 and 4.4.7 are satisfied and in estimate (4.4.15) the matrix C is negative semi-definite.*

Then the solution $x(t, 0, 0)$ of system (4.3.1) is unique for all $t \in [0, T]$.

Proof. Under conditions (A) it is reasonable to apply the algorithm of the matrix-valued function $U(t, x)$ construction discussed in Section 4.4. Under condition (3) the function $v(t, x, \eta)$ satisfies the conditions

$$(4.5.1) \qquad \begin{aligned} v(t, 0, \eta) &= 0 \quad \text{and} \quad v(t, x, \eta) > 0 \\ &\text{for} \quad x \neq 0 \quad \text{and} \quad t \in [0, T]. \end{aligned}$$

By condition (4) of Theorem 4.5.1 function $D^+ v(t, x, \eta)$ satisfies the estimate

$$(4.5.2) \qquad D^+ v(t, x, \eta) \leq 0 \quad \text{for all} \quad (t, x) \in \bigcup_{k=1}^{\infty} G_k \cap \mathcal{D}.$$

If condition (5) is satisfied, it can be easily seen that

$$(4.5.3) \qquad v(t + 0, \, x + I_k(x), \, \eta) \leq v(t, x, \eta) \quad \text{for all} \quad (t, x) \in \sigma_k \cap \mathcal{D}.$$

Further we suppose that $x(t; 0, 0) \neq 0$ for all $t \in [0, T]$. In addition, there exist the values t_1, t_2 on interval $[0, T]$ such that $x(t_1; 0, 0) = 0$ and $x(t; 0, 0) \neq 0$ for all $t \in (t_1, t_2)$. Then we find from (4.5.2) and (4.5.3)

$$(4.5.4) \qquad v(t_2, \, x(t_2; 0, 0), \, \eta) \leq v(t_1, \, x(t_1; 0, 0), \, \eta) = 0.$$

This inequality contradicts condition (4.5.1) and proves Theorem 4.5.1.

Theorem 4.5.2. *Assume that*

(1) *vector function $f(t, x)$ in system (4.3.1) is continuous and bounded on $[0, T] \times R^n$;*

(2) *vector function $I^k(x)$, $k \in \mathcal{Z}$, is definite for all $x \in R^n$;*

(3) *for all $x \in R^n$ the correlations*

$$\tau_k(x) \equiv t_k, \quad t_k \in (0, \infty), \quad k \in \mathcal{Z}, \quad \lim_{k \to \infty} t_k = \infty, \quad t_{k+1} > t_k$$

are satisfied.

Then any solution $x(t; t_0, x_0)$ of system (4.3.1) is continuable up to $t = T$.

Proof. In view of Theorem 3.4 by Yoshizawa [1] and the fact that there is only a finite number of points t_k, $k = 1, 2, \ldots, p$, on the finite interval $[0, T]$, the assertion of the theorem is obvious.

Proposition 4.5.1. *Let*

(1) *conditions A_1 and A_2 be satisfied;*
(2) *the sequence $\{\tau_k(x)\} \in S(\Omega)$ for all $k \in \mathcal{Z}$;*
(3) *system (4.3.1) have the solution $x(t; t_0, x_0)$ defined on $J(t_0, x_0) \in R_+$ such that for $t \in [t_0, a) \subset J(t_0, x_0)$ the inequality $\|x(t; t_0, x_0)\| < \beta$ $(0 < \beta < H)$ is satisfied.*

Then the solution $x(t; t_0, x_0)$ for $t \in [t_0, a)$ intersects the hypersurfaces σ_k, $k \in \mathcal{Z}$, a finite number of times.

Proof. Assume that the solution $x(t; t_0, x_0)$ for $t \in [t_0, a)$ intersects the hypersurfaces $\sigma_{k_1}, \sigma_{k_2}, \ldots$ at points $\xi_1 < \xi_2 < \ldots$. Moreover, $\xi_k \in [t_0, a)$, i.e. $|\xi_k| < a$, $k \in \mathcal{Z}$. Conditions (1) and (2) of Proposition 4.5.1 imply that for any $\varepsilon > a$ there exists a $N > 0$ such that for $n > N$ the inequality $\tau_n(x) > \varepsilon$ holds for all $x \in \Omega$. Therefore, for $n > N$: $\xi_n = \tau_{k_n}(x(\xi_n)) > \varepsilon > a$. The contradiction obtained proves Proposition 4.5.1.

Theorem 4.5.3. *Assume that*

(1) *conditions $A_1 - A_4$ are satisfied;*
(2) *vector function $f(t, x)$ is continuous and bounded on the set $[0, T] \times \Omega$;*
(3) *the sequence $\{\tau_k(x)\} \in S(\Omega)$ for all $k \in \mathcal{Z}$;*
(4) *any solution $x(t; t_0, x_0)$ of system (4.3.1) is strictly bounded by constant β $(\beta \leq \rho)$ for any $t \in J(t_0, x_0)$, where $J(t_0, x_0)$ is a maximal existence interval for the solution $x(t; t_0, x_0)$ of system (4.3.1);*
(5) *vector function $I_k(x)$, $k \in \mathcal{Z}$, is definite on Ω and for any $x \in \Omega$, $x + I_k(x) \in \Omega$.*

Then any solution of system (4.3.1) is continuable up to $t = T$.

Proof. Let the solution $x(t; t_0, x_0)$ be definite and exist on $J(t_0, x_0) = [t_0, a)$, $a \leq T$. By Proposition 4.5.1 the solution $x(t; t_0, x_0)$ intersects hypersurfaces $\sigma_{k_1}, \sigma_{k_2}, \ldots, \sigma_{k_p}$ at points $\xi_1 < \xi_2 < \cdots < \xi_p < a$.

We consider the system of equations

$$\frac{dx}{dt} = f(t, x)$$

with the initial condition

$$x(\xi_p) = x(\xi_p; t_0, x_0) + I_{k_p}(x(\xi_p; t_0, x_0)).$$

Conditions (4) and (5) of Theorem 4.5.3 imply that any solution $x^*(t)$ of the initial problem satisfies the condition $\|x^*(t)\| < \beta$ for any $t \geq t_p$ for which $x^*(t)$ exists. Besides, there exists a solution $\bar{x}(t) = x(t; t_0, x_0)$ for $t \in (\xi_p, a)$. According to Yoshizawa [1, Theorem 3.3] the function $\bar{x}(t)$ is definite for $t \in [t_p, a)$. Then $\lim_{t \to a^-} x(t; t_0, x_0) = \lim_{t \to a-0} \bar{x}(t) = \bar{x}(a)$. This contradicts the choice of number a and proves the theorem.

Theorem 4.5.4. *Assume that*

(1) *conditions (1)–(4) of Theorem 4.5.2 are satisfied for $\Omega = R^n$;*
(2) *functions $I_k(x)$, $k \in \mathcal{Z}$, are definite for all $x \in R^n$.*

Then any solution of system (4.3.1) is continuable up to $t = T$.

The Proof is similar to that of Theorem 4.5.2.

Corollary 4.5.1. *Let the conditions of Theorem 4.5.4 be satisfied for $\Omega = R^n$ and $[0, T] \times \Omega \equiv R_+ \times R^n$. Then any solution of system (4.3.1) with the initial conditions $(t_0, x_0) \in \text{int}\,(R_+ \times R^n)$ is definite for all $t \geq t_0$.*

Proposition 4.5.2. *Let sequence $\{\tau_k(x)\} \in S(R^n)$ and function $v(t, x, \eta) \in (W_0([0, T] \times R^n \times R_+^m, R)$.*

Then for any $\alpha > 0$ and $t_0 \in [0, T]$ there exists a number $K(t_0, \alpha) > 0$ such that for $\|x\| < \alpha$ and $(t_0, x) \in \bigcup_{k=1}^{\infty} G_k$ the inequality

$$(4.5.5) \qquad v(t_0, x, \eta) \leq K(t_0, \alpha)$$

is satisfied.

Proof. Assume that conditions of Proposition 4.5.2 are satisfied, and there exist $\alpha > 0$ and point $(t_0, x_n) \in \bigcup_{k=1}^{\infty} G_k$, $\|x_n\| \leq \alpha$, such that

$$(4.5.6) \qquad v(t_0, x_n, \eta) \geq n.$$

Since the sequence $\{x_n\}$ is bounded, there exists a convergent subsequence x_{n_k}. We designate

$$\beta = \lim_{n_k \to \infty} x_{n_k}$$

and, moreover, $\|\beta\| \leq \alpha$. Further let us consider two cases.

Case 1. Let $(t_0, \beta) \in G_k$ for all $k \in \mathcal{Z}$. For any n there exists a $N > 0$ such that for $n > N$ the inclusion $(t_0, x_{n_k}) \in G_k$ is satisfied. Since the function $v(t, x, \eta)$ is continuous on sets G_k, then $v(t_0, x_{n_k}, \eta) \to v(t_0, \beta, \eta)$ as $n_k \to \infty$, that contradicts inequality (4.5.6).

Case 2. Let $(t_0, \beta) \in \sigma_j$, i.e. $t_0 = \tau_j(\beta)$, $j \in \mathcal{Z}$. The continuity of $\tau_j(x)$ implies that $\tau_j(\beta) = \lim\limits_{n_k \to \infty} \tau_j(x_{n_k})$. Besides, there exist a number $N > 0$ and points t_{n_k}: $\lim\limits_{n_k \to \infty} t_{n_k} = t_0$, such that for $n_k > N$ the inclusion $(t_{n_k}, x_{n_k}) \in G_j$, $j \in \mathcal{Z}$, holds.

Then

$$v(t_0, \beta, \eta) = v(t_0 - 0, \beta, \eta) = \lim\limits_{n_k \to \infty} v(t_{n_k}, x_{n_k}, \eta).$$

The contradiction obtained shows that the assumption made is not correct. This proves Proposition 4.5.2.

Theorem 4.5.5. *Assume that*

(1) *conditions (A) are satisfied;*

(2) *vector function $f(t, x)$ is definite and continuous on $[0, T] \times R^n$;*

(3) *functions $I^k(x)$, $k \in \mathcal{Z}$, are definite for $x \in R^n$ and such that for $\|x\| > p$ inequality $\|x + I^k(x)\| \geq p$, $p = \text{const}$, holds;*

(4) *conditions of Assumptions 4.4.1 and 4.4.2 are satisfied for $\mathcal{N}_i = R^{n_i}$, $i = 1, 2, \ldots, N$, and functions $\varphi_{ip} \in KR$;*

(5) *conditions of Assumptions 4.4.3 and 4.4.6 are satisfied for $\mathcal{N}_i = R^{n_i}$ and $\mathcal{N} = R^n$ and for functions $\beta_{ip} \in KR$;*

(6) *conditions of Assumptions 4.4.4, 4.4.5 and 4.4.7 are satisfied for $\mathcal{N}_i = R^{n_i}$ and $\mathcal{N} = R^n$;*

(7) *in the estimates (4.4.6), (4.4.13) and (4.4.15)*

    (a) *the matrix A is positive definite;*

    (b) *the matrix S is negative semi-definite;*

    (c) *the matrix C is negative semi-definite.*

Then any solution $x(t; t_0, x_0)$ of (4.3.1) for which $(t_0, x_0) \in [0, T] \times \Omega^c(\rho)$ is continuable up to $t = T$.

Proof. Let $\alpha > \rho$ be an arbitrary constant and $x(t; t_0, x_0)$ be a solution of (4.3.1) for which $(t_0, x_0) \in \bigcup\limits_{k=1}^{\infty} G_k \cap \mathcal{D}_1$, $\mathcal{D}_1 = [0, T] \times \{x \colon \|x\| \geq \rho\}$, and moreover $\|x_0\| \leq \alpha$.

By Proposition 4.5.2 there exists a constant $K(t_0, \alpha)$ such that

$$(4.5.7) \qquad\qquad v(t_0, x, \eta) \leq K(t_0, \alpha)$$

for $(t_0, x) \in \bigcup\limits_{k=1}^{\infty} G_k \cap \mathcal{D}_1$ and $\|x\| \leq \alpha$.

Under conditions (4) and (6a) of Theorem 4.5.5 the function $v(t, x, \eta)$ is positive definite, and such that $\varphi_{ip} \in KR$ then there exists a function

$a(r) \to \infty$ as $r \to \infty$, being of class KR such that in estimate (4.4.6)

(4.5.8) $$a(\|x\|) \leq w^{\mathrm{T}} H^{\mathrm{T}} A H w \quad \text{for} \quad \|x\| \geq \rho$$

and therefore,

$$a(\|x\|) \leq v(t, x, \eta) \quad \text{for} \quad \|x\| \geq \rho, \quad t \in [0, T].$$

For function $a(r)$ there exists a number $\gamma > 0$ such that

(4.5.9) $$a(\gamma) > K(t_0, \alpha).$$

If $\|x(t; t_0, x_0)\| < \gamma$ for $t \in J(t_0, x_0)$, where $J(t_0, x_0)$ is a maximal interval, where the solution $x(t; t_0, x_0)$ is definite, then it can be easily shown that $J(t_0, x_0) = [t_0, T]$. Assume that there exists a value t^* such that $\|x(t^*; t_0, x_0)\| \geq \gamma$ and introduce designation

$$\xi = \inf\{t^* \in J(t_0, x_0): \ \|x(t^*; t_0, x_0)\| \geq \gamma\}.$$

The solution $x(t; t_0, x_0)$ intersects the hypersurfaces $\sigma_{k_1}, \sigma_{k_2}, \ldots, \sigma_{k_p}$ at points $t_1 < t_2 < \cdots < t_p$ respectively.

Further two cases of solution behavior are considered.

Case 1. Let $(\xi, x(\xi; t_0, x_0)) \in \bigcup\limits_{k=1}^{\infty} G_k$. The fact that the solution $x(t; t_0, x_0)$ is continuous at point $t = \xi$ implies that there exists a point $\zeta \in [t_0, \xi)$ such that $\rho < \|x(\zeta; t_0, x_0)\| \leq \alpha$, $\|x(\xi; t_0, x_0)\| \geq \gamma$ and $\rho < \|x(\zeta; t_0, x_0)\| < \gamma$ for all $t \in (\zeta, \xi)$.

Under conditions (5) and (7b) of Theorem 4.5.5 estimate

(4.5.10) $$D^+ v(t, x, \eta) \leq 0 \quad \text{for} \quad (t, x) \in \bigcup\limits_{k=1}^{\infty} G_k \cap \mathcal{D}_1$$

is valid, and under conditions (6) and (7c) of Theorem 4.5.5 we have

(4.5.11) $$v(t - 0, x + I_k(x), \eta) \leq v(t, x, \eta) \quad \text{for} \quad (t, x) \in \sigma_k \cap \mathcal{D}_1.$$

In view of (4.5.7), (4.5.10) and (4.5.11) we get

(4.5.12) $$v(\xi, x(\xi; t_0, x_0), \eta) \leq v(\zeta, x(\zeta; t_0, x_0), \eta) \leq K(t_0, \alpha),$$

that contradicts the inequality (4.5.9).

Case 2. Let $(\xi, x(\xi; t_0, x_0)) \in \sigma_{k_{p+1}}$. For the time ξ defined above two possibilities occur:

(a) $\|x(\xi; t_0, x_0)\| = \gamma$ and $\|x(t; t_0, x_0)\| < \gamma$ for $t < \xi$.

In this case inequality (4.5.12) is satisfied, that contradicts inequality (4.5.9).

(b) $\rho \leq \|x(\xi; t_0, x_0)\| < \gamma$ and $\|x(t; t_0, x_0)\| > \gamma$ for $t \in (\xi, \xi + \varepsilon)$, where $\varepsilon > 0$ is sufficiently small.

Moreover, the inequality

$$\|x(\xi + 0; t_0, x_0)\| \geq \gamma$$

together with conditions (4.5.11) and (4.5.12) yield the inequalities

$$v(\xi + 0, x(\xi + 0; t_0, x_0), \eta) \leq v(\xi, x(\xi; t_0, x_0), \eta) \leq v(t_0, x_0, \eta) \leq K(t_0, \alpha).$$

However, this contradicts inequality (4.5.9).

Summing up the analysis one can see that $\|x(t; t_0, x_0)\| < \gamma$ for all $t \in J(t_0, x_0)$. This proves Theorem 4.5.5.

Corollary 4.5.2. *Let conditions (1) – (6) of Theorem 4.5.5 be satisfied for $t \in R_+$, $x \in R^n$ and $[0, T] \times \{\|x\| \geq \rho\} = R_+ \times \{\|x\| \geq \rho\}$. Then any solution $x(t; t_0, x_0)$ of (4.3.1), for which $(t_0, x_0) \in R_+ \times \{\|x\| \geq \rho\}$, is definite for $t \geq t_0$.*

We return ourselves to the case when $\tau_k(x) = t_k$, $k \in \mathcal{Z}$. The following result holds true.

Theorem 4.5.6. *Assume that*

(1) *vector function $f(t, x)$ is continuous on $[0, T] \times R^n$;*

(2) *vector functions $I_k(x)$, $k \in \mathcal{Z}$, are continuous for all $x \in R^n$;*

(3) *points $t_k \in R_+$, $k \in \mathcal{Z}$ and $0 < t_1 < t_2 < \ldots$, $\lim t_k = \infty$, $t_k = \tau_k(x)$, $k \in \mathcal{Z}$;*

(4) *every solution of (4.3.1) is continuable up to $t = T$.*

Then, for any $\alpha > 0$ there exists a $\beta(\alpha) > 0$ such that if $\|x_0\| \leq \alpha$, then $\|x(t; t_0, x_0)\| < \beta$ for all $t \in [t_0, T]$.

Proof. Condition (3) implies that $t_i \in (0, T)$, $i = 1, 2, \ldots, p$. According to Yoshizawa [1, Theorem 3.6] a number $\beta(\alpha) > 0$ exists, such that for $t \in (t_0, t_1)$ one gets $\|x(t; t_0, x_0)\| < \beta$. Therefore, $\|x(t; t_0, x_0)\| \leq \beta$. Condition (2) implies that there exists a constant $K_1(\beta) > 0$ such that, if $\|x\| \leq \beta$, then $\|I_1(x)\| \leq K_1(\beta)$. Then $\|x(t_1 + 0)\| \leq \beta + K_1(\beta)$. Applying Theorem 3.6 by Yoshizawa [1] once again we conclude that a $\bar{\beta}(\beta + K_1(\beta))$ exists such that $\|x(t; t_1, x_1)\| < \bar{\beta}$ for all $t \in (t_1, t_2]$. Preceding as above with the same arguments for every next interval $(t_i, t_{i+1}]$, $i = 1, 2, \ldots, p-1$, and $(t_p, T]$ in turn, we prove the theorem.

4.6 On Boundedness of the Solutions

As it has been noted earlier the construction of Liapunov functions in a finite form in terms of dynamical properties of subsystems makes possible the progress in qualitative analysis of an impulsive system (4.3.1). We present some results dealing with the boundedness conditions of the solutions.

Definition 4.6.1. The *solution* $x(t; t_0, x_0)$ of system (4.3.1) is

(a) *bounded*, if for any $\alpha > 0$ and any $t_0 \in R_+$, there exists a constant $\beta(t_0, \alpha) > 0$ such that $(t_0, x_0) \in \bigcup_{k=1}^{\infty} G_k$, $\|x_0\| \leq \alpha$ or $(t_0, x_0) \in \sigma_k$, $\|x_0 + I_k(x_0)\| \leq \alpha$, $k \in \mathcal{Z}$, then $\|x(t; t_0, x_0)\| < \beta$ holds for $t \geq t_0$;

(b) *uniformly bounded*, if for any $\alpha > 0$ and any $t_0 \in R_+$, there exists a constant $\beta(\alpha) > 0$ such that $(t_0, x_0) \in \bigcup_{k=1}^{\infty} G_k$ and $\|x_0\| \leq \alpha$ or $(t_0, x_0) \in \sigma_k$ and $\|x_0 + I^k(x_0)\| \leq \alpha$, $k \in \mathcal{Z}$, then $\|x(t; t_0, x_0)\| < \beta$ holds for $t \geq t_0$;

(c) *ultimately bounded for bound B*, if there exists a $T > 0$, such that for every solution $x(t; t_0, x_0)$ of (4.3.1) $\|x(t; t_0, x_0)\| < B$ for all $t \geq t_0 + T$, where T may depend on each solution;

(d) *equi-ultimately bounded for bound B*, if for any $\alpha > 0$ and any $t_0 \in R_+$, a number $T(t_0, \alpha) > 0$ exists, such that if $(t_0, x_0) \in \bigcup_{k=1}^{\infty} G_k$ and $\|x_0\| \leq \alpha$ or $(t_0, x_0) \in \sigma_k$ and $\|x_0 + I_k(x_0)\| \leq \alpha$, $k \in \mathcal{Z}$, then $\|x(t; t_0, x_0)\| < B$ holds for $t \geq t_0 + T$;

(e) *uniformly ultimately bounded*, if a number $B > 0$ exists, such that for any $\alpha > 0$ and $t_0 \in R_+$, a number $T(\alpha) > 0$ exists, such that if $(t_0, x_0) \in \bigcup_{k=1}^{\infty} G_k$ and $\|x_0\| \leq \alpha$ or $(t_0, x_0) \in \sigma_k$ and $\|x_0 + I_k(x_0)\| \leq \alpha$, then $\|x(t; t_0, x_0)\| < B$ holds for $t \geq t_0 + T$.

Definition 4.6.2. We will say that *system* (4.3.1) is *T-periodic*, if numbers $T > 0$ and $p > 0$ (p is an integer) exist, such that $f(t+T, x) = f(t, x)$ for $(t, x) \in R_+ \times R^n$ and $I_{k+p}(x) = I_k(x)$, $\tau_{k+p}(x) = \tau_k(x) + T$, for $x \in R$.

Further we consider system (4.3.1) assuming that the impulsive perturbation takes place at fixed times, i.e. $\tau_k = t_k$, $k = 1, 2, \ldots$.

With regard to the results by Yoshizawa [1] for systems of ordinary differential equations we formulate the following assertions.

Theorem 4.6.1. *Assume that*

(1) *for all* $x \in R^n$, $\tau_k(x) = t_k$, $k \in \mathcal{Z}$ *and* $0 < t_1 < t_2 < \ldots$, $\lim\limits_{k \to \infty} t_k = \infty$;

(2) *system (4.3.1) is T-periodic;*

(3) *the solutions of system (4.3.1) are bounded.*

Then the solutions of (4.3.1) are uniformly bounded.

Proof. Let $\alpha > 0$ and $t_0 \in [0, T]$ be arbitrary numbers. Let also $x_0 \in R^n$ and $\|x_0\| \leq \alpha$ be some fixed value of x_0. In view that any solution $x(t; t_0, x_0)$ of (4.3.1) is continuable up to $t = T$, then by Theorem 4.5.6 there exists a number $\beta(\alpha) > 0$ such that $\|x(t; t_0, x_0)\| < \beta$ for all $t \in [t_0, T]$. According to condition (3) of Theorem 4.6.1 there exists a number $\gamma(\alpha) > 0$ such that, if $\|x_0\| \leq \beta$, then $\|x(t; t_0, x_0)\| < \gamma$ for all $t \geq T$. However, if $0 \leq t_0 < T$ and $\|x_0\| < \alpha$, $\|x_0 + I_k(x_0)\| < \alpha$, $k \in \mathcal{Z}$, then $\|x(t; t_0, x_0)\| < \gamma$ for all $t \geq t_0$. Since by condition (2) the system (4.3.1) is T-periodic, for any $t_0 \in R_+$ and $\|x_0\| \leq \alpha$ we get $\|x(t; t_0, x_0)\| < \gamma$ for all $t > t_0$. Theorem 4.6.1 is proved.

Theorem 4.6.2. *Assume that*

(1) *condition (1) of Theorem 4.6.1 holds;*

(2) *vector function $f(t, x)$ satisfies Lipschitz condition with respect to the second argument;*

(3) *solutions of system (4.3.1) are uniformly bounded and ultimately bounded.*

Then the solutions of system (4.3.1) are equi-ultimately bounded.

The Proof of this theorem is similar to that of Theorem 8.6 by Yoshizawa [1].

Further we incorporate Liapunov function (4.4.3) constructed for system (4.3.1) to investigate the boundedness.

Assumption 4.6.1. *Let*

(a) *conditions of Assumptions 4.4.1 and 4.4.2 be satisfied with functions φ_i, $\varphi_{ip} \in KR$ when $\mathcal{N}_i = R^{n_i}$ and $\mathcal{N}_{ip} = R^{n_{ip}}$;*

(b) *conditions of Assumption 4.4.3 be satisfied with functions $\beta_{ip} \in KR$ when $\mathcal{N}_{ip} = R^{n_{ip}}$;*

(c) *conditions of Assumption 4.4.4 be satisfied with functions $\Psi_{ip} \in KR$ when $\mathcal{N}_{ip} = R^{n_{ip}}$ and $\Omega = R^n$.*

Theorem 4.6.3. *Assume that*

(1) *conditions* $A_1 - A_5$ *are satisfied;*

(2) *sequence* $\{\tau_k(x)\} \in S(R^n)$, $k \in \mathcal{Z}$;

(3) *vector function* $f(t, x)$ *is definite and continuous on* $R_+ \times R^n$;

(4) *vector functions* $I_k(x)$, $k \in \mathcal{Z}$, *are definite on* R^n;

(5) *all conditions of Assumption 4.6.1 hold and*

      (a) *in estimate (4.4.6) matrix* A *is positive definite;*

      (b) *in estimate (4.4.13) matrix* S *is negative semi-definite;*

      (c) *in estimate (4.4.15) matrix* C *is negative semi-definite.*

Then the solutions of system (4.3.1) are bounded.

Proof. Under conditions (1) – (4) of Theorem 4.6.3 it is reasonable to construct function (4.4.3) for which estimates (4.4.6), (4.4.13) and (4.4.15) are valid on R^n. Condition (5a) implies that the function $v(t, x, \eta)$ is positive definite and, therefore, there exists a function $a \in KR$ such that

$$(4.6.1) \qquad a(\|x\|) \le v(t, x, \eta) \quad \text{for} \quad (t, x) \in R_+ \times R^n.$$

Condition (5b) yields

$$(4.6.2) \qquad D^+ v(t, x, \eta) \le 0 \quad \text{for} \quad (t, x) \in \bigcup_{k=1}^{\infty} G_k \cap (R_+ \times R^n)$$

and under condition (5c) we have

$$(4.6.3) \qquad \begin{aligned} v(t+0, x + I^k(x), \eta) &\le v(t, x, \eta) \\ \text{for} \quad (t, x) &\in \sigma_k \cap (R_+ \times R^n). \end{aligned}$$

Consider the solution $x(t; t_0, x_0)$ of system (4.3.1) for which $(t, x) \in \bigcup_{k=1}^{\infty} G_k$, $\|x_0\| \le \alpha$.

According to Proposition 4.5.2 for the function (4.6.1) a constant $K(t_0, \alpha)$ exists such that $v(t_0 x_0, \eta) \le K(t_0, \alpha)$. We take $\beta > 0$ so that

$$(4.6.4) \qquad\qquad a(\beta) > K(t_0, \alpha).$$

It is clear that $\beta = \beta(t_0, \alpha)$.

Assume that conditions of Theorem 4.6.3 hold true and a value $t^* \in J(t_0, x)$ exists for which $\|x(t^*; t_0, x_0)\| \ge \beta$. We use designation

$$\xi = \inf\{t \in R_+ : \|x(t; t_0, x_0)\| \ge \beta\}.$$

By Proposition 4.5.1 the solution $x(t; t_0, x_0)$ for all $t \in [t_0, \xi)$ intersects the hypersurfaces $\sigma_{k1}, \ldots, \sigma_{kp}$ at times $t_1 < t_2 < \cdots < t_p$.
Consider two possible cases of the solution behavior.

Case 1. Let $(\xi, x(\xi; t_0, x_0)) \in \bigcup\limits_{k=1}^{\infty} G_k$. The continuity of the solution $x(t; t_0, x_0)$ at point ξ provides that $\|x(\xi; t_0, x_0)\| = \beta$ and $\|x(t; t_0, x_0)\| < \beta$ for all $t \in (t_0, \xi)$. Inequalities (4.6.1) and (4.6.2) imply

$$(4.6.5) \qquad a(\beta) \leq v(\xi, x(\xi; t_0, x_0), \eta) \leq v(t_0, x_0, \eta) \leq K(t_0, \alpha).$$

This contradicts the choice of value of β satisfying inequality (4.6.4).

Case 2. Let $(\xi, x(\xi; t_0, x_0)) \in \sigma_k$, $k \in \mathcal{Z}$. Then for $\|x(\xi; t_0, x_0)\| = \beta$ inequality (4.6.5) holds, and for $\|x(\xi; t_0, x_0)\| < \beta$ and $\|x(\xi + 0; t_0, x_0)\| \geq \beta$ we have

$$a(\beta) \leq v(\xi + 0, x(\xi + 0; t_0, x_0), \eta) \leq v(\xi, x(\xi; t_0, x_0), \eta)$$
$$\leq v(t_0, x_0, \eta) \leq K(t_0, \alpha).$$

This leads to the contradiction with the choice of value of β.
Therefore, $\|x(t; t_0, x_0)\| < \beta$ for $t \in J(t_0, x)$. Since $J(t_0, x) = [t_0, \infty)$, Theorem 4.6.3 is proved.

Theorem 4.6.4. *Assume that*

(1) *conditions $A_1 - A_5$ are satisfied;*
(2) *sequence $\{\tau_k(x)\} \in S(R^n)$, $k \in \mathcal{Z}$;*
(3) *vector function $f(t, x)$ is definite and continuous on $R_+ \times R^n$;*
(4) *vector functions $I_k(x)$, $k \in \mathcal{Z}$, are definite on R^n and for $\|x\| < \mu < +\infty$ inequalities $\|x + I_k(x)\| < \mu$, $k \in \mathcal{Z}$, are satisfied;*
(5) *all conditions of Assumption 4.6.1 are satisfied and*
    (a) *in estimate (4.4.6) matrices A and B are positive definite;*
    (b) *in estimate (4.4.13) matrix S is negative semi-definite;*
    (c) *in estimate (4.4.15) matrix C is negative semi-definite.*

Then the solutions of system (4.3.1) are uniformly bounded.

Proof. Let $x(t; t_0, x_0)$ be a solution of system (4.3.1), definite for all $t \in J(t_0, x)$ for which $\mu \leq \|x_0\| \leq \alpha$, $t_0 \in R_+$, $\alpha > \mu$ is an arbitrary constant. Under conditions $A_1 - A_5$ for system (4.3.1) the Liapunov function (4.4.3) can be constructed. If condition (5a) is satisfied, then for function

$v(t, x, \eta) \in W_2(R_+ \times R^n \times R_+^m, R_+)$ there exist functions $a, b \in KR$ such that

$$(4.6.6) \qquad a(\|x\|) \leq v(t, x, \eta) \leq b(\|x\|) \quad \text{for all} \quad (t, x) \in R_+ \times \mathcal{D}_1,$$

where $\mathcal{D}_1 = \{x \in R^n: \|x\| \geq \mu\}$. We take the number $\beta(\alpha) > 0$ so that $b(\alpha) < a(\beta)$, and assume that there exists a number $\gamma \in J(t_0, x_0)$ such that $\|x(\gamma; t_0, x_0)\| \geq \beta$.

We designate

$$\xi = \inf\{t \in J(t_0, x_0): \|x(t; t_0, x_0)\| \geq \beta\}.$$

Condition (3) of Theorem 4.6.4 implies the existence of point $\sigma < \xi$, $\sigma \in J(t_0, x_0)$, such that $\mu \leq \|x(\sigma; t_0, x_0)\| < \alpha$ and $\|x(\sigma; t_0, x_0)\| \geq \mu$ for all $t \in [\sigma, \xi]$.

According to Proposition 4.5.1 the solution $x(t; t_0, x_0)$ of system (4.3.1) for all $t \in [t_0, \xi)$ intersects the hypersurfaces $\sigma_{k_1}, \sigma_{k_2}, \ldots, \sigma_{k_p}$ at points $0 < t_1 < t_2 < \cdots < t_p < \xi$ respectively. As in the proof of Theorem 4.6.3 we deal with two cases.

Case 1. Let $(\xi, x(\xi; t_0, x_0)) \in \bigcup\limits_{k=1}^{\infty} G_k$. Since the solution $x(t; t_0, x_0)$ is continuous at point ξ, then $\|x(\xi; t_0, x_0)\| = \beta$ and $\|x(t; t_0, x_0)\| < \beta$ for all $t \in (t_0, \xi)$. We have under condition (5b)

$$(4.6.7) \qquad D^+ v(t, x, \eta) \leq 0 \quad \text{for} \quad (t, x) \in \bigcup\limits_{k=1}^{\infty} G_k \cap \mathcal{D}_1.$$

In view of (4.6.6) and (4.6.7) we have for $t = \xi$

$$(4.6.8) \quad a(\beta) = a(\|x(\xi; t_0, x_0)\|) \leq v(\xi, x(\xi; t_0, x_0), \eta) \leq v(t_0, x_0, \eta) \leq b(\alpha).$$

Inequality (4.6.8) contradicts the choice of β made earlier.

Case 2. Let $(\xi, x(\xi; t_0, x_0)) \in \sigma_{k_p}$. Under conditions (4) and (5c) of Theorem 4.6.4 we obtain from estimate (4.4.15) the inequality

$$(4.6.9) \quad v(t + 0, x + I_k(x), \eta) \leq v(t, x, \eta) \quad \text{for all} \quad (t, x) \in \sigma_k \cap \mathcal{D}_1.$$

For $\|x(\xi; t_0, x_0)\| = \beta$ inequality (4.6.8) holds, while $\|x(\xi; t_0, x_0)\| < \beta$ and $\|x(\xi + 0; t_0, x_0)\| \geq \beta$,

$$(4.6.10) \quad \begin{aligned} a(\beta) &\leq v(t + 0, x(\xi + 0; t_0, x_0), \eta) \leq v(\xi, x(\xi; t_0, x_0), \eta) \\ &\leq v(t_0, x_0, \eta) \leq b(\alpha). \end{aligned}$$

Similar to Case 1 the inequality (4.6.10) contradicts the choice of β.

The obtained contradiction shows that $\|x(t;t_0,x_0)\| < \beta$ for all $t \in J(t_0,x_0)$. Since $J(t,x_0) = [t_0,+\infty)$ (see Corollary 4.5.1), then Theorem 4.6.4 is proved.

Theorem 4.6.5. *Assume that*

(1) *conditions (1), (2), (3) and (4) of Theorem 4.6.4 are satisfied;*
(2) *all conditions of Assumption 4.6.1 hold and*
    (a) *in estimate (4.4.6) matrices A and B are positive definite,*
    (b) *in estimate (4.4.13) matrix S is negative definite,*
    (c) *in estimate (4.4.15) matrix C is negative semi-definite.*

Then the solutions of system (4.3.1) are uniformly ultimately bounded.

Proof. Under conditions (2a) for system (4.3.1) the function $v(t,x,\eta)$ can be constructed, that satisfies estimate (4.6.6).

Condition (2b) and estimates (4.4.13) imply that there exists a function $c \in KR$ such that

$$(4.6.11) \qquad D^+v(t,x,\eta) \le -c(\|x\|) \quad \text{for all} \quad (t,x) \in \bigcup_{k=1}^{\infty} G_k \cap \mathcal{D}_1.$$

Finally, condition (2c) yields estimate (4.6.9). Under conditions of Theorem 4.6.5 all conditions of Theorem 4.6.4 are satisfied, and, hence, the solutions $x(t;t_0,x_0)$ of system (4.3.1) are uniformly bounded. Hence, it follows that a number $\beta(\alpha)$ exists such that estimate $\|x(t;t_0,x_0)\| < \beta(\alpha)$ holds for all $t \ge t_0$ whenever $\|x_0\| < \alpha$, $\beta > \alpha \ge \mu$.

The uniform boundedness of the solution $x(t;t_0,x_0)$ implies that there exists a number $B > 0$ such that for $\|x_0\| < \mu$ when all $t \ge t_0$ the estimate $\|x(t;t_0,x_0)\| < B$ holds. Let $x_0 \in R^n$ and $\|x_0\| \le \alpha$. We shall show that there exists a time $t_1 > t_0$ such that $\|x(t;t_0,x_0)\| < \mu$. Assume on the contrary that estimate $\|x(t;t_0,x_0)\| \ge \mu$ is true for all $t \ge t_0$. By condition (4.6.11) for $\mu \le \|x\| \le \beta$ a $\delta(\alpha) > 0$ can be taken so that

$$D^+v(t,x,\eta) \le -\delta(\alpha), \quad (t,x) \in \bigcup_{k=1}^{\infty} G_k \cap \mathcal{D}_1.$$

The inequality

$$(4.6.12) \quad a(\mu) \le v(t,x(t;t_0,x_0),\eta) \le v(t_0,x_0,\eta) - \delta(\alpha)(t-t_0) < a(\mu)$$

is satisfied for all $t > t_0 + T_1(a)$, where

$$T_1(\alpha) = [b(\alpha) - a(\mu)]/\delta(\alpha) > 0.$$

Inequality (4.6.12) provides that $t_1 \notin J(t_0, x_0)$. Then for all $t > t_1$ or $t > t_0 + T_1(\alpha)$ the estimate $\|x(t; t_0, x_0)\| < B$ is satisfied. Theorem 4.6.5 is proved.

4.7 Novel Methodology for Stability

Along with the boundedness property of system (4.3.1) it is of interest to investigate stability of solutions to this system. In terms of the Liapunov function constructed in Section 4.4 some sufficient stability conditions are formulated for the solutions of system (4.3.1).

Consider the system of differential equations (4.3.1):

$$\frac{dx}{dt} = f(t, x), \quad t \neq \tau_k(x),$$
$$\Delta x = I_k(x), \quad t = \tau_k(x), \quad k \in \mathcal{Z},$$
$$x(t_0^+) = x_0.$$

4.7.1 Stability conditions

Stability analysis of the equilibrium state $x = 0$ of the system (4.3.1) can be made in terms of the general theorems presented below. We recall that system (4.3.1) is considered in domain $R_+ \times \Omega(\rho)$, $\Omega(\rho) \subseteq R^n$.

Theorem 4.7.1. *Assume that*

(1) *conditions $A_1 - A_5$ are satisfied;*

(2) *all conditions of Assumptions 4.4.1 – 4.4.4, 4.4.6 and 4.4.7 are satisfied;*

(3) *in estimate (4.4.6)*

    (a) *matrix A is positive definite;*

    (b) *matrix B is positive definite;*

(4) *in estimate (4.4.13) matrix S is negative semi-definite or equal to zero;*

(5) *matrix C in estimate (4.4.15) is*

    (a) *negative semi-definite or equal to zero;*

    (b) *negative definite.*

Then, respectively,

(a) *conditions (1) – (3a) and (5a) are sufficient for the equilibrium state $x = 0$ of system (4.3.1) to be stable;*
(b) *conditions (1) – (4) and (5a) are sufficient for the equilibrium state $x = 0$ of system (4.3.1) to be uniformly stable;*
(c) *conditions (1) – (3a) and (5b) are sufficient for the equilibrium state $x = 0$ of system (4.3.1) to be asymptotically stable;*
(d) *conditions (1) – (4) and (5b) are sufficient for the equilibrium state $x = 0$ of system (4.3.1) to be uniformly asymptotically stable.*

Proof. We start with assertion (a). Under condition (1) of Theorem 4.7.1 system (4.3.1) can be represented in the form (4.3.4) with further reduction of subsystems to the form (4.3.7). If conditions of Assumptions 4.4.1 and 4.4.2 are satisfied, then Proposition 4.4.2 is valid and for function $v(t, x, \eta)$ estimate (4.4.6) holds, which ensures under condition (3) of Theorem 4.7.1 some definite positiveness of function $v(t, x, \eta)$ provided all $t \in R_+$. Under conditions of Assumptions 4.4.3 and 4.4.6 for function $D^+ v(t, x, \eta)$ estimate (4.4.13) is valid, which yields, under condition (4) of Theorem 4.7.1

$$(4.7.1) \qquad D^+ v(t, x, \eta) \leq 0, \quad t \neq \tau_k(x), \quad k \in \mathcal{Z}.$$

According to conditions of Assumptions 4.4.4 and 4.4.7 estimate (4.4.15) is satisfied, that yields under condition (5a) of Theorem 4.7.1

$$(4.7.2) \qquad v(t, x + I_k(x), \eta) \leq v(t, x, \eta),$$

provided $t = \tau_k(x)$, $k \in \mathcal{Z}$.

Since function $v(t, x, \eta)$ is positive definite for all $(t, x) \in R_+ \times \Omega(\rho)$, there exists a function $a \in K$ such that

$$(4.7.3) \qquad a(\|x\|) \leq v(t, x, \eta) \quad \text{for all} \quad (t, x, \eta) \in R_+ \times \Omega(\rho) \times R_+^s.$$

Let any initial time $t_0 \in R_+$ and $\varepsilon > 0$ be given. The properties of the function $v(t, x, \eta)$ imply the existence of $\delta = \delta(t_0, \varepsilon) > 0$ such that

$$(4.7.4) \qquad \sup_{\|x\| < \delta} v(t_0 + 0, \, x, \, \eta) < \min\left(a(\xi), \, a(\mu)\right).$$

Let $x_0 \in \Omega(\rho)$, $\|x_0\| < \delta$, and $x(t) = x(t; t_0, x_0)$ be a solution of system (4.3.1). Conditions (4.7.1) and (4.7.2) imply that the function $v(t, x(t), \eta)$

does not increase on interval $J^+(t_0, x_0)$, where the solution of system (4.3.1) is continuable to the right. We find from conditions (4.7.1) – (4.7.3)

(4.7.5)
$$a(\|x(t; t_0, x_0)\|) \leq v(t, x(t), \eta)$$
$$\leq v(t_0 + 0, x_0, \eta) < \min(a(\varepsilon), a(\mu))$$

for all $t \in J^+(t_0, x_0)$. Hence, it follows

$$\|x(t; t_0, x_0)\| < \min(\varepsilon, \mu) \quad \text{for all} \quad t \in J^+(t_0, x_0).$$

If $J^+(t_0, x_0) = (t_0, +\infty)$, then the solution $x = 0$ of the system (4.3.1) is stable in the sense of Liapunov.

Assertion (b) of Theorem 4.7.1 is proved similarly, though in view of condition (3b) there exists a function $b \in K$ such that

(4.7.6) $\quad v(t, x, \eta) \leq b(\|x\|) \quad$ for all $\quad (t, x, \eta) \in R_+ \times \Omega(\rho) \times R_+^s$.

Therefore, for any $\varepsilon > 0$ the value δ can be taken independently of t_0 setting, for instance, $\delta(\varepsilon) = \min b^{-1}(a(\varepsilon), a(\mu))$. Then, if $\|x_0\| < \delta$, we find from conditions (4.7.1), (4.7.2) and (4.7.6)

$$a(\|x(t; t_0, x_0)\|) \leq v(t, x(t), \eta) \leq v(t_0 + 0, x_0, \eta)$$
$$\leq b(\|x_0\|) < \min b(b^{-1}(a(\varepsilon), a(\mu))) = \min(a(\varepsilon), a(\mu)).$$

Hence for $J^+(t_0, x_0) = (t_0, +\infty)$ the solution $x = 0$ of the system (4.3.1) is uniformly stable in the sense of Liapunov.

Now we prove assertion (c). Under Assumption 4.4.7 we have estimate (4.4.15) from Proposition 4.4.8 instead of estimate (4.7.2). If condition (5b) is satisfied, then (4.4.15) implies

(4.7.7) $\quad v(\tau_k(x), x + I_k(x), \eta) - v(\tau_k(x), x, \eta) \leq -\lambda_M(C)\psi^T\psi,$

where $\lambda_M(C)$ is the maximal eigenvalue of matrix C.

Since under conditions (1) – (3a) and (5a) the solution $x = 0$ is stable, then in order that to prove assertion (c) it is sufficient to show that $\lim v(t, x(t; t_0, x_0), \eta) = 0$ as $t \to \infty$.

By condition (4.7.1) function $v(t, x, \eta)$ does not increase, and since it satisfies condition (4.7.2), there exists a $\lim_{t \to \infty} v(t, x, \eta) = \alpha$ for all $(t, x, \eta) \in R_+ \times \Omega(\rho) \cap \Omega(\mu) \times R_+^s$. Let $\alpha > 0$. We designate

$$\Delta = \min_{\alpha \leq s \leq v(t_0, x_0, \eta)} \psi^T(s)\psi(s).$$

If the solution $x(t)$ intersects the surfaces $t = \tau_k(x)$, $k \in \mathcal{Z}$, at the points $(\tau_k(x_k), x_k)$, then we have by estimate (4.7.7) the inequality

$$v(\tau_k(x_k) + 0, \cdot) - v(\tau_k(x_k), \cdot) \leq -\lambda_M(C)\psi^T(\tau_k(x_k))\psi(\tau_k(x_k))$$

for all $k \in \mathcal{Z}$.

Since $\alpha \leq \psi^T(\tau_k(x_k))\psi(\tau_k(x_k)) \leq v(t_0, x_0, \eta)$, then $-\lambda_M(C)\psi^T(\tau_k(x_k))$ $\times\psi(\tau_k(x_k)) \leq -\Delta$. Therefore,

$$v(\tau_k(x_k) + 0, \cdot) - v(\tau_k(x_k), \cdot) \leq -\lambda_M(C)\Delta.$$

Thus, for the decreasing function $v(t, x, \eta)$ we have on every continuity interval

$$(\tau_k(x_k) + 0, \cdot) \geq v(\tau_k(x_k), \cdot).$$

Then the inequality

(4.7.8)
$$\begin{aligned}
v(\tau_k(x_k) + 0, \cdot) &\leq v(\tau_k(x_k) + 0, \cdot) \\
&+ \sum_{k=0}^{l_1} \left(v(\tau_k(x_k) + 0, \cdot) - v(\tau_{k+1}(x_{k+1}), \cdot)\right) \\
&= v(t_0, x_0, \eta) + \sum_{k=0}^{l-1} \left(v(\tau_k(x_k) + 0, \cdot) - v(\tau_k(x_k), \cdot)\right) \\
&\leq v(t_0, x_0, \eta) - l\lambda(C)\Delta
\end{aligned}$$

holds for all $l \in \mathcal{Z}$.

For the large values of l the right-hand part of the inequality (4.7.8) becomes negative, and this contradicts positive definiteness of function $v(t, x, \eta)$ and assumption that $\alpha \geq 0$.

This proves assertion (c).

Assertion (d) of Theorem 4.7.1 is proved similarly to assertion (c) with regard to the fact that the value δ can be taken independently of t_0 in the way mentioned in the proof of assertion (b).

Remark 4.7.1. The analysis of condition (4.7.1) and conditions of Assumption 4.4.7 together with condition (5b) of Theorem 4.7.1 shows that the impulsive perturbations of certain type (condition (5b)) change the property of solution $x(t)$ from stability up to asymptotic stability.

Theorem 4.7.2. *Let the differential equations (4.3.1) of a perturbed motion be such that in the domain* $R_+ \times \Omega(\rho)$ *conditions* $A_1 - A_5$ *hold and*

(1) *system (4.3.1) admits the first and second level decompositions;*
(2) *all conditions of Assumptions 4.4.1 – 4.4.3, 4.4.5, 4.4.6 and 4.4.8 are satisfied;*
(3) *matrix* A *in estimate (4.4.6) is positive definite;*
(4) *matrix* B *in estimate (4.4.6) is positive definite;*
(5) *matrix* S *in estimate (4.4.13) is negative definite;*
(6) *matrix* C^* *in estimate (4.4.16) is positive definite;*
(7) *there exists a constant* $\vartheta > 0$ *for which*

$$\sup_k \left(\min_{\Omega(\rho)} \tau_{k+1}(x) - \max_{\Omega(\rho)} \tau_k(x) \right) = \vartheta > 0;$$

(8) *there exists a constant* $a_0 > 0$ *such that*

$$-\frac{1}{\lambda_M(S)} \int\limits_{a}^{\lambda_M(C^*)Q_2(a)} \frac{dy}{Q_1(y)} \leq \vartheta$$

for all $a \in (0, a_0);$
(9) *there exist a constant* $a_0 > 0$ *and a number* $\gamma > 0$ *such that*

$$-\frac{1}{\lambda_M(S)} \int\limits_{a}^{\lambda_M(C^*)Q_2(a)} \frac{dy}{Q_1(y)} \leq \vartheta - \gamma.$$

Then, correspondingly,

(a) *conditions (1) – (3), (5), (8) are sufficient for the stability of equilibrium state* $x = 0$ *of the system (4.3.1);*
(b) *conditions (1) – (8) are sufficient for uniform stability of equilibrium state* $x = 0$ *of the system (4.3.1);*
(c) *conditions (1) – (3), (5) – (7) and (9) are sufficient for asymptotic stability of equilibrium state* $x = 0$ *of the system (4.3.1);*
(d) *conditions (1) – (7) and (9) are sufficient for uniform asymptotic stability of equilibrium state* $x = 0$ *of the system (4.3.1).*

Proof. We begin with assertion (a). Conditions $A_1 - A_5$ provide the application of general approach based on hierarchical matrix function. Under conditions of Assumptions 4.4.1 and 4.4.2 the function $v(t, x, \eta)$ is positive definite. If conditions of Assumptions 4.4.3 and 4.4.6 are satisfied, then for the function $D^+v(t, x, \eta)$ the estimate from Proposition 4.4.6

$$(4.7.9) \qquad D^+v(t, x, \eta) \leq \beta^{\mathrm{T}} S \beta \quad \text{for} \quad t \neq \tau_k(x), \quad \text{and} \quad k \in \mathcal{Z}$$

takes place.

Conditions of Assumptions 4.4.5 and 4.4.8 yield that estimate from Proposition 4.4.10

$$(4.7.10) \qquad \begin{aligned} &v(\tau_k(x),\, x + I_k(x), \eta) \leq \psi^{\mathrm{T}} C^* \psi \\ &\text{for} \quad t = \tau_k(x) \quad \text{and} \quad k \in \mathcal{Z}, \end{aligned}$$

is valid.

Since by conditions (5) of Theorem 4.7.2 the matrix S is negative definite, then $\lambda_M(S) < 0$ and according to Proposition 4.4.7 we have

$$(4.7.11) \qquad \begin{aligned} &D^+v(t, x, \eta) \leq \lambda_M(S) H_1(v(t, x, \eta)) \\ &\text{for} \quad t \neq \tau_k(x), \quad \text{and} \quad k \in \mathcal{Z}. \end{aligned}$$

By condition (6) of Theorem 4.7.2 the matrix C^* is positive definite, i.e. $\lambda_M(C^*) > 0$, but by Proposition 4.4.11 we have

$$(4.7.12) \qquad v(\tau_k(x),\, x + I_k(x), \eta) \leq \lambda_M(C^*) Q_2(v(\tau_k(x), x, \eta))$$

for all $k \in \mathcal{Z}$.

Let $t_0 \in R_+$ and $\varepsilon \in (0, \rho)$. We compute a greatest lower bound of the function $v(t, x, \eta)$

$$(4.7.13) \qquad l = \inf_{t \geq t_0,\, \|x\| \leq \varepsilon} v(t, x, \eta)$$

and by given t_0 and ε take $\delta = \delta(t_0, \varepsilon) > 0$ so that inequality

$$(4.7.14) \qquad m = \sup_{\|x\| < \delta} v(t_0, x, \eta) < l$$

holds.

Assume that the solution $x(t) = x(t; t_0, x_0)$ initiated in $\Omega(\delta)$. Assertion (a) is proved, if any solution $x(t)$ does not leave the domain $\Omega(\varepsilon)$. We designate by $v(t) = v(t, x(t), \eta)$ the value of the function $v(t, x, \eta)$ along the solution $x(t; t_0, x_0)$. In view of (4.7.13) and (4.7.14) the assertion (a) of Theorem 4.7.2 is proved, if $v(t) < l$ for all $t \geq t_0$. Assume the contrary, i.e. there exists a $t^* > t_0$ such that $x(t^*) \notin \Omega(\varepsilon)$ without reaching surface $t = \tau_1(x)$. Then

$$(4.7.15) \qquad v(t^*) = v(t^*, x(t^*), \eta) \geq l$$

and, on the other hand, according to inequality (4.7.11) function $v(t)$ does not increase for all $x \in \overline{\Omega}(\varepsilon)$ and $v(t^*) \leq m < l$. Therefore, the solution $x(t)$ reaches surface $t = \tau_1(x)$. We designate the reaching point by $(\tau_1(\tilde{x}_1), \tilde{x}_1)$, where $\tilde{x}$ denotes a fixed point on hypersurface $t = \tau_1(x)$. For the values $t \in [t_0, \tau_1(\tilde{x}_1)]$ we get from inequality (4.7.11)

$$D^+ v(t) \leq -\lambda_M(S) H_1(v(t))$$

and, therefore,

$$(4.7.16) \qquad -\frac{1}{\lambda_M(S)} \int\limits_{t_0}^{\tau_1(\tilde{x})} \frac{v'(t)\, dt}{H_1(v(t))} \geq \tau_1(\tilde{x}) - t_0.$$

Setting $v(t) = y$ and in view of condition (7) of Theorem 4.7.2 we obtain

$$(4.7.17) \qquad \frac{1}{\lambda_M(S)} \int\limits_{v(\tau_1(\tilde{x}))}^{v(t_0)} \frac{dy}{H_1(y)} \geq \tau_1(\tilde{x}) - t_0 \geq \vartheta.$$

Under condition (8) of Theorem 4.7.2, and having designated $a = v(\tau_1(\tilde{x}))$ we arrive at the estimate

$$(4.7.18) \qquad \int\limits_{v(\tau_1(\tilde{x}))}^{v(\tau_1(\tilde{x})+0)} \frac{dy}{H_1(y)} \leq \int\limits_{v(\tau_1(\tilde{x}))}^{\lambda_M(C^*)Q_2(v(\tau_1(\tilde{x})))} \frac{dy}{H_1(y)} \leq \vartheta.$$

Inequalities (4.7.17) and (4.7.18) yield

$$(4.7.19) \qquad \int\limits_{v(\tau_1(\tilde{x})+0)}^{v(t_0)} \frac{ds}{H_1(s)} = \int\limits_{v(\tau_1(\tilde{x}))}^{v(t_0)} \frac{dy}{H_1(y)} - \int\limits_{v(\tau_1(\tilde{x}))}^{v(\tau_1(\tilde{x})+0)} \frac{dy}{H_1(y)} \geq 0.$$

Inequality (4.7.19) implies that $v(\tau_1(\tilde{x}_1) + 0) \leq v(t_0)$. Incorporating the method of mathematical induction we get

$$v(\tau_k(\tilde{x}_k) + 0) \leq v(t_0), \quad k \in \mathcal{Z}.$$

This proves assertion (a) of Theorem 4.7.2.

Now let us prove assertion (b). To this end we assume that conditions (1)– (7) and (9) of Theorem 4.7.2 are satisfied. Now we analyze condition (9). Let the solution $x(t)$ of the system (4.3.1) intersect surfaces $t = \tau_k(x)$ at points $(\tau_k(\tilde{x}_k), \tilde{x}_k)$. By inequality (4.7.11) we have

$$-\frac{1}{\lambda_M(S)} \int\limits_{\tau_k(\tilde{x}_k)}^{\tau_{k+1}(\tilde{x}_{k+1})} \frac{v'(t)dt}{H_1(v(t))} \geq \tau_{k+1}(x_{k+1}) - \tau_k(x_k) \geq \vartheta.$$

If in the inequality of condition (9) we take $a = v(\tau_{k+1}(\tilde{x}_{k+1}))$, then in view of (4.7.12) we get

$$\int\limits_{v(\tau_{k+1}(\tilde{x}_{k+1}))}^{v(\tau_{k+1}(\tilde{x}_{k+1})+0)} \frac{ds}{H_1(s)} \leq \int\limits_{v(\tau_{k+1}(\tilde{x}_{k+1}))}^{\lambda_M(C^*)Q_2(v(\tau_{k+1}(\tilde{x}_{k+1})))} \frac{ds}{H_1(s)} \leq \vartheta - \gamma.$$

Let us designate $a_k^+ = v(\tau_k(\tilde{x}_k) + 0)$, $k \in Z$. Then we have

$$\int\limits_{a_{k+1}^+}^{a_k^+} \frac{ds}{H_1(s)} \leq \int\limits_{a_{k+1}}^{a_k^+} \frac{ds}{H_1(s)} - \int\limits_{a_{k+1}}^{a_{k+1}^+} \frac{ds}{H_1(s)} \geq \gamma.$$

Hence, it follows that for sequence $\{a_k^+\}$ the inequality

$$(4.7.20) \qquad \int\limits_{a_{k+1}^+}^{a_k^+} \frac{ds}{H_1(s)} \geq \gamma, \quad k \in Z$$

is valid. Inequality (4.7.20) implies that the sequence $\{a_k^+\}$ decreases for $k \to \infty$, and, therefore, $\lim\limits_{k\to\infty} v(\tau_k(x_k) + 0) = 0$.

Let this be not correct, i.e. $\lim\limits_{k\to\infty} v(\tau_k(x_k) + 0) = \alpha > 0$. We designate $c = \min\limits_{a\leq y\leq v(t_0)} \lambda_M(S)H_1(y)$. Then we get from inequality (4.7.20)

$$\gamma \leq \int\limits_{a_{k+1}^+}^{a_k^+} \frac{ds}{H_1(s)} \leq \frac{1}{c}(v(\tau_k(x_k) + 0) - v(\tau_{k+1}(x_{k+1}) + 0)),$$

i.e.

$$v(\tau_k(x_k) + 0) - v(\tau_{k+1}(x_{k+1}) + 0) \geq \gamma, \quad c = \text{const}.$$

The obtained inequality contradicts the convergence of sequence $v(\tau_k(x_k) + 0)$. Therefore

$$v(\tau_k(x_k) + 0) \to 0 \quad \text{as} \quad k \to \infty.$$

Further, in view of the fact that for all $t \neq \tau_k(x)$, $k \in \mathcal{Z}$, the function $v(t)$ decreases, and, therefore

$$\sup_{\tau_k(x_k) < t < \tau_{k+1}(x_{k+1})} v(t) = v(\tau_k(x_k) + 0).$$

Alongside the inequality

$$v(\tau_k(x_k) + 0) > v(\tau_{k+1}(x_{k+1}) + 0),$$

that holds for all $k \in \mathcal{Z}$, we get

$$v(t) < v(\tau_k(x_k) + 0) \quad \text{for all} \quad t > \tau_k(x_k).$$

Thus, condition $v(\tau_k(x_k) + 0) \to 0$ as $k \to \infty$ implies

(4.7.21) $$v(t) \to 0 \quad \text{as} \quad t \to \infty.$$

Since $v(t) = v(t, x(t; t_0, x_0), \eta)$, then (4.7.21) yields $\|x(t; t_0, x_0)\| \to 0$ as $t \to \infty$. This proves assertion (c).

The proof of assertions (b) and (d) is made in the same way, and, moreover, by virtue of condition (4) the value δ can be taken independent of t_0.

Example 4.7.1. Consider an impulsive forth order system consisting of two subsystems of the second order, that are described by the equations

(4.7.22)
$$\frac{dx_i}{dt} = -x_i^3 + 0,5\,x_j^3, \quad i, j = 1, 2, \quad t \neq \tau_k(x), \quad k \in \mathcal{Z},$$
$$\Delta x_i = -x_i + \sigma x_j, \quad t = \tau_k(x), \quad k \in \mathcal{Z},$$

where $x_i^{\mathrm{T}} = (x_{i1}, x_{i2}) \in R^2$, $x \in R^4$.

Here the independent subsystems of the first level decomposition are

$$\frac{dx_i}{dt} = -x_i^3, \quad t \neq \tau_k(x), \quad k \in \mathcal{Z},$$
$$\Delta x_i = -x_i, \quad t = \tau_k(x), \quad k \in \mathcal{Z}.$$

The second level decomposition yields

$$\frac{dx_{ij}}{dt} = -x_{ij}^3 + h_{ij}(x_i), \quad i, j = 1, 2, \quad t \neq \tau_k(x),$$

$$\Delta x_{ij} = -x_{ij}, \quad t = \tau_k(x), \quad k \in \mathcal{Z},$$

where $x_{ij} \in R$,

$$h_{11}(x_1) = -x_{11}x_{12}^2, \quad h_{12}(x_1) = -x_{11}^2 x_{12},$$
$$h_{21}(x_2) = -x_{21}x_{22}^2, \quad h_{22}(x_2) = -x_{21}^2 x_{22}.$$

In the matrices $B_i(t, \cdot)$, $i = 1, 2$, the elements u_{ij} are taken as follows

$$u_{11}^{(1)} = x_{11}^2, \quad u_{22}^{(1)} = x_{12}^2, \quad u_{12}^{(1)} = u_{21}^{(1)} = 0,$$
$$u_{11}^{(2)} = x_{21}^2, \quad u_{22}^{(2)} = x_{22}^2, \quad u_{12}^{(2)} = u_{21}^{(2)} = 0.$$

Functions U_{12} and U_{21} are taken in the form

$$U_{12} = U_{21} = x_1^{\mathrm{T}} \operatorname{diag}[0.1; \, 0.1] \, x_2.$$

For the vector $\eta = (1, 1)^{\mathrm{T}} \in R_+^2$ the matrix A in estimate (4.4.2) is

$$A = \begin{pmatrix} 1 & 0,1 \\ 0,1 & 1 \end{pmatrix}.$$

For the vectors $\xi_i = (1, 1)^{\mathrm{T}} \in R_+^2$, $i = 1, 2$, the matrices S and C are of the form

$$S = \begin{pmatrix} -1.3 & 0.6 \\ 0.6 & -1.3 \end{pmatrix}, \quad C = \begin{pmatrix} \sigma^2 - 1 & 0.1 \, |\sigma^2 - 1| \\ 0.1 \, |\sigma^2 - 1| & \sigma^2 - 1 \end{pmatrix}.$$

It can be easily verified that the matrix A is positive definite, the matrix S is negative definite, and the matrix C for $\sigma = \pm 1$ equals to zero, and for $|\sigma| < 1$ is negative definite. By Theorem 4.7.1 the solution $(x = 0) \in R^4$ is stable.

Stability analysis of unsteady motions of impulsive system (4.3.1) in terms of the hierarchical Liapunov functions proposed here is distinguished by simplicity and general character. Liapunov functions composed of matrix-valued functions for subsystems of first and second level decompositions and their interconnection functions are versatile tools in qualitative analysis of systems modeling various technology and engineering processes. It is also natural that the present construction algorithm for Liapunov functions is applicable for autonomous linear and nonlinear large scale impulsive systems as well.

4.8 Notes

4.1. To describe mathematically an evolution of a real world phenomenon with a short-term perturbation, it is sometimes convenient to neglect the duration of the perturbation and to consider these perturbations to be "instantaneous." For such an idealization, it becomes necessary to study dynamical systems with discontinuous trajectories or, as they might be called, differential equations with impulses, i.e. impulsive differential equations. The state of investigations in this area is reflected in the monographs Bainov and Simeonov [1], Halanay and Wexler [1], Lakshmikantham, Bainov, *et al.* [1], Larin [1], Pandit and Deo [1], Samoilenko and Perestyuk [1], and many papers.

4.2. Construction of auxiliary Liapunov function for large scale impulsive systems (4.2.2) is an important problem for the theory of these systems. The application of matrix-valued function $U(t,x)$ in construction of scalar function $v(t,x,\eta)$ diminishes some difficulties due to weakening of requirements to components $v_{j,l}$, $j, l = 1, 2, \ldots, m$, what in its turn allows better account of interactions between independent subsystems. All established sufficient conditions for stability, asymptotic stability and instability are formulated in terms of restrictions on maximal or minimal eigenvalues of special matrices due to Martynyuk and Miladzhanov [1] (cf. Samoilenko and Perestyuk [1]).

4.3. The theorems like Theorem 4.3.1 can be found in Lakshmikantham, Bainov, *et al.* [1], Samoilenko and Perestyuk [1], etc.

4.4. Hierarchical Liapunov matrix-valued function for impulsive systems are constructed due to some results of the paper by Martynyuk and Begmuratov [2].

4.5. Theorems 4.5.1 and 4.5.5 are new, while Theorems 4.5.2, 4.5.3 and 4.5.4 are due to Hristova and Bainov [1].

4.6. Theorems 4.6.1 and 4.6.2 are due to Hristova and Bainov [1]. Theorems 4.6.3, 4.6.4 and 4.6.5 are due to Martynyuk and Chernetskaya [1].

4.7. Theorems 4.7.1 and 4.7.2 are new.

For the use of impulsive systems in other situations see Bainov and Dishliev [1], Bainov and Kulev [1], Bainov and Simeonov [1], Barbashin [1], Blaquiere [1], Carvalho and Ferreira [1], Das and Sharma [1], Larin [2], Lella [1], Liu and Willms [1], Pavlidis [1], Sree Hari Rao [1], etc.

5

APPLICATIONS

5.1 Introduction

This chapter shows some applications of the general results presented in the previous chapters for solution of the problems of mechanics, theoretical electrodynamics and theory of automatic control.

In Section 5.2 the original Zubov's result is set forth and a new algorithm is established for the asymptotic stability domain of estimation for nonlinear time-invariant systems via Liapunov matrix-valued functions method.

In Section 5.3 a new algorithm is set out to estimate the domain of asymptotic stability for the equations modeling the dynamics of a three-machine power system. The result is compared with those obtained before in terms of vector Liapunov function.

Section 5.4 deals with the method of constructing the matrix-valued function for a three-mass system, which occurs frequently in mechanics and engineering. Here we set out the method of constructing the matrix-valued Liapunov function which is applied in stability investigation of two non-stationary connected oscillators.

In the final Section 5.5 motion stability conditions are established for a discrete-time system. As an example the Lur'e-Postnikov system admitting a homogeneous hierarchical decomposition is considered.

5.2 Estimations of Asymptotic Stability Domains in General

5.2.1 A fundamental Zubov's result

Let $\mathbb{R}$ be a metric space and X and Y be two sets contained in $\mathbb{R}$. The functional V is given on the set X, if the law is prescribed by which every element $p \in X$ is associated with the real number $V(p)$. Let the set $\mathcal{M} \subset \mathbb{R}$ be compact in $\mathbb{R}$.

Let the autonomous system of n differential equations

(5.2.1) $$\frac{dx}{dt} = f(x), \quad x(t_0) = x_0$$

be determined in R^n and the components f_s of the function $f(x) = (f_1(x), \ldots, f_n(x))^\mathrm{T}$ be continuous in R^n for $-\infty < x_s < +\infty$. Besides, we assume that these functions satisfy the Lipschitz condition in any finite domain of the space R^n. Assume also that (5.2.1) induces the dynamical system $f(p, t)$ (see Birkhoff [1], and Nemytskii and Stepanov [1]).

We give the following definitions according to Zubov [3].

If the closed invariant set $\mathcal{M}$ of the dynamical system $f(p, t)$ is asymptotically stable, then the *totality* $\mathcal{A}$ of all points $p \in \mathbb{R}$ and $p \notin \mathcal{M}$ possessing the property

(5.2.2) $$\rho(f(p, t), \mathcal{M}) \to 0 \quad \text{as} \quad t \to +\infty$$

is called the *domain of asymptotic stability* of this invariant set. Here $\rho(p, X) = \inf\limits_{q \in X} q(p, q)$ and $\rho(p, q)$ is the metric distance between the elements p and q of the space $\mathbb{R}$.

The nonempty set of all the points $q \in \bar{\mathcal{A}} \setminus \mathcal{A}$ and $q \notin \mathcal{M}$ is called the *boundary of the asymptotic stability domain*.

It is proved that the boundary of the asymptotic stability domain is also the invariant set.

Zubov's theorem presented below is the fundamental result solution of the problem of estimating the domain of asymptotic stability and constructing its boundary for the dynamical system $f(p, t)$ induced, in particular, by the system (5.2.1).

Theorem 5.2.1. *For the given system $f(p, t)$ let two functionals $V(p)$ and $W(p)$ exist such that*

(1) *the functional $V(p)$ is given and continuous in $\mathcal{A}$, and $-1 < V(p) < 0$ for $p \in \mathcal{A}$;*

(2) *the functional $W(p)$ is given and continuous in $\mathbb{R}$ and $W(p) > 0$ for $p \in \mathbb{R}$, $\rho(p, \mathcal{M} \neq 0$ and $W(p) = 0$ for $p \in \mathcal{M}$;*

(3) *for sufficiently small $\gamma_2 > 0$ the values γ_1 and α_1 can be defined such that*

(5.2.3)
$$V(p) < -\gamma_1 \quad \text{for} \quad \rho(p, \mathcal{M}) \geq \gamma_2,$$
$$W(p) > \alpha_1 \quad \text{for} \quad \rho(p, \mathcal{M}) \geq \gamma_2;$$

(4) *the functionals $V(p)$ and $W(p)$ vanish as $\rho(p, \mathcal{M}) \to 0$;*

(5) *if there exists a point $q \notin \mathcal{M}$, $q \in \bar{A} \setminus A$, then $\lim(V(p):$ $\rho(p, q) \to 0) = -1$;*

(6) $\left.\dfrac{dV(p)}{dt}\right|_{f(p,t)} = W(p)(1 + V(p)).$

Then and only then, the open invariant set $A \subset \mathbb{R}$ which contains a neighborhood of a closed invariant set $\mathcal{M} \subset \mathbb{R}$ is the domain of asymptotic stability of the uniformly asymptotically stable and uniformly attractive set $\mathcal{M}$.

For the Proof of this theorem see Zubov [1, 3].
We set out some comments to Theorem 5.2.1.

Remark 5.2.1. For any $\lambda \in (0, 1)$ the equation $1 + V(p) = \lambda$ gives a condition for a set of points to be a cross-section of the open invariant set A.

Remark 5.2.2. The boundary of the asymptotic stability domain is composed of the totality S of points q (if any) such that $p \to q$ as $V(p) \to -1$.

Remark 5.2.3. The modification of Theorem 5.2.1 remains valid, if instead of functionals $V(p)$ and $W(p)$ the functions $v(x)$ and $w(x)$ are used with the same properties $(1) - (6)$ from Theorem 5.2.1.

Hence, it follows that using the functional $V(p)$ (functions $v(x, \eta)$ constructed in terms of matrix-valued function) one can always solve the problem of determining the boundary of the asymptotic stability domain.

Further, Theorem 5.2.1 is importance in the development of the estimation algorithm for the asymptotic stability domain of the class of systems (5.2.1) in terms of quadratic matrix-valued Liapunov functions.

5.2.2 Some estimates for quadratic matrix-valued functions

Consider a time-invariant large-scale system

(5.2.4) $\qquad \dfrac{dx_i}{dt} = f_i(x_i) + g_i(x_1, \ldots, x_m), \quad i = 1, 2, \ldots, m,$

where $x_i \in R^{n_i}$; $f_i \in C(R^{n_i}, R^{n_i})$, $g_i \in C(R^{n_1} \times \ldots \times R^{n_m}, R^{n_i})$, $f_i(x_i) = 0$ if and only if $x_i = 0$, $i = 1, 2, \ldots, m$, and $g_i(x_1, \ldots, x_m) = 0$ if and only if $x_1 = \cdots = x_m = 0$, $i = 1, 2, \ldots, m$. For system (5.2.4) the free subsystems

(5.2.5) $\qquad \dfrac{dx_i}{dt} = f_i(x_i), \quad i = 1, 2, \ldots, m$

and (i, j)-pair of the free subsystems

(5.2.6)
$$\frac{dx_i}{dt} = f_i(x_i, x_j),$$
$$\frac{dx_j}{dt} = f_j(x_j, x_i), \quad \text{for all} \quad (i \neq j)$$

will be a basis for construction of hierarchical Liapunov matrix-valued function.

We associate with free subsystems (5.2.5) and (i, j)-couples (5.2.6) the elements v_{ii} and v_{ij} for all $(i \neq j)$ of the matrix-valued function $U(x, K_{ij})$. Let it be quadratic forms

(5.2.7) $\quad v_{ij} = \begin{cases} x_i^T K_{ii} x_i, & K_{ii} > 0, \quad x_i \in R^{n_i}, & \text{for} \quad i = j; \\ x_{ij}^T K_{ij} x_{ij}, & K_{ij} > 0, \quad x_{ij} \in R^{n_i} \times R^{n_j}, & \text{for} \quad i < j; \\ x_{ji}^T K_{ji} x_{ji}, & & \text{for} \quad i > j. \end{cases}$

Definition 5.2.2. *Matrix-valued function $U(x, K_{ij})$ belongs to the class of quadratic matrix-valued function, if its elements are of the form of (5.2.7).*

Proposition 5.2.1. *If the matrix-valued function $U \in C(R^n, R^{s \times s})$ belongs to the class of quadratic matrix-valued function, then there exists a $n \times n$ matrix C such that*

(5.2.8) $\qquad \eta^T U(x, K_{ij}) \eta = x^T C x, \qquad x = (x_1^T, \ldots, x_s^T)^T \in R^n,$

where $C = C(K_{ij}, \eta)$, $\eta \in R_+^s$, $\eta > 0$.

Proof. The function $\eta^T U(x, K_{ij}) \eta$, that takes into account expressions (5.2.7) is presented as

$$v(x, \eta, K_{ij}) = \eta^T U(x, K_{ij}) \eta = \sum_{i=1}^{s} \eta_i^2 x_i^T K_{ii} x_i + 2 \sum_{i=1}^{s} \sum_{j=i+1}^{s} \eta_i \eta_j x_{ij}^T K_{ij} x_{ij}$$

$$= \sum_{i=1}^{s} x_i^T \left[\eta_i^2 K_{ii} + 2 \sum_{j=i+1}^{s} \eta_i \eta_j K_{ij}^i + 2 \sum_{j=1}^{i-1} \eta_i \eta_j K_{ij}^j \right] x_i$$

$$+ 4 \sum_{i=1}^{s} \sum_{j=i+1}^{s} x_i^T [\eta_i \eta_j \bar{K}_{ij}] x_j = x^T C x,$$

where the blocks of the matrix C are of the form

$$C_{ij} = C_{ji}^{\mathrm{T}} = \begin{cases} \eta_i^2 K_{ii} + 2\eta_i \left[\sum\limits_{j=1}^{i-1} \eta_j K_{ij}^j + \sum\limits_{j=i+1}^{s} \eta_j K_{ij}^i \right], & \text{for } i = j; \\ 2\eta_i \eta_j \bar{K}_{ij}, & \text{for } i < j. \end{cases}$$

We recall some well-known facts necessary for our presentation (see Michel, Sarabudla, et al. [1]). If n by n matrix $C = C^{\mathrm{T}} = [c_{ij}]$ is positive definite, then for a fixed $m > 0$ the equation

(5.2.9) $v(x, \eta, K_{ij}) = x^{\mathrm{T}} C x = m$

defines the ellipsoid in R^n and all the eigenvalues of the matrix C: $\lambda_1(C)$, $\ldots, \lambda_n(C)$ are real and positive. The main ellipsoid (5.2.9) axes are defined by the expressions

(5.2.10) $\left(\dfrac{m}{\lambda_k(C)} \right)^{\frac{1}{2}}, \quad k = 1, 2, \ldots, n,$

and its hypervolume is proportional to the value

(5.2.11) $1 \Big/ \left(\prod\limits_{k=1}^{n} \lambda_k(C) \right)^{\frac{1}{2}}.$

It is clear that by approximate choice of the block matrices C_{ij}^l one can make the ellipsoids

(5.2.12) $v^l(x, \eta, K_{ij}) = x^{\mathrm{T}} C_l x = m_l, \quad l = 1, 2, \ldots$

embedded into each other. If for some fixed l hypervolume of the ellipsoid (5.2.12) is taken for the maximal estimate of the asymptotic stability domain (set E), then the domain E_l is defined by

(5.2.13) $E_l = \{ x \in R^n : v_l(x, \eta, K_{ij}) = x^{\mathrm{T}} C_l x = m_l \},$

and $Dv_l(x, \eta, K_{ij}) < 0$, $l = 1, 2, \ldots,$ must satisfy the condition

(5.2.14) $E_1 \subseteq E_2 \subseteq \ldots \subseteq E.$

Naturally, $E_1, E_2, \ldots$ satisfy conditions (i) – (iii) of Definition 7 from Grujić, et al. [1]:

 (i) E is a neighborhood of $x = 0$;
 (ii) $E \subseteq \mathcal{D}$, $\mathcal{D} \subseteq R^n$ is the domain of asymptotic stability of $x = 0$ of the system (5.2.4);
 (iii) E is positively invariant set of the system (5.2.4), that is, that $x_0 \in E$ implies $x(t, 0, x_0) \in E$ for every $t \in R_+$.

Let

$$\alpha_1(C) = \sum_{i=1}^{n} \lambda_i(C) = \operatorname{tr} C,$$

and

$$\alpha_2(C) = \prod_{i=1}^{n} \lambda_i(C) = \det C,$$

$$\det C_i > 0, \quad i = 1, 2, \ldots, n,$$

where $\det C_i$ is the ith principal minor of the matrix C. Thus, the problem of estimation of the asymptotic stability domain is reduced to the problem of conventional maximization of the domain E_l at the expense of parameters of the matrix C_l or conventional minimization of functions $\alpha_1(C)$ or $\alpha_2(C)$. As it was mentioned by Michel, Sarabudla, *et al.* [1] the minimization of function $\alpha_1(C)$ is preferable in view of computation, since it means uniform minimization of all eigenvalues of matrix C, while the minimization of the function $\alpha_2(C)$ is reduced to that of the smallest eigenvalue of the matrix C.

Remark 5.2.3. Problem of the function $\alpha_1(C)$ minimization in the space of parameters (η, K_{ij}) can be reduced to a sequence of problems of smaller dimensions. Since the equation

$$\operatorname{tr} C = \sum_{i=1}^{s} \operatorname{tr} C_{ii} = \sum_{i=1}^{s} \left[\eta_i^2 \operatorname{tr} K_{ii} + 2\eta_i \sum_{j=i+1}^{s} \eta_j (\operatorname{tr} K_{ij}^i + \operatorname{tr} K_{ij}^j) \right]$$

$$= \sum_{i=1}^{s} \eta_i^2 \operatorname{tr} K_{ii} + 2 \sum_{i=1}^{s} \sum_{j=i+1}^{s} \eta_i \eta_j \operatorname{tr} K_{ij}$$

takes place, the minimization of $\alpha_1(C)$ for a fixed $\eta \in R_+^s$ can be reduced to a gradual minimization of the functions $\operatorname{tr} K_{ij}$ for all $(i, j) \in [1, s]$. If $\min \operatorname{tr} K_{ij}$ is reached for a fixed value of diagonal elements of the matrices K_{ij} for all $(i, j) \in [1, s]$, then $\min \alpha_2(C)$ can be obtained by an approximate choice of nondiagonal elements of the matrix K_{ij}.

On each step of computations, when inclusions (5.2.14) are constructed, it is necessary to verify the condition

$$(5.2.15) \qquad Dv_l(x, \eta, K_{ij}) < 0, \quad l = 1, 2, \ldots.$$

If the right side part of the system (5.2.1) is smooth enough, it is sufficient to verify the condition (5.2.15) on the network covering boundary of the sets ∂E_{l-1} and ∂E_l.

The network of points from R^n forms some an m-pointwise set L, $L \in \partial E$. Discrete m-pointwise set L can be constructed so that for $m \to \infty$ the set L covers the entire boundary ∂E of the set E.

5.2.3 Algorithm of constructing a point network covering boundary of domain E

It is sufficient to verify the condition (5.2.15) at points of the network covering the boundary E. We present an algorithm for construction of such a network. The location of a point $x \in R^n$ is determined by its coordinates $(x_1, \ldots, x_n)^{\mathrm{T}}$. To find out the coordinate with a fixed value we use the upper index. For arbitrary real constant $a_1^0 > 0$ and some positive integer $N_1 \geq 2p$, $p = 2, 3, 4$, we specify N_1-pointwise set L_1^0 as

$$L_1^0 = \{x_1^{i_1} \in R \colon x_1^{i_1} = a_1^0 \cos \alpha_{i_1}\},$$

where $\alpha_{i_1} = \pi(i_1 - 1)/(N_1 - 1)$, $i_1 \in [1, N_1]$.

It is clear that L_1^0 consists of N_1 fixed values $x_1^{i_1}$, $i_1 \in [1, N_1]$ of the first coordinate x_1 of the point $x \in R^n$, and moreover $x_1^1 = a_1^0$, $x_1^{N_1} = -a_1^0$.

For every fixed value $x_k^{i_k} \in L_k^{i_{k-1}}$, $k \in [1, n-2]$, $i_k \in [1, N_k]$ for all $k \in [1, n-2]$ and $i_0 = 0$, where $N_k \geq 2p$ ($p = 2, 3, \ldots$, $k \in [1, n-2]$) we define sets $L_{k+1}^{i_k}$ as

$$(5.2.16) \qquad \begin{array}{ll} \{x_{k+1}^1 \in R \colon x_{k+1}^1 = 0\}, & i_k = 1, \quad i_k = N_k; \\[2mm] \{x_{k+1}^{i_{k+1}} \in R \colon x_{k+1}^{i_{k+1}} = a_{k+1}^{i_k} \cos \alpha_{i_{k+1}}\}, & i_k \in [2, N_k - 1], \end{array}$$

where $\alpha_{i_{k+1}} = \pi(i_{k+1} - 1)/(N_{k+1} - 1)$, for all $i_{k+1} \in [1, N_{k+1}]$, and

$$(5.2.17) \qquad a_{k+1}^{i_k} = \left[(a_k^{i_{k-1}})^2 - (x_k^{i_k})^2\right]^{\frac{1}{2}}.$$

It is obvious that $x_{k+1}^1 = a_{k+1}^{i_k}$ and $x_{k+1}^{N_{k+1}} = -a_{k+1}^{i_k}$ for all $i_k \in [2, N_{k-1}]$.

The sets $L_{k+1}^{i_k}$ are a totality of fixed values of $(k+1)$-th coordinate x_{k+1} of the point $x \in R^n$ generated by every fixed value $x_k^{i_k}$ of k-th coordinate of the point $x \in R^n$. Thus, every fixed value $x_k^{i_k}$, $k \in [1, n-2]$, $i_k \in [1, N_k]$ of coordinate of the point $x \in R^n$ having index k, generates $N_{k+1} - 2$ values of $(k+1)$-th nonzero coordinate of the point $x \in R^n$ and two equal

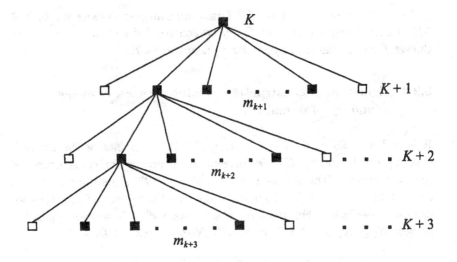

Figure 5.2.1.

to zero values of this coordinate. This is shown on Figure 5.2.1 in the form
a graph.

Symbols □ and ■ denote nodes of the graph corresponding to the
fixed values of the coordinates, which generate only zero and, respectively,
nonzero values of the consequent coordinates of the point $x \in R^n$, and
the edges connecting nodes show the succession of this values' formation.
Further, for every fixed value $x_{n-1}^{i_{n-1}} \in L_{n-1}^{i_{n-2}}$ for all $i_n \in [1, N_{n-1}]$ we define
the set $L_n^{i_{n-1}}$ as

$$
L_n^{i_{n-1}} = \begin{cases} \{x_n^1 \in R : x_n^1 = 0\}, & i_{n-1} = 1, \quad i_{n-1} = N_{n-1}; \\ \{(x_n^1, x_n^2) \in R : x_n^1 = a_{n-1}^{i_{n-1}} \sin \alpha_{i_{n-1}}, \quad x_n^2 = -x_n^1\}, \\ \quad\quad\quad\quad i_{n-1} \in [2, N_{n-1} - 1], \end{cases}
$$

where $a_{n-1}^{i_{n-1}}$ are specified according to formulas (5.2.17).

Herewith, the graph generated by kth fixed value x_1^k, $|x_1^k| \neq a_1^0$ of the
first coordinate x_1, $k \in [2, N_1 - 1]$ of the point $x \in R^n$ can be shown
according to Figure 5.2.2, where every level k, $k \in [2, n]$ corresponds to
the fixed values of kth coordinate of the point $x \in R^n$. Thus, the set L_1^0
generates $(N_1 - 2)$ different and two equal to zero values of coordinate
x_2 of the point $x \in R^n$, $(N_1 - 2)(N_2 - 2)$ nonzero and $2 + 2(N_1 - 2)$
equal to zero values of the coordinate x_3 of the point $x \in R^n$ etc. Finally,

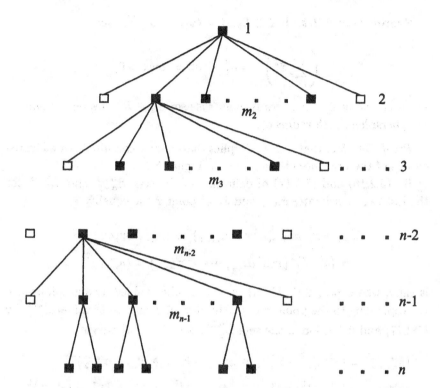

Figure 5.2.2.

we get that the set l_1^0 generates $N^* = 2 \prod\limits_{i=1}^{n-1} (N_i - 2)$ sets of values of the coordinates $x_1, x_2, \ldots, x_n$ of the point $x \in R^n$, such that $x_i \neq 0$, $i \in [1, n]$ and $N^0 = 2 + \sum\limits_{i=1}^{n-2} \prod\limits_{j=1}^{i} (N_j - 2)$ sets of coordinates $x_1, x_2, \ldots, x_n$ of point x such that there exists at least one index $k \in [1, n]$ such that $x_k = 0$.

By definition of the sets $L_k^{i_k-1}$ we get that the set $L^0 - 1$ generates N-pointwise set $L_s \in R^n$, $N = N^* + N^0$ of points of a set in R^n that differ from each other by value of at least one coordinate. The set L_s can be presented in the form of totality M_1, of graphs shown on Figure 5.2.2, each branch of which denotes a fixed set of values of coordinates of points $x \in L_s$, and, besides, for the branch that ends with a node designated by $\square$ and all subsequent coordinates equal to zero.

For the points of set L_s the following assertion holds true.

Proposition 5.2.2. *If $\tilde{x} \in L_s$, $\tilde{x} = (\tilde{x}_1, \ldots, \tilde{x}_n)^T$, then*

$$\left(\sum_{i=1}^{n} \tilde{x}_i^2 \right)^{\frac{1}{2}} = a_1^0, \quad a_1^0 = \text{const} > 0,$$

i.e. every point $\tilde{x}$ of discrete N-pointwise set $L_s \in R^n$ lies on the surface of hypersphere with radius a_1^0.

Proof. The fact that $\tilde{x} \in L_s$ implies that point $\tilde{x}$ coordinates are located on one of the branches of the graph on Figure 5.2.2.

By (5.2.16) and (5.2.17) of definition of the sets $L_{n-1}^{i_{n-2}}$ and $L_n^{i_{n-1}}$ for the last two coordinates $\tilde{x}_{n-1}$ and $\tilde{x}_n$ of point $\tilde{x}$ the equality

$$\tilde{x}_{n-1}^2 + \tilde{x}_n^2 = \left(a_{n-1}^{i_{n-2}} \cos \alpha_{i_{n-1}} \right)^2 + \left(a_{n-1}^{i_{n-2}} \sin \alpha_{i_{n-1}} \right)^2$$
$$= \left(a_{n-1}^{i_{n-2}} \right)^2 \left(\cos^2 \alpha_{i_{n-1}} + \sin^2 \alpha_{i_{n-1}} \right) = \left(a_{n-1}^{i_{n-1}} \right)^2,$$

is valid, where $i_{n-1} \in [1, M_{n-1}]$ is a fixed value defined by a graph branch corresponding to the point $\tilde{x} \in L_s$. In view of definition of value $a_{n-1}^{i_{n-2}}$ by (5.2.17) and definition of the sets $L_k^{i_k-1}$, $k \in [1, n-1]$ we get

$$\left(a_{n-1}^{i_{n-2}} \right)^2 = \left(a_{n-2}^{i_{n-3}} \right)^2 - \tilde{x}_{n-2}^2 \Rightarrow \tilde{x}_{n-2}^2 + \tilde{x}_{n-1}^2 + \tilde{x}_n^2 = \left(a_{n-2}^{i_{n-3}} \right)^2$$
$$\left(a_{n-2}^{i_{n-3}} \right)^2 = \left(a_{n-3}^{i_{n-4}} \right)^2 - \tilde{x}_{n-3}^2 \Rightarrow \tilde{x}_{n-3}^2 + \tilde{x}_{n-2}^2 + \tilde{x}_{n-1}^2 + \tilde{x}_n^2 = \left(a_{n-3}^{i_{n-4}} \right)^2,$$

$\ldots$

$$\left(a_2^{i_1} \right)^2 = \left(a_1^{i_0} \right)^2 - \tilde{x}_1^2 \Rightarrow \sum_{i=1}^{n} \tilde{x}_i^2 = \left(a_1^0 \right)^2, \quad i_0 = 0.$$

This proves Proposition 5.2.2.

Let $\tilde{x} \in L_s$ and a hyperplane Q pass through k, $k \in [1, n-2]$ of the first coordinates of point $\tilde{x}$. Then the intersection obtained is a discrete set, every point of which lies on the hypersphere of a surface in R^{n-k} with radius $a_{k+1}^{i_k}$. Figure 5.2.3 illustrates an example of point network formulation on sphere in R^3 (in one orthant) with radius $a_1^0 = 1$ for $N_1 = N_2 = 10$.

Then we return to the boundary of the set E:

$$\partial E = \{ x \in R^n : v(x, \eta, K_{ij}) = m(\eta, K_{ij}) \},$$

where

$$m(\eta, K_{ij}) = \min_{x \in D} v(x, \eta, K_{ij}),$$

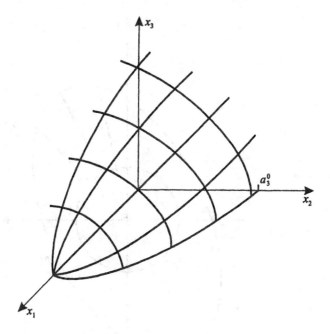

Figure 5.2.3.

and

$$\mathcal{D} = \{x \in R^n \colon Dv(x, \eta, K_{ij}) = 0\}.$$

In order to cover with a control point network the boundary ∂E of the set E for a fixed l, we use the following property of the function $v(x, \eta, K_{ij})$ of the quadratic matrix-valued function class.

For an arbitrary constant $c \geq 0$ for all $x \in R^n$ in view of Proposition 5.2.1 we have

$$v(cx, \eta, K_{ij}) = (cx)^{\mathrm{T}} C(cx) = c^2 x^{\mathrm{T}} C x = c^2 v(x, \eta, K_{ij}).$$

Now, if we take a point $\tilde{x} \in L_s$, then the corresponding point x on the boundary ∂E can be found by formula $x = c\tilde{x}$ (component-wise), where the constant c is defined by

$$c = \left(\frac{m(\eta, K_{ij})}{v(\tilde{x}, \eta, K_{ij})} \right)^{\frac{1}{2}}.$$

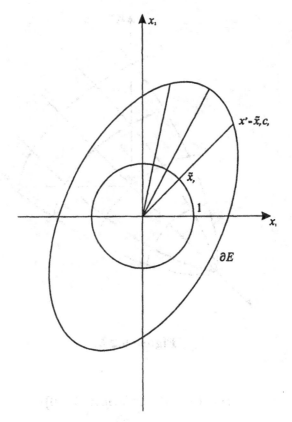

Figure 5.2.4.

Thus, the discrete m-pointwise set $L \in \partial E$ can be obtained from points $\tilde{x}^p \in L_s$ for all $p \in [1, m]$ by formula $x^p = c_p \tilde{x}^p$, where

$$c_p = \left(\frac{m(\eta, K_{ij})}{v(\tilde{x}^p, \eta, K_{ij})} \right)^{\frac{1}{2}}, \quad x_p \in L \subset \partial E.$$

Figure 5.2.4 shows the formation of points $x^p \in L \subset \partial E$ in R^2.

5.2.4 Numerical realization and discussion of the algorithm

In view of the above presented results the algorithm of construction of the initial estimate of asymptotic stability domain and its further refining can be reduced to a sequence of the steps.

Step 1. Compute the values of the Jacobians

$$J_i(x_i) = \left(\frac{\partial f_i}{\partial x_i}\right)\bigg|_{x_i=0}, \quad i = 1, 2, \ldots, m,$$

$$J_{ij}(x_{ij}) = \left(\frac{\partial f_{ij}}{\partial x_{ij}}\right)\bigg|_{x_{ij}=0}, \quad (i < j) \in [1, m]$$

and solve the Liapunov equations

$$J_i^{\mathrm{T}}(0)K_{ii}^0 + K_{ii}^0 J_i(0) = -I_{ii}, \quad i \in [1, m],$$

and

$$J_{ij}^{\mathrm{T}}(0)K_{ij}^0 + K_{ij}^0 J_{ij}(0) = -I_{ij}, \quad (i < j) \in [1, m].$$

For the stability of independent subsystems (5.2.5) and (5.2.6) put $K_{ii}^0 = I_{ii}$, $K_{ij}^0 = I_{ij}$, where $I_{ij} = \mathrm{diag}\,(1, \ldots, 1)$ for all $(i \le j) \in [1, m]$. Choose a vector $\eta \in R_+^m$, $\eta > 0$ and construct initial quadratic matrix-valued function $U(x, K_{ij}^0)$ and scalar function

$$v_0(x, \eta, K_{ij}^0) = x^{\mathrm{T}} C^0 x.$$

Step 2. Using the Rodden technique (see Rodden [1]) find out all points in R^n satisfying the conditions

(i) $Dv_0(x, \eta, K_{ij}^0) = 0,$

(ii) $\dfrac{\nabla v_0(x, \cdot)}{|\nabla v_0(x, \cdot)|} - \left(\dfrac{\nabla v_0(x, \cdot)^{\mathrm{T}}}{|\nabla v_0(x, \cdot)|}\dfrac{\nabla Dv_0(x, \cdot)}{|\nabla Dv_0(x, \cdot)|}\right)\dfrac{\nabla Dv_0(x, \cdot)}{|\nabla Dv_0(x, \cdot)|} = 0,$

where $\nabla v_0(x, \cdot)$ denotes the gradient vector of $v_0(x, \cdot)$, and $|\cdot|$ denote a norm in R^m.

Let $\mathcal{D}$ be all such points. Now we compute

$$m_0(\eta, K_{ij}^0) = \min_{x \in \mathcal{D}} v_0(x, \eta, K_{ij}^0)$$

and define the set

$$E_0 = \{x \in R^n \colon v_0(x, \eta, K_{ij}^0) < m_0(\eta, K_{ij}^0)\}$$

so that

$$Dv_0(x, \eta, K_{ij}^0) < 0 \quad \text{for all} \quad (x \ne 0) \in E_0.$$

Step 3. Cover the boundary ∂E_0 of the set E_0:

$$\partial E_0 = \{x \in R^n : v_0(x, \eta, K_{ij}) = m_0(\eta, K_{ij}^0)\},$$

with N-pointwise network $L_0 \in \partial E_0$.

Step 4. Take the block matrix K_{ij}^0 for all $(i \le j) \in [1, m]$ and

(4.1). Using the optimization algorithm proposed by Rosenbrock [1] construct a sequence of matrices K_{ij}^σ, $\sigma = 1, 2, \ldots, Q$, so that

$$\text{tr}\,(K_{ij}^\sigma) \to \min$$

under the restrictions

    (i) $K_{ij}^\sigma = (K_{ij}^\sigma)^{\text{T}}, \quad K_{ij}^\sigma > 0;$

    (ii) $Dv_0(x, \eta, K_{ij}^\sigma) < 0$ for all $(C_l x^l) \in L_\sigma \subset \partial E_\sigma$,

where $x^l \in L_{\sigma-1} \subset \partial E_{\sigma-1}$, $l = 1, 2, \ldots, M$. Here ∂E_σ is the boundary of the set E_σ:

$$E_\sigma = \{x \in R^n : v_\sigma(x, \eta, K_{ij}^\sigma) < m_0(\eta, K_{ij}^0)\},$$

and the constants c_l satisfy the condition

$$c_l = \frac{m_0(\eta, K_{ij}^0)}{v_\sigma(x^l, \eta, K_{ij}^\sigma)} \ge 1 \quad \text{for all} \quad l \in [1, N].$$

(4.2). Construct the scalar function

$$v_M(x, \eta, K_{ij}^M) = x^{\text{T}} C_M x, \quad x \in R^n$$

and define the set

$$E_M = \{x \in R^n : v_M(x, \eta, K_{ij}^M) < m_M(\eta, K_{ij}^M)\}$$

from the conditions

    (i) $Dv_M(x, \eta, K_{ij}^M) < 0$ for all $(x \ne 0) \in E_M$;

    (ii) $m_M(\eta, K_{ij}^M) = \min\limits_{x \in \mathcal{D}_M} v_M(x, \eta, K_{ij}^M),$

where

$$\mathcal{D}_M = \{x \in R^n : Dv_M(x, \eta, K_{ij}^M) = 0\}.$$

Step 5. Cover the boundary ∂E_M of the set E_M with a point network $L_M \subset \partial E_M$ according to the relation

$$\tilde{x}^l = c_l x^l,$$

where $\tilde{x}^l \in L_M$ and $x^l \in L_0$,

$$c_l = \left(\frac{m_M(\eta, K_{ij}^M)}{v_M(x^l, \eta, K_{ij}^M)} \right)^{\frac{1}{2}}.$$

Step 6. Take the matrix K_{ij}^M for the initial, and using optimization algorithm (see, for example, Rosenbrock [1]) define the final matrix K_{ij}^F so that

$$\det(K_{ij}^P) \to \min \quad \text{for} \quad P \to F, \quad P \in [M+1, \ldots, F]$$

under the restrictions

(i) $K_{ij}^P = (K_{ij}^P)^{\mathrm{T}}, \quad K_{ij}^P > 0,$

(ii) $\operatorname{tr} K_{ij}^p = \operatorname{tr} K_{ij}^M \quad$ for all $\quad p \in [M+1, \ldots, F],$

(iii) $D v_p(x_l x^l, \eta, K_{ij}^p) < 0 \quad$ for all $\quad c_l x^l \in L_p \subset \partial E_p,$

where $x^l \in L_{p-1} \subset \partial E_{p-1}$, $l \in [1, N]$. Here ∂E_p is a boundary of the set E_p

$$E_p = \{x \in R^n : v_p(x, \eta, K_{ij}^P) < m_p(\eta, K_{ij}^P)\}$$

and the constants c_l satisfy the condition

$$c_l = \left(\frac{m_p(\eta, K_{ij}^p)}{v_p(x^l, \eta, K_{ij}^p)} \right)^{\frac{1}{2}} \geq 1$$

for all $l \in [1, N]$.

The presented algorithm of constructing an estimate of the asymptotic stability domain of system (5.2.1) admits application of structural programming principles. This is possible since separate steps of the algorithm are isolated problems. We discuss some peculiarities of the Steps 1–6.

The main problem that arises on Step 1 is the solution of the matrix Liapunov equations. In view that the dimensions of the system of linear equations is connected with the decomposition procedure for system (5.2.1), it is important to make the system (5.2.1) decomposition so that the first level independent subsystems be of the lowest possible order.

On Step 2 and 4.2 it is reasonable to apply the method proposed by Rodden [1] to construct an attraction domain for solutions of system (1) via the Liapunov function obtained in the result of the numerical solution of the Zubov equation (see Zubov [3]).

The application of the matrix quadratic Liapunov function in the framework of Rodden's method ensures an effective initial estimation of the domain E_0.

Algorithm of the Steps 3 and 5 realization was described earlier.

On stages 4.2 and 6 for each set of indexes $(i < j) \in [1, m]$, the partial problem of mathematical programming with restrictions given by a system of inequalities is solved. Here the method of random search of the best trial in combination with the method of penalty functions are perspective for application. An essential condition on stage 4.2 is inequality $c_l \geq 1$ for all $l \in [1, N]$. It is easy to see that if for any point $x^* \in \partial E_p$, where ∂E_p is the surface of an ellipsoid, a constant $c^* \geq 1$ is found so that $c^* x^* \in \partial E_{p+1}$, then $\partial E_p \subseteq \partial E_{p+1}$.

In result of the Step 6 the estimates of asymptotic stability domains E_p, $p = M + 1, \ldots, F$ are constructed such that

$$v_F(x, \eta, K_{ij}^F) \geq v_k(x, \eta, K_{ij}^k) \quad \text{for all} \quad k \in [M, F - 1]$$

i.e. the set E_F is a maximal estimate of asymptotic stability domain of the equilibrium state $x = 0$ of the system (5.2.1). We note that the proposed algorithm does not presuppose the optimization with respect to parameter $\eta \in R_+^s$, $\eta > 0$ though the reasonability of such an optimization is undoubtful.

5.2.5 Illustrative examples

In order to demonstrate the effectiveness of the proposed algorithm of estimating asymptotic stability domains we cite some examples from the paper by Michel, Sarabudla, et al. [1].

For all examples mentioned below the domains of asymptotic stability were constructed and, moreover, for Example 5.2.8 on Figure 5.2.12 the intersection of the domain E_F estimate by plane $x_3 = 0$ was shown. The numbers $\odot$ on Figures 5.2.5 – 5.2.12 denote:

  1. The points lying on the boundary of strict the domain E, obtained by direct integration of the system by Runge-Kutta method.

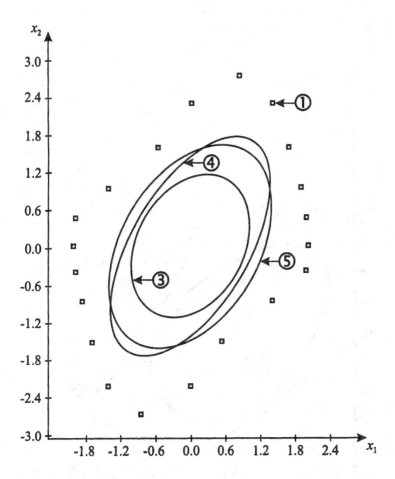

Figure 5.2.5. $\dot{x}_1 = -x_2, \ \dot{x}_2 = x_1 - x_2(1 - x_1^2).$

2. The points lying out of the domain E, obtained by Runge-Kutta method.

3. The initial estimation of the domain E obtained by algorithm of Section 5.2.

4. The maximal estimation of the domain E obtained via the quadratic matrix-valued function (see Section 5.2).

5. The estimation of the domain E, obtained by Michel, Sarabudla, *et al.* [1] via quadratic Liapunov function for the first approximation of the systems without decomposition of the systems.

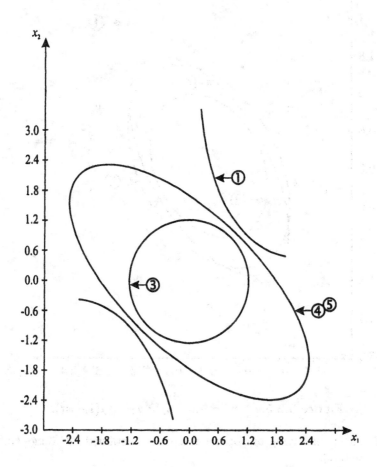

Figure 5.2.6. $\dot{x}_1 = -x_1 + 2x_1^2 x_2, \ \dot{x}_2 = -x_2.$

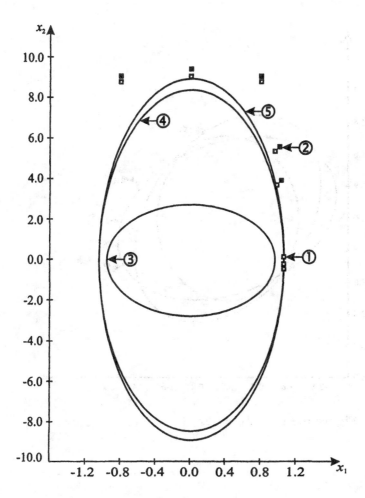

Figure 5.2.7. $\dot{x}_1 = -2x_1(1-x_1) + 0.1x_1x_2,$
$\dot{x}_2 = -2x_2(9-x_2) + 0.1(x_1+x_2).$

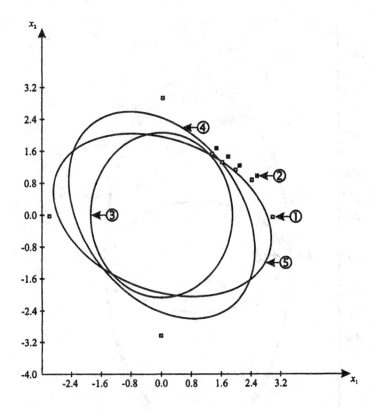

Figure 5.2.8. $\dot{x}_1 = -2x_1 + x_1x_2,\ \dot{x}_2 = -x_2 + x_1x_2.$

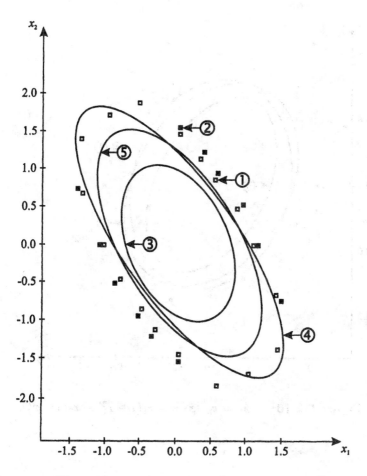

Figure 5.2.9. $\dot{x}_1 = x_2, \ \dot{x}_2 = -x_1 + x_2 + x_1^3.$

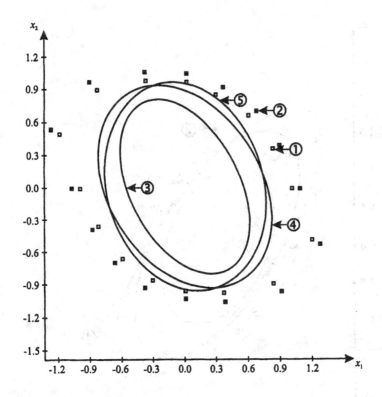

Figure 5.2.10. $\dot{x}_1 = x_2, \ \dot{x}_2 = -x_1(1 - x_1^2) - x_2(1 - x_2^2).$

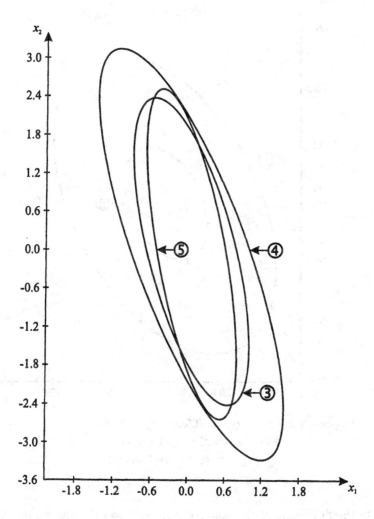

Figure 5.2.11. $\dot{x}_1 = x_2,$
$\dot{x}_2 = -x_1 - 4x_2 + 0.25(x_2 - 0.5x_1)$
$\times (x_2 - 2x_1)(x_2 + 2x_1)(x_2 + x_1).$

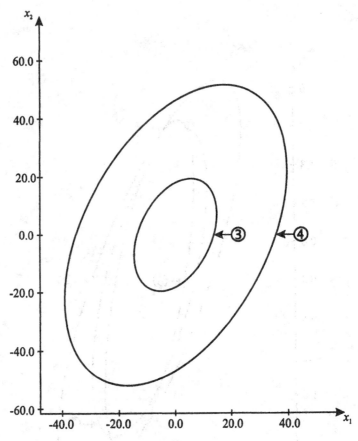

Figure 5.2.12. $\quad \dot{x}_1 = -0.5x_1 - x_2 - 0.5x_3,$
$\qquad\qquad\quad \dot{x}_2 = x_1 - x_2(1 + 0.1x_2^2),$
$\qquad\qquad\quad \dot{x}_3 = x_1 - x_3(1 + 0.1x_3^2).$

In the discussion of the examples the first level decomposition with two
subsystems and the second level decomposition with formation of one couple
$(1, 2)$ that coincides with the whole system were made. In Example 5.2.8
the first level decomposition into three subsystems and the second level
decomposition into three couples of subsystems in view of nonlinearities
were made.

5.3 Construction of Estimate for the Domain E of Power System

According to Šiljak [1] consider some *power system* where absolute motion of ith machine is described by

$$(5.3.1) \qquad M_i \ddot{\delta}_i + D_i \dot{\delta}_i = P_{mi} - P_{ei}, \quad i = 1, 2, 3,$$

where $P_{ei} = \sum\limits_{j=1}^{n} E_i E_j Y_{ij} \cos(\delta_i - \delta_j - \theta ij)$ and n is the number of system generators; δ_i is the absolute rotor angle of the ith machine; M_i is the inertia coefficient of the ith machine; D_i is the damping coefficient of the ith machine; P_{mi} is the mechanical power delivered to the ith machine; P_{ei} is the electrical power delivered by the ith machine; E_i is the inertial voltage; Y_{ij} is mutual conductivities of the machine $(i \neq j) = 1, 2, 3$; θ_{ij} is the phase angle of transfer admittance between the ith and jth machines.

Assume that M_i, P_{mi} and E_i are constant for all generators and $D_i M_i^{-1} = \lambda$, $i = 1, 2, 3$.

System (5.3.1) can be presented as

$$(5.3.2) \qquad \frac{dx_i}{dt} = y_{i3} \quad (i \neq j) = 1, 2,$$

$$\frac{dy_{i3}}{dt} = -\lambda_i y_{i3} - \mu_i \phi_1(x_i) + \nu_i \phi_1(x_j) - \beta_i \phi_2(x_i, x_j).$$

Here x_i is a deviation of rotor of the ith generator from the rotor of a standard generator, y_{i3} is a velocity of x_i change, functions ϕ_1 and ϕ_2 are defined by

$$\phi_1(x_i) = \cos(x_i - \theta_{i3}) - \cos\theta_{i3},$$

$$\phi_2(x_i, x_j) = \cos(x_i - x_j - \theta_{ij}) - \cos\theta_{ij},$$

and constants μ_i, ν_i and β_i are defined as

$$\mu_i = E_i E_3 Y_{i3}(M_i^{-1} - M_3^{-1}),$$

$$\nu_i = E_j E_3 Y_{j3} M_3^{-1},$$

$$\beta_i = E_1 E_2 Y_{12} M_i^{-1}, \quad (i \neq j) \in [1, 2].$$

We admit the numerical parameters of the system (5.3.1) as

$$M_1 = 0.01, \qquad M_2 = 0.01, \quad M_3 = 2.00;$$

$$E_1 = 1.017, \qquad E_2 = 1.005, \quad E_3 = 1.033;$$

$$Y_{12} = 0.98 \times 10^{-3}, \quad Y_{13} = 0.114, \quad Y_{23} = 0.106;$$

$$\theta_{13} = 90°, \quad \theta_{23} = 92°, \quad \theta_{12} = 87°, \quad \theta_{21} = 85°;$$

$$\lambda_1 = \lambda_2 = 100.$$

For independent subsystems of first level decomposition of the system (5.3.2)

$$\frac{dx_i}{dt} = y_{i3},$$

$$\frac{dy_{i3}}{dt} = -\lambda_i y_{i3} - \mu_i \phi_1(x_i), \quad i = 1, 2;$$

the elements $v_{ii}(x_i)$ of the initial matrix-valued function $U_0(x)$ are taken according to Step 1 in the form (see Krapivnyi [1])

$$v_{ii}(u_i) = u_i^T K_{ii}^0 u_i, \quad u_i = (x_i, y_{i3})^T,$$

where K_{ii}^0 are defined from the Liapunov equations

$$J_{ii}^{0T} K_{ii}^0 + K_{ii}^0 J_{ii}^0 = -I_{ii}.$$

Here

$$J_{ii}^0 = \begin{pmatrix} 0 & 1 \\ \mu_i \sin(-\theta_{i3}) & -\lambda_i \end{pmatrix}, \quad i = 1, 2.$$

We have for the numerical values of parameters

$$J_{11}^0 = \begin{pmatrix} 0.00 & 1.00 \\ -11.917 & -100 \end{pmatrix}; \quad K_{11}^0 = \begin{pmatrix} 4.259 & 0.419 \\ 0.0419 & 0.00542 \end{pmatrix};$$

$$J_{22}^0 = \begin{pmatrix} 0.00 & 1.00 \\ -10.939 & -100 \end{pmatrix}; \quad K_{22}^0 = \begin{pmatrix} 4.63 & 0.0457 \\ 0.0457 & 0.00546 \end{pmatrix}.$$

The interconnected second level decomposition subsystem coincides with the system (5.3.2) and for it

$$J_{12}^0 = \begin{pmatrix} 0.0 & 1.0 & 0.0 & 0.0 \\ -12.036 & -100 & 0.055 & 0.0 \\ 0.0 & 0.0 & 0.0 & 1.0 \\ 0.06 & 0.0 & -11.05 & -100 \end{pmatrix}.$$

Since matrix J_{12}^0 differs insignificantly from the matrix

$$\begin{pmatrix} J_{11}^0 & 0 \\ 0 & J_{22}^0 \end{pmatrix},$$

the matrix K_{12}^0 is given as

$$K_{12}^0 = \begin{pmatrix} K_{11}^0 & 0 \\ 0 & K_{22}^0 \end{pmatrix}.$$

In this case the matrix $J_{12}^{0T} K_{12}^0 + K_{12}^0 J_{12}^0$ is negative definite and the elements $v_{12}(x_{12}) = v_{21}(x_{21})$ of the matrix-valued function $U_0(x)$ can be taken in the form

$$v_{12}(x_{12}) = v_{21}(x_{21}) = x_{12}^T K_{12}^0 x_{12}, \qquad x_{12} = (x_1^T, x_2^T)^T.$$

According to the above algorithm we find

$$K_{11}^F = \begin{pmatrix} 4.228 & 0.0607 \\ 0.0607 & 0.00133 \end{pmatrix}, \qquad K_{22}^F = \begin{pmatrix} 4.63 & 0.0457 \\ 0.0457 & 0.00547 \end{pmatrix},$$

$$K_{12}^F = \begin{pmatrix} 4.259 & 0.0342 & 0.00724 & -0.00181 \\ 0.0342 & 0.0542 & 0.00013 & -0.00196 \\ 0.00724 & 0.00013 & 4.63 & 0.0517 \\ -0.00181 & -0.00196 & 0.0517 & 0.00546 \end{pmatrix}.$$

Figure 5.3.1 shows the intersection of the estimate of the domain E_F by the plane $y_{13} = y_{23} = 0$.

The numbers $\odot$ on Figure 5.3.1 denote:

1. The initial estimate of the domain E_0;
2. The maximal E_F estimate of the domain E;
3. The estimate of the domain E, obtained for the system (5.3.1) by Abdullin, Anapolskii, *et al.* [1] in the result of the vector Liapunov function application with components in the form of the linear forms moduli.

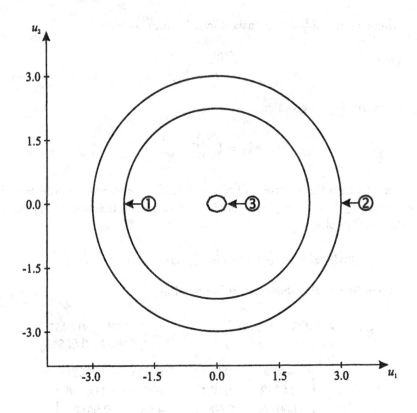

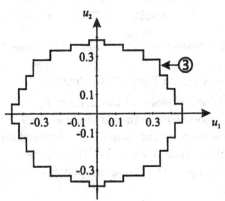

Figure 5.3.1

5.4 Oscillations and Stability of Some Mechanical Systems

5.4.1 Three-mass systems

Following Djordjević [3] and Krapivnyi [1] we consider a mechanical system consisting of three pendulums with equal mass m and the length l (see Figure 5.4.1). The pendulums are connected with each other by springs with rigidity k, that are attached to the pendulum rods at the distance h from the point of fixation. The points of the pendulum fixation are located in horizontal plane. The rods and spring masses are neglected. We take the deviation angulars ϕ_i of the pendulums from vertical position as generalized coordinates and we assume that the equilibrium state of the system is defined as $\phi_i = 0$, $i = 1, 2, 3$. Oscillation amplitudes of the system are supposed to be small, i.e., $\phi_i \approx \sin\phi_i$, $i = 1, 2, 3$.

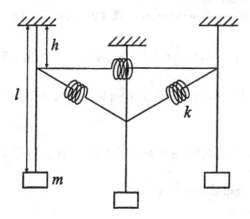

Figure 5.4.1

Motion equations of such system are

(5.4.1)
$$ml^2\ddot{\phi}_i + mgl\phi_i + h^2k(\phi_i - \phi_j) + h^2k(\phi_i - \phi_s) = 0,$$
$$(i, j, s) \in [1, 3], \quad i \neq j, \quad j \neq s, \quad s \neq i.$$

First level decomposition of the system (5.4.1) is reduced to the designation of free subsystems

(5.4.2)
$$ml^2\ddot{\phi}_i + mgl\phi_i = 0, \quad i = 1, 2, 3$$

and the interconnection functions between them

(5.4.3) $h_i = h^2 k(\phi_i - \phi_j) + h^2 k(\phi_i - \phi_s), \quad i = 1, 2, 3.$

Second level decomposition is determined by three couples of independent subsystems

$$ml^2 \ddot{\phi}_i + mgl\phi_i + h^2 k(\phi_i - \phi_j) = 0,$$

(5.4.4) $ml^2 \ddot{\phi}_j + mgl\phi_j + h^2 k(\phi_j - \phi_i) = 0,$

$$(i < j) \in [1, 3].$$

These equations describe the oscillations of two interconnected pendulums without taking into account the influence of the third pendulum on the system oscillations.

The energy of ith free subsystem (5.4.2) is defined by

(5.4.5) $E_{ii}(\phi_i) = \dfrac{1}{2} ml\dot{\phi}_i^2 + \dfrac{1}{2} mgl\phi_i^2, \quad i = 1, 2, 3.$

Interconnection energy of (i, j) couples of the subsystems (5.4.4) is defined by

(5.4.6) $E_{ij}(\phi_i, \phi_j) = \dfrac{1}{2} h^2 k(\phi_i - \phi_j)^2, \quad (i \neq j) \in [1, 3].$

The matrix-valued function

(5.4.7) $U(\phi_i, \phi_j) = [E_{ij}(\cdot)], \quad i, j = 1, 2, 3,$

together with vector $\eta = (1, 1, 1)^{\mathrm{T}}$ can be applied to construct the Liapunov's function

(5.4.8) $v(\phi, \eta) = \eta^{\mathrm{T}} U(\phi_i, \phi_j)\eta.$

The function (5.4.8) is positive definite and its total derivative $Dv(\phi, \eta)$ by virtue of the system (5.4.1) is equal to zero.

Thus, elements (5.4.5) and (5.4.6) are suitable for construction of the Liapunov's matrix-valued function (5.4.8). This example shows the energy nature of the matrix-valued function (5.4.7).

5.4.2 Nonautonomous oscillator

We shall study the motion of two nonautonomously connected oscillators whose behavior is described by the equations

(5.4.9)

$$\frac{dx_1}{dt} = \gamma_1 x_2 + v \cos \omega t y_1 - v \sin \omega t y_2,$$

$$\frac{dx_2}{dt} = -\gamma_1 x_1 + v \sin \omega t y_1 + v \cos \omega t y_2,$$

$$\frac{dy_1}{dt} = \gamma_2 y_2 + v \cos \omega t x_1 + v \sin \omega t x_2,$$

$$\frac{dy_2}{dt} = -\gamma_2 y_2 + v \cos \omega t x_2 - v \sin \omega t x_1,$$

where γ_1, γ_2, v, ω, $\omega + \gamma_1 - \gamma_2 \neq 0$ are some constants.

For the independent subsystems

(5.4.10)

$$\frac{dx_1}{dt} = \gamma_1 x_2, \qquad \frac{dx_2}{dt} = -\gamma_1 x_1,$$

$$\frac{dy_1}{dt} = \gamma_2 y_2, \qquad \frac{dy_2}{dt} = -\gamma_2 y_1,$$

the auxiliary functions v_{ii}, $i = 1, 2$, are taken in the form

(5.4.11)

$$v_{11}(x) = x^T x, \qquad x = (x_1, x_2)^T,$$

$$v_{22}(y) = y^T y, \qquad y = (y_1, y_2)^T.$$

We use the equation (2.4.5) (see Assumption 2.4.2) to determine the nondiagonal element $v_{12}(x, y)$ of the matrix-valued function $U(t, x, y) = [v_{ij}(\cdot)]$, $i, j = 1, 2$. To this end set $\eta = (1, 1)^T$ and $v_{12}(x, y) = x^T P_{12} y$, where $P_{12} \in C^1(\mathcal{T}_\tau, R^{2 \times 2})$. For the equation

(5.4.12)

$$\frac{dP_{12}}{dt} + \begin{pmatrix} 0 & -\gamma_1 \\ \gamma_1 & 0 \end{pmatrix} P_{12}$$

$$+ P_{12} \begin{pmatrix} 0 & \gamma_2 \\ -\gamma_2 & 0 \end{pmatrix} + 2v \begin{pmatrix} \cos \omega t & -\sin \omega t \\ \sin \omega t & \cos \omega t \end{pmatrix} = 0,$$

the matrix

$$P_{12} = -\frac{2v}{\omega + \gamma_1 - \gamma_2} \begin{pmatrix} \sin \omega t & \cos \omega t \\ -\cos \omega t & \sin \omega t \end{pmatrix}$$

is a partial solution bounded for all $t \in \mathcal{T}_\tau$.

Thus, for the function $v(t, x, y) = \eta^{\mathrm{T}} U(t, x, y)\eta$ it is easy to establish the estimate of (2.4.7) type with matrices $\mathcal{C}$ and $\bar{C}$ in the form

$$\mathcal{C} = \begin{pmatrix} \varsigma_{11} & \varsigma_{12} \\ \varsigma_{12} & \varsigma_{22} \end{pmatrix}, \qquad \bar{C} = \begin{pmatrix} \bar{c}_{11} & \bar{c}_{12} \\ \bar{c}_{12} & \bar{c}_{22} \end{pmatrix},$$

where $\bar{c}_{11} = \varsigma_{11} = 1$; $\bar{c}_{22} = \varsigma_{22} = 1$, $\bar{c}_{12} = -\varsigma_{12} = \frac{|2v|}{|\omega + \gamma_1 - \gamma_2|}$. Besides, the vector $u_1{}^{\mathrm{T}} = (\|x\|, \|y\|) = u_2{}^{\mathrm{T}}$ since the system (5.1) is linear.

For system (5.4.9) the estimate (2.4.9) becomes

$$Dv(t, x, y)\big|_{(5.1)} = 0$$

for all $(x, y) \in R^2 \times R^2$ because $M = 0$.

Due to (5.4.12) the motion stability conditions for system (5.4.9) are established based on the analysis of matrices $\mathcal{C}$ and $\bar{C}$ property of having fixed sign.

It is easy to verify that the matrices $\mathcal{C}$ and $\bar{C}$ are positive definite, if

$$1 - \frac{4v^2}{(\omega + \gamma_1 - \gamma_2)^2} > 0.$$

Consequently, the motion of nonautonomously connected oscillators is uniformly stable in the whole, if

$$|v| < \frac{1}{2}|\omega + \gamma_1 - \gamma_2|.$$

5.5 Absolute Stability of Discrete Systems

As noted before (see Chapter 1) the vector Liapunov function is a partial case of the matrix-valued function. The development of the method of constructing the vector Liapunov function associated with the employment of hierarchical structure of the system under consideration yields the refinement of results obtained in terms of simple vector function. To illustrate this statement we shall consider the application of Theorem 3.2.1 in the Lur'e-Postnikov problem for large scale discrete system.

We now consider systems described by difference equations of the form

$$x_{11}(\tau+1) = 0.2\,x_{11}(\tau) + 0.1\,f_{11}(\tau,\sigma_1)$$
$$+ p\,h_{11}^{(1)}(\tau,\sigma_1) + p\,h_{12}^{(1)}(\tau,\sigma_2),$$
$$x_{12}(\tau+1) = 0.5\,x_{12}(\tau) + 0.1\,f_{12}(\tau,\sigma_1)$$
$$+ p\,h_{21}^{(1)}(\tau,\sigma_1) + p\,h_{22}^{(1)}(\tau,\sigma_2),$$

(5.5.1)

$$x_{21}(\tau+1) = 0.5\,x_{21}(\tau) + 0.2\,f_{21}(\tau,\sigma_2)$$
$$+ p\,h_{11}^{(2)}(\tau,\sigma_1) + p\,h_{12}^{(2)}(\tau,\sigma_2),$$
$$x_{22}(\tau+1) = 0.8\,x_{22}(\tau) + 0.2\,f_{22}(\tau,\sigma_2)$$
$$+ p\,h_{21}^{(2)}(\tau,\sigma_1) + p\,h_{22}^{(2)}(\tau,\sigma_2),$$

where $\tau \in \mathcal{T}$, $x_{ij} \in R$, $i=1,2$, $\sigma_1 = x_{11} + x_{12}$, $\sigma_2 = x_{21} + x_{22}$, p is a positive constant, functions $f_{ij}\colon \mathcal{T} \in R \to R$ are such that $f_{ij}[\tau,0]=0$ for all $\tau \in \mathcal{T}$ and the inequalities

$$0 \leqslant \sigma_1 f_{1j}(\tau,\sigma_1) \leqslant \sigma_1^2, \quad 0 \leqslant \sigma_2 f_{2j}(\tau,\sigma_2) \leqslant 0.5\,\sigma_1^2,$$

are satisfied for $i,j=1,2$.

The functions $h_{ij}^{(k)}\colon \mathcal{T} \times R \to R$ such that

$$|h_{ij}^{(k)}(\tau,\sigma_j)| \leqslant |\sigma_j|, \quad \text{for all}\quad \tau \in \mathcal{T}, \quad i,j,k=1,2.$$

We decompose system (5.5.1) into two interconnected subsystems

$$x_{11}(\tau+1) = 0.2\,x_{11}(\tau) + 0.1\,f_{11}(\tau,\sigma_1)$$
$$+ p\,h_{11}^{(1)}(\tau,\sigma_1) + p\,h_{12}^{(1)}(\tau,\sigma_2),$$
$$x_{12}(\tau+1) = 0.5\,x_{12}(\tau) + 0.1\,f_{12}(\tau,\sigma_1)$$
$$+ p\,h_{21}^{(1)}(\tau,\sigma_1) + p\,h_{22}^{(1)}(\tau,\sigma_2),$$
$$x_{21}(\tau+1) = 0.5\,x_{21}(\tau) + 0.2\,f_{21}(\tau,\sigma_2)$$
$$+ p\,h_{11}^{(2)}(\tau,\sigma_1) + p\,h_{12}^{(2)}(\tau,\sigma_2),$$
$$x_{22}(\tau+1) = 0.8\,x_{22}(\tau) + 0.2\,f_{22}(\tau,\sigma_2)$$
$$+ p\,h_{21}^{(2)}(\tau,\sigma_1) + p\,h_{22}^{(2)}(\tau,\sigma_2).$$

Each of the obtained independent subsystems

(5.5.2)
$$x_{11}(\tau+1) = 0.2\,x_{11}(\tau) + 0.1\,f_{11}(\tau,\sigma_1),$$
$$x_{12}(\tau+1) = 0.5\,x_{12}(\tau) + 0.1\,f_{12}(\tau,\sigma_1),$$

$$(5.5.3) \qquad \begin{aligned} x_{21}(\tau+1) &= 0.5\,x_{21}(\tau) + 0.2\,f_{21}(\tau,\sigma_2), \\ x_{22}(\tau+1) &= 0.8\,x_{22}(\tau) + 0.2\,f_{22}(\tau,\sigma_2), \end{aligned}$$

is decomposed into interconnected components

$$\begin{aligned} x_{11}(\tau+1) &= 0.2\,x_{11}(\tau) + 0.1\,f_{11}(\tau,\sigma_1), \\ x_{12}(\tau+1) &= 0.5\,x_{12}(\tau) + 0.1\,f_{12}(\tau,\sigma_1), \\ x_{21}(\tau+1) &= 0.5\,x_{21}(\tau) + 0.2\,f_{21}(\tau,\sigma_2), \\ x_{22}(\tau+1) &= 0.8\,x_{22}(\tau) + 0.2\,f_{22}(\tau,\sigma_2). \end{aligned}$$

In results we get four independent components

$$\begin{aligned} x_{11}(\tau+1) &= 0.2\,x_{11}(\tau), \\ x_{12}(\tau+1) &= 0.5\,x_{12}(\tau), \\ x_{21}(\tau+1) &= 0.5\,x_{21}(\tau), \\ x_{22}(\tau+1) &= 0.8\,x_{22}(\tau). \end{aligned}$$

Take the functions

$$v_{ij} = |x_{ij}|, \quad \psi_{ij} = |x_{ij}|, \quad i,j = 1,2.$$

Compute the constants

$$\pi_{11} = 0.8, \quad \pi_{12} = 0.5, \quad \xi_{11}^1 = \xi_{12}^1 = \xi_{21}^1 = \xi_{22}^1 = 0.1$$

and the matrix

$$W_1 = \begin{pmatrix} 0.7 & -0.1 \\ -0.1 & 0.4 \end{pmatrix},$$

which is the M-matrix. Choose the constants $d_{11} = 1$ and $d_{12} = 5$, then

$$a_1{}^{\mathrm{T}}W_1 = (1;5)\begin{pmatrix} 0.7 & -0.11 \\ -0.1 & 0.4 \end{pmatrix} = (0.2;\,1.9).$$

For the function

$$v_1(x_1) = |x_{11}| + 5\,|x_{12}|$$

the estimate

$$\Delta v_1(x_1)\big|_{(5.5.2)} \leqslant -0.2\,|x_{11}| - 1.9\,|x_{12}|$$

is valid.

Similarly we have

$$\pi_{21} = 0.5, \quad \pi_{22} = 0.2, \quad \xi_{11}^2 = \xi_{12}^2 = \xi_{21}^2 = \xi_{22}^2 = 0.1,$$

$$W_2 = \begin{pmatrix} 0.4 & -0.1 \\ -0.1 & 0.1 \end{pmatrix}.$$

The matrix W_2 is also the M-matrix. We take $d_{21} = 1$ and $d_{22} = 2$. Then we get

$$a_2{}^T W_2 = (1; 2) \begin{pmatrix} 0.4 & -0.11 \\ -0.1 & 0.1 \end{pmatrix} = (0.2; 0.1).$$

For the function

$$v_2(x_2) = |x_{21}| + 2|x_{22}|$$

we obtain the inequality

$$\Delta v_2(x_2)\big|_{(5.5.3)} \leqslant -0.2|x_{21}| - 0.1|x_{22}|.$$

Take the functions

$$\psi_1 = |x_{11}| + |x_{12}|, \quad \psi_2 = |x_{21}| + |x_{22}|$$

and compute the constants

$$\pi_1 = 0.2, \quad \pi_2 = 0.1, \quad \xi_{11} = \xi_{12} = 6p, \quad \xi_{21} = \xi_{22} = 3p.$$

The matrix

$$W = \begin{pmatrix} 0.2 - 6p & -6p \\ -3p & 0.1 - 3p \end{pmatrix}$$

is the M-matrix, if $p < 1/60$. We take

$$d_1 = 18p^2 - 0.6p + 0.001, \quad d_2 = -36p^2 + 1.2p.$$

Then

$$a^T W = (18p^2 - 0.6p + 0.001; \ -36p^2 + 1.2p) \begin{pmatrix} 0.2 - 6p & -6p \\ -3p & 0.1 - 3p \end{pmatrix} =$$

$$= (3.6p^2 - 0.18p + 0.002; \ 0.48p - 3.6p^2).$$

For $0 < p < 1/60$ the vector a has positive components. Since all conditions of Theorem 3.7.1 are satisfied, the equilibrium state $x = 0$ of system (5.5.1) is asymptotically stable in the whole. The function

$$v(x) = (18p^2 - 0.6p + 0.01)(|x_{11}| + 5|x_{12}|) + (-36p^2 + 1.2p)(|x_{21}| + |x_{22}|)$$

is the hierarchical Liapunov function for system (5.5.1).

The study of a discrete system as hierarchical structure is adequate for analyzing the complex evolution of a real system in animate or inanimate nature (see Bronowski [1], Levins [1], Šiljak [1] and Simon [1]). The Liapunov function corresponding to the structure of such the system is also hierarchical. It is natural to expect that stability conditions established via such a function prove to be more close to the required ones and the estimates of stability domains are more precise as compared with those obtained in terms of an ordinary vector Liapunov function.

5.6 Notes

5.1. The paper by Zubov [1] (the development of the idea of this work is presented in Zubov [3]) is the first attempt to solve the problem of constructing the Liapunov function which determines a complete domain of asymptotic stability or its boundary.

Aulbach [1] proved the analogue to Zubov's method in the case of equilibria of nonautonomous differential systems where he had to restrict time dependence to almost periodicity. In the second part of the paper Aulbach [2] established the extension of Zubov's method to autonomous systems with asymptotically stable limit cycles.

Various aspects of Zubov's method are developed by Abu Hassan and Storey [1], Bertoni [1], Casti [1], Fallside, Patel, *et al.* [1], Kirin, Nelepin, *et al.* [1], Knight [1], Prabhakara, El-Abiad, *et al.* [1], etc.

5.2. The results of this section are due to Krapivnyi [1], Krapivnyi and Martynyuk [1], and Martynyuk [16, 17]. See also Michel, Sarabudla, *et al.* [1], Rodden [1], and Rosenbrock [1] are used.

5.3. The equations of the three-machine power system are due to Šiljak [1]. The obtained result is compared with that by Abdullin, Anapolskii, *et al.* [1].

5.4. Three-mass systems model satisfactorily the processes in dynamics of machines with elastic links and other real objects. Our presentation is based on the results by Djordjević [3], Krapivnyi [1], and Martynyuk and Slyn'ko [1].

5.5. Discrete-time systems are the object of investigation of many papers (see Martynyuk [12], Michel, Wang, *et al.* [1], etc.). This section is based on the results by Lukyanova and Martynyuk [1].

For the beginner investigators interested in various approaches that were applied before to solve the problems of estimating the stability and attraction domains we recall the works by Burland and Sarlos [1], Chiang and Thorp [1], Dikin, Shelkunova, *et al.* [1], Foster and Davies [1], Garg and Rabins [1], Genesio, Gartaglia, *et al.* [1], Genesio and Vicino [1], Loparo and Blankenship [1], Martynyuk and Radzishewski [1], Noldus, Galle, *et al.* [1], Sastry [1], Walker and McClamroch [1], Weissenberger [1], Willems [1], etc.

REFERENCES

Abdullin, R. Z., Anapolskii, L. Yu., et al.
[1] *Method of Vector Lyapunov Functions in Stability Theory*. Moscow: Nauka, 1987. [Russian].

Abu Hassan, M. and Storey, C.
[1] Numerical determination of domains of attraction for electrical power systems using the method of Zubov. *Int. J. Control* **34** (1981) 371–381.

Akinyele, O.
[1] Cone valued Lyapunov functions and Lipschitz stability of certain nonlinear perturbed systems. *An. Stiint. Univ. Al. I. Cuza Iasi, Ser. Noua, Mat.* **42**(1) (1996) 65–70.

Akpan, E. P.
[1] Stability of large scale systems and the method of cone-valued Lyapunov functions. *Nonlin. Anal.* **26** (1996) 1613–1620.

Aliev, F. A. and Larin, V. B.
[1] *Optimization of Linear Control Systems. Analytical Methods and Computational Algorithms*. Amsterdam: Gordon and Breach Publishers, 1998.

Araki, M., Ando, K. and Kondo, B.
[1] Stability of sampled-data composite systems with many nonlinearities. *IEEE Trans. Automat. Control* **AC-16** (1971) 22–27.

Aulbach, B.
[1] Asymptotic stability regions via extensions of Zubov's method. I. *Nonlin. Anal.* **7** (1983) 1431–1440.
[2] Asymptotic stability regions via extensions of Zubov's method. II. *Nonlin. Anal.* **7** (1983) 1441–1454.

Azbelev, N. V.
[1] On the Chaplygin problem. About a method of estimations. *Author's abstract of doctor dissertation in physicomathematical sciences.* Izhevsk, 1962.

Bailey, F. N.
[1] The application of Lyapunov's second method to interconnected systems. *J. Soc. Ind. Appl. Math. Ser. A* **3** (1965) 443–462.

Bainov, D. D. and Dishliev, A. B.
[1] Sufficient conditions for absence of "beating" in systems of differential equations with impulses. *Appl. Anal.* **18** (1984) 67–73.

Bainov, D. D. and Kulev, G.
[1] Application of Lyapunov's direct method to the investigation of global stability of solutions of systems with impulse effect. *Appl. Anal.* **26** (1988) 255–270.

Bainov, D. D. and Simeonov, P. S.
[1] *Systems with Impulse Effect Stability, Theory and Applications.* New York, etc.: Halsted Press, 1989.

Barbashin, Ye. A.
[1] On stability with respect to impulsive perturbations. *Diff. Uravn.* **2** (1966) 863–871.
[2] *The Liapunov Functions.* Moscow: Nauka, 1967. [Russian].

Barbashin, Ye. A. and Krasovskii, N. N.
[1] On the stability of motion in the large. *Dokl. Akad. Nauk SSSR* **86** (1952) 453–456. [Russian].
[2] On the existence of Liapunov functions in the case of asymptotic stability in the whole. *Prikl. Math. Mekh.* **18** (1954) 345–350. [Russian].

Barnett, S. and Storey, C.
[1] *Matrix Methods in Stability Theory.* London: Nelson, 1970.

Basson, M. and Fogarty Michael, J.
[1] Harvesting in discrete-time predator-prey systems. *Math. Biosci.* **141**(1) (1997) 41–74.

Begmuratov, K. A. and Martynyuk, A. A.
[1] A hierarchical method for constructing Lyapunov functions in the theory of the stability of unperturbed motion. *Prikl. Mekh.* **29**(5) (1993) 80–87. [Russian].

Begmuratov, K., Martynyuk, A. A. and Miladzhanov, V. G.

[1] A method of constructing Lyapunov's matrix functions for Lur'e-Postnikov nonautonomous systems. *Engineering Simulation* **13** (1995) 1–12.

Bellman, R.

[1] *Stability Theory of Differential Equations.* New York: Academic Press, 1953.

[2] Vector Lyapunov functions. *SIAM Journ. Contr., Ser. A1* (1962) 32–34.

Bertoni, G.

[1] An application of Zubov's method by means of series expansion. *Atti Accad. Sci. Ist. Bologna Rc* **12**(1–2) (1967/68) 121–134. smallskip

Bhatia, N. P. and Szegö, G. P.

[1] *Dynamical Systems: Stability Theory and Applications.* Berlin: Springer-Verlag, 1967.

Birkhoff, G. G.

[1] *Dynamical Systems.* Moscow-Leningrad: GITTL, 1941. [Russian].

Blaquiere, A.

[1] Differential games with piecewise continuous trajectories. In: *Lecture Notes in Control and Information Sciences*, New York, Springer-Verlag, (1977) 34–69.

Bromberg, P. V.

[1] *Matrix Methods in the Theory of Impulsive Control.* Moscow: Nauka, 1967. [Russian].

Bronowski, J.

[1] New concepts in the evolution of complexity. *Syntheses* **21** (1970) 228–246.

Bulgakov, N. G.

[1] *Semi-Definite Functions in Stability Theory.* Minsk: Izdat. Universiteta, 1984. [Russian].

Burland, G. and Sarlos, G.

[1] Determination of the domain of stability. *J. Mathematical Analysis and Applications* **23** (1968) 714–722.

Carvalho, L. A. V. and Ferreira, R. R.

[1] On a new extension of Liapunov's direct method to discrete equations. *Quarterly of Appl. Math.* **XLVI**(4) (1988) 779–788.

Casti, J.

[1] Zubov procedures for estimating the domain of attraction. *Anal. Comput. Equil. Reg. Stab.* (1975) 154–160.

Chaplygin, S. A.

[1] *New Method of Approximate Integration of Differential Equations. Selected Works.* Moscow: Nauka, 1976. [Russian].

Chen, W.

[1] Lipschitz stability in dynamical systems. *Acta Math. Sin.* **38**(5) (1995) 621–627.

[2] Lipschitz stability and almost periodicity in dynamical systems. *Nonlin. Anal.* **26**(11) (1996) 1811–1821.

Chetaev, N. G.

[1] *Stability of Motion.* Moscow: Nauka, 1990. [Russian].

Chiang, H. D. and Thorp, J. S.

[1] Stability regions of nonlinear dynamical systems: A constructive methodology. *IEEE Trans. Aut. Control* **34** (1989) 1229–1241.

Coppel, W. A.

[1] *Stability and Asymptotic Behaviour of Differential Equations.* Boston: Heath and Co., 1965.

Dannan, F. M. and Elaydi, S.

[1] Lipschitz stability of nonlinear systems of differential equations. *J. Math. Anal. Appl.* **113** (1986) 562–577.

[2] Lipschitz stability of nonlinear systems of differential equations. II. Liapunov functions. *J. Math. Anal. Appl.* **143** (1989) 517–529.

Das, P. C. and Sharma, R. R.

[1] Existence and stability of measure differential equations. *Czech. Math. J.* **22**(97) (1972) 145-158.

Dash, A. T. and Cressman, R.

[1] Polygamy in human and animal species. *Math. Biosci.* **88**(1) (1988) 49–66.

Demidovich, B. P.

[1] *Lectures on the Mathematical Theory of Stability.* Moscow: Nauka, 1967. [Russian].

Dieudonne, J.

[1] *Calcul Infinitesimal.* Paris: Hermann, 1968.

Dikin, I. I., Shelkunova, L. V., Skibenko, V. P. and Voropai, N. I.
[1] A new approach to construction of optimal stability regions of electric power systems on the base of quadratic Lyapunov functions. Energy System Institute, Irkutsk, 1999.

Djordjević M. Z.
[1] Stability analysis of interconnected systems with possibly unstable subsystems. *Systems and Control Letters* 3 (1983) 165–169.
[2] Stability analysis of nonlinear systems by the matrix Lyapunov method. *Large Scale Systems* 5 (1983) 252–262.
[3] Zur Stabilitat Nichtlinearer Gekoppelter Systeme mit der Matrix-Ljapunov-Methode. *Diss. ETH 7690*, Zurich, 1984.

Elaydi, S. and Peterson, A.
[1] Stability of difference equations. Differential equations and applications. *Proc. Int. Conf.* (Columbus/OH (USA)) I (1988) 235–238.

El-Sheikh, M. M. A. and Soliman, A. A.
[1] On Lipschitz stability for nonlinear systems of ordinary differential equations. *Differ. Equ. Dyn. Syst.* 3(3) (1995) 235–250.

Fallside, F., Patel, M. R., Etherton, M., Margolis, S. G. and Vogt, W. G.
[1] Control engineering applications of V.I. Zubov's construction procedure for Liapunov functions. *IEEE Trans. Automatic Control* 10 (1965) 220–222.

Fausett, D. W. and Koksal, S.
[1] Variation of parameters formula and Lipschitz stability of nonlinear matrix differential equations. *World Congress of Nonlinear Analysts 92. Proceedings of the first world congress*, (Ed.: V. Lakshmikantham), Tampa, FL, USA, August 19-26, 1992. 4 volumes. Berlin: de Gruyter, 1996, 1415–1426.

Foster, W. R. and Davies, M. L.
[1] Estimating the domain of attraction for systems with multiple nonlinearities. *Int. J. Control* 15 (1972) 1001–1003.

Fu, J.-H. and Abed, E. H.
[1] Families of Lyapunov functions for nonlinear systems in critical cases. *IEEE Trans. Autom. Control* 38(1) (1993) 3–16.

Furasov, V. D.
[1] *Stability and Stabilization of Discrete Processes*. Moscow: Nauka, 1982. [Russian].

Galperin, E. A. and Skowronski, J. M.
[1] Geometry of V-functions and the Liapunov stability theory. *Nonlin. Anal.* 11(2) (1987) 183-197.

Gantmacher, F. R.
[1] *The Theory of Matrices.* New York: Chelsea Publ. Co, 1960.

Garg, D. P. and Rabins, M. J.
[1] Stability bounds for nonlinear systems designed via frequency domain stability criteria. *Trans. ASME J. Dynamic Systems, Measurement and Control* 104 (1972) 262-265.

Genesio, R., Gartaglia, M. and Vicino, A.
[1] On the estimation of asymptotic stability regions: state of the art and new proposals. *IEEE Trans. Aut. Control* AC-30 (1985) 747-755.

Genesio, R. and Vicino, A.
[1] New techniques for constructing asymptotic stability regions for nonlinear systems. *IEEE Trans. Circ. Systems* CAS-31 (1984) 574-581.

Grujić, Lj. T.
[1] Novel development of Liapunov stability of motion. *Int. J. Control* 22 (1975) 525-549.
[2] On large scale systems stability. *Proc. 12th World Congress IMACS*, Vol. 1, (1988) 224-229.

Grujić, Lj. T., Martynyuk, A. A. and Ribbens-Pavella, M.
[1] *Large Scale Systems Stability under Structural and Singular Perturbations.* Berlin: Springer Verlag, 1987.

Hahn, W.
[1] *Theorie and Anwendung der Direkten Methode von Liapunov.* Berlin, etc.: Springer-Verlag, 1959.
[2] *Stability of Motion.* Berlin: Springer-Verlag, 1967.

Halanay, A. and Wexler, D.
[1] *Qualitative Theory of Impulsive Systems.* Moscow: Mir, 1971.

Hale, J. K.
[1] *Ordinary Differential Equations.* New York: Wiley, 1969.

He, J. X. and Wang, M. S.
[1] Remarks on exponential stability by comparison functions of the same order of magnitude. *Ann. Diff. Eqs.* 7(4) (1991) 409-414.

Hirsch, M. W. and Smale, S.
[1] *Differential Equations, Dynamical Systems, and Linear Algebra.* New York: Academic Press, 1974.

Hristova, S. G. and Bainov, D. D.

[1] Application of Lyapunov's functions for studying the boundedness of solutions of systems with impulses. *COMPEL* **5** (1986) 23–40.

Hsieh, Y.

[1] The phenomenon of unstable oscillation in population models. *Math. Comput. Modelling* **10**(6) (1988) 429–435.

Ikeda, M. and Šiljak, D. D.

[1] Generalized decomposition of dynamic systems and vector Lyapunov functions. *IEEE Trans. Autom. Control* **AC-26**(5) (1981) 1118–1125.

[2] Hierarchical Liapunov functions. *Journal of Mathematical Analysis and Applications* **112** (1985) 110–128.

Jin, G.

[1] Lipschitz stability of general control systems. *Ann. Diff. Eqs.* **12**(4) (1996) 432–440.

Kalman, R. E. and Bertram, J. E.

[1] Control system analysis and design via the "second method" of Liapunov. *I. Trans. of ASME: J. Basic Enf.* **82** (1960) 371–393.

Kameneva, T. A.

[1] On a way of stability analysis of discrete system. *Dokl. Nats. Acad. Nauk of Ukraine* **5** (2001) 45–50. [Russian].

Kamke, E.

[1] Zur theory der systeme gewohnlicher differentialgleichungen. *Acta Math.* **58** (1932) 57–85.

Kim, J.-M., Kye, Y.-H. and Lee, K.-H.

[1] Orbital Lipschitz stability and exponential asymptotic stability in dynamical systems. *J. Korean Math. Soc.* **35**(2) (1998) 449–463.

Kinnen, E. and Chen, C. S.

[1] Liapunov functions derived from auxiliary exact differential equations. *Automatica* **4** (1968) 195–204.

Kirin, N. E.; Nelepin, R. A. and Bajdaev, V. N.

[1] Construction of the attraction region by Zubov's method. *Diff. Eqns.* **17** (1982) 871–880.

Knight, R. A.

[1] Zubov's condition revisited. *Proc. Edinb. Math. Soc., II. Ser.* **26** (1983) 253–257.

Kotelyanskii, D. M.

[1] On some properties of matrices with positive elements. *Matem. Sborn.*
     31 (1952) 497–506. [Russian].

Krapivnyi, Yu. N.

[1] Methods for construction of matrix Liapunov functions and estimation
     of the domain of asymptotic stability for large-scale systems. *Author's
     Abstract of Kandidat Dissertation in Physikomathematical Sciences*,
     Kiev, 1988. [Russian].

Krapivnyi, Yu. N. and Martynyuk, A. A.

[1] Decomposition of large-scale systems in the investigation of stability
     of motion. *Prikl. Mekh.* **24**(8) (1988) 91–98. [Russian].
[2] The Lyapunov matrix functions and stability of large-scale discrete
     systems. *Electron. Model.* **13**(4) (1991) 3–7. [Russian].

Krasovskii, N. N.

[1] *Certain Problems of the Theory of Stability of Motion.* Moscow: Fiz-
     matgiz, 1959. [Russian].

Krylov, N. M. and Bogolyubov, N. N.

[1] *Introduction to Nonlinear Mechanics.* Kiev: Publ. Acad. Nauk of
     Ukraine, 1937. [Russian].

Kudo, M.

[1] On integral stability and the uniform integral stability of nonlinear
     differential equations by using comparison principle. *Research reports
     of Akita National College of Technology* **23** (1988) 67–72.
[2] On the uniformly Lipschitz stability of nonlinear differential equations
     by the comparison principle. *Research reports of Akita National Col-
     lege of Technology* **26** (1991) 41–44.

Kulev, G. K.

[1] Uniform asymptotic stability in impulsive perturbed systems of dif-
     ferential equations. *Journal of Computational and Applied Mathema-
     tics* **41** (1992) 49–55.

Kuo, B. C.

[1] *Digital Control Systems.* Illinois, SRL Publishing Company, Cham-
     paign, 1977.

Kurzweil, J. and Papaschinopoulos, G.

[1] Structural stability of linear discrete systems via the exponential di-
     chotomy. *Czech. Math. J.* **38**(113) No. 2. (1988) 280–284.

Ladde, G. S.

[1] Cellular systems I: stability of chemical systems. *Math. Biosci.* **29** (1976) 309–330.

Lakshmikantham, V., Bainov, D. D. and Simeonov, P. S.

[1] *Theory of Impulsive Differential Equations.* Singapore: World Scientific, 1989.

Lakshmikantham, V. and Leela, S.

[1] *Differential and Integral Inequalities.* New York: Academic Press, 1969.

[2] Cone-valued Lyapunov functions. *Nonlin. Anal.* **1** (1977) 215–222.

Lakshmikantham, V., Leela, S. and Martynyuk, A. A.

[1] *Stability Analysis of Nonlinear Systems.* New York: Marcel Dekker, 1988.

Lakshmikantham, V., Matrosov, V. M. and Sivasundaram, S.

[1] *Vector Lyapunov Functions and Stability Analysis of Nonlinear Systems.* Amsterdam: Kluwer Academic Publishers, 1991.

Lakshmikantham, V. and Papageorgiou, N. S.

[1] Cone-valued Lyapunov functions and stability theory. *Nonlin. Anal.* **22** (1994) 381–390.

Lakshmikantham, V. and Salvadori, L.

[1] On Massera type converse theorem in terms of two different measures. *Bull. U.M.* **13** (1976) 293–301.

LaSalle, J. P.

[1] *The Stability of Dynamical Systems.* Philadelphia: SIAM, 1976.

[2] *The Stability and Control of Discrete Processes.* Berlin: Springer-Verlag, 1986.

Lankaster, P.

[1] *Matrix Theory.* Moscow: Nauka, 1978. [Russian].

Larin, V. B.

[1] *Control of Walking Machines.* Kiev: Naukova Dumka, 1980. [Russian].

[2] Problem of control of a hopping apparatus. *J. Franclin Institute* **335B** (1998) 579–593.

Leela, S.

[1] Stability of differential systems with impulsive perturbations in terms of two measures. *Nonlinear Analysis* **1** (1977) 667-677.

Leitmann, G., Udwadia, F. E. and Kryazhimskii, A. V. (Eds.)
[1] *Dynamics and Control.* Amsterdam: Gordon and Breach Science Publishers, 1999.

Levin, A. G.
[1] Lyapunov function construction for systems connected by a common coordinate. *J. Comput. Syst. Sci. Int.* **31**(1) (1993) 36–42.

Levins, R.
[1] Complex systems. In: *Towards a Theoretical Biology.* (Ed.: C.H. Waddington). Edinburgh: Edinburg Univ. Press, 1969, 73–88.

Liapunov, A. M.
[1] *General Problem of Stability of Motion.* Harkov: Math. Soc., 1892. (Published in: *Collected Papers.* Moscow-Leningrad: Ac. Sci. USSR. **2** (1956) 5–263. [Russian].

Liu, X. and Willms, A.
[1] Stability analysis and applications to large scale impulsive systems: A new approach. *Canadian Applied Mathematics Quarterly* **3** (1995) 419–444.

Liu, Y. and Zhang, C.
[1] The problem about construction of the Lyapunov function in nonlinear systems. *Adv. Model. Simul.* **26**(2) (1991) 33–42.

Loparo, K. A. and Blankenship, G. L.
[1] Estimating the domain of attraction of nonlinear feedback systems. *IEEE Trans. Aut. Control* **AC-23** (1978) 602–608.

Lukyanova, T. A. and Martynyuk, A. A.
[1] Connective stability of discrete systems via hierarchical Liapunov function method. *Int. Appl. Mekh.*, (to appear).

Martynyuk, A. A.
[1] *Stability of Motion of Complex Systems.* Kiev: Naukova Dumka, 1975. [Russian].
[2] The Lyapunov matrix function. *Nonlin. Anal.* **8** (1984) 1223–1226.
[3] Extension of the state space of dynamical systems and the problem of stability. *Colloquia Mathematica Societaties Janos Bolyai 47. Differential Equationa: Qualitative Theory*, Szeged (Hungary), 1984, 711–749.
[4] On matrix Liapunov function and stability of motion. *Dokl. AN USSR* **249**(5) (1985) 59–63. [Russian].

[5] Matrix-function of Liapunov's and stability of hybrid systems. *Prikl. Mekh.* **21**(4) (1985) 89–96. [Russian].

[6] The Lyapunov matrix function and stability of hybrid systems. *Nonlin. Anal.* **10** (1986) 1449–1457.

[7] On application of Liapunov matrix-function in the investigation of motion of systems with distributed and lamped parameters. *Teorijska i primenjena mehanika* **14** (1988) 73–83. [Russian].

[8] Hierarchical matrix Lyapunov function. *Differential and Integral Equations* **2**(4) (1989) 411–417.

[9] A theorem on polystability. *Dokl. Akad. Nauk SSSR* **318**(4) (1991) 808–811. [Russian].

[10] Analysis of the stability of nonlinear systems on the basis of Lyapunov matrix functions (a survey). *Prikl. Mekh.* **27**(8) (1991) 3–15. [Russian].

[11] On multistability of motion with respect to some of the variables. *Russian Dokl. Akad. Nauk* **324**(1) (1992) 39–41. [Russian].

[12] Matrix method of comparison in the theory of the stability of motion. *Int. Appl. Mech.* **29** (1993) 861–867.

[13] On exponential polystability of motion. *Teorijska i primenjena mehanika* **20** (1994) 143–151. [Russian].

[14] On exponential multistability of separating motions. *Russian Dokl. Akad. Nauk* **336** (1994) 446–447. [Russian].

[15] Qualitative analysis of nonlinear systems by the method of matrix Lyapunov functions. *Second Geoffrey J. Butler Memorial Conference in Differential Equations and Mathematical Biology (Edmonton, AB, 1992).* Rocky Mountain J. Math. **25**(1) (1995) 397–415.

[16] Forms of aggregation in the study of the stability of motion of large-scale systems. Criteria of stability (Review). *Int. Appl. Mech.* **31** (1995) 683–694.

[17] Forms of aggregation of nonlinear systems. Domains of asymptotic stability (Review). *Int. Appl. Mech.* **32** (1996) 241–255.

[18] Exponential polystability of separating motions. *Ukr. Mat. Zh.* **48**(5) (1996) 642–649. [Russian].

[19] On integral stability and Lipschitz stability of motion. *Ukr. Mat. Zh.* **49** (1997) 84–92.

[20] *Stability by Liapunov's Matrix Functions Method with Applications.* New York: Marcel Dekker, Inc., 1998.

[21] Stability and Liapunov's matrix functions method in dynamical systems. *Prikl. Mekh.* **34**(10) (1998) 144–152. [Russian].

288   REFERENCES

[22] Stability analysis of discrete systems (Survey). *International Applied Mechanics* **36**(7) (2000) 3–35.

Martynyuk, A. A. and Begmuratov, K.
[1]   Hierarchical Lyapunov matrix function and its application. *Electron. Modelling* **13**(5) (1991) 3–9. [Russian].
[2]   Analytical construction of the hierarchical matrix Lyapunov function for impulsive systems. *Ukr. Mat. Zh.* **49** (1997) 548–557. [Russian].

Martynyuk, A. A. and Chernetskaya, L. N.
[1]   On the polystability of linear systems with periodic coefficients. *Dop. Akad. Nauk Ukraine* **11** (1993) 61–65. [Russian].
[2]   On the polystability of linear autonomous systems. *Dop. Akad. Nauk Ukraine* **8** (1993) 17–20. [Russian].
[3]   On boundedness of the solutions to impulsive systems. *Prikl. Mekh.* **33**(7) (1997) 89–94. [Russian].

Martynyuk, A. A. and Krapivnyi, Yu. N.
[1]   *An Application of Lyapunov Matrix Functions in Stability Theory of Discrete Large-Scale Systems.* Kiev: Preprint Inst. of Math. AN Ukr.SSR, 1988. [Russian].

Martynyuk, A. A. and Miladzhanov, V. G.
[1]   *Stability Analysis in Whole of Dynamical System via Matrix Lyapunov Function.* Kiev: Preprint 87.62, Institute of Mathematics, 1987. [Russian].
[2]   Stability analysis of solutions of large-scale impulsive systems. *Electronnoe Modelirovanie*, (to appear).

Martynyuk, A. A., Miladzhanov, V. G. and Begmuratov, K.
[1]   Construction of hierarchical matrix Lyapunov functions. *Differ. Equ. Dyn. Syst.* **1**(1) (1993) 3–21.
[2]   Construction of hierarchical matrix Lyapunov functions. *J. Math. Anal. Appl.* **185** (1994) 129–145.

Martynyuk, A. A. and Obolenskii, A. Yu.
[1]   Stability of autonomous Wazewski systems. *Diff. Uravn.* **16** (1980) 1392–1407. [Russian].
[2]   On the theory of one-side models in spaces with arbitrary cones. *J. of Applied Mathematics and Stochastic Analysis* **3** (1990) 85–97.

Martynyuk, A. A. and Radzishevski, B.
[1]   To the theory of stability of motion. *Matem. Fiz.* **22** (1977) 14–22. [Russian].

Martynyuk, A. A. and Slyn'ko, V. I.
[1] Solution of the problem of constructing Liapunov matrix functions for
     a class of large scale systems. *Nonlinear Dynamics and Systems Theory*
     1(2) (2001) 193-203.
[2] Matrix-valued Lyapunov function and overlapping decomposition of
     dynamical system. *Int. Appl. Mekh.*, (to appear).

Massera, J. L.
[1] On Liapunov's conditions of stability. *Ann. of Math.* **50** (1949)
     705-721.
[2] Contribution to stability theory. *Ann. of Math.* **64** (1956) 182-206.

Matrosov, V. M.
[1] To the theory of stability of motion. *Prikl. Math. Mekh.* **26** (1962)
     992-1002. [Russian].

McShane, E. L.
[1] *Integration.* Princeton: Princeton University Press, 1944.

Mejlakhs, A. M.
[1] Construction of a Lyapunov function for parametrically perturbed
     linear systems. *Autom. Remote Control* **52**(10) pt. 2 (1991) 1479-
     1481.

Mel'nikov, G. I.
[1] Nonlinear stability of motion of the ship on a course. *Vestn. Leningrad.
     Gos. Un-ta* **13**(3) (1962) 90-98. [Russian].

Michel, A. N.
[1] Stability analysis of interconnected systems. *J. SIAM Control* **12**
     (1974) 554-579.

Michel, A. N. and Miller, R. K.
[1] *Qualitative Analysis of Large Scale Dynamical Systems.* New York,
     etc.: Academic Press, 1977.

Michel, A. N., Sarabudla, N. R. and Miller, R. K.
[1] Stability analysis of complex dynamical systems: Some computational
     methods. *Circuits Systems Signal Process* **1** (1982) 171-202.

Michel, A. N., Wang, K., and Hu, B.
[1] *Qualitative Theory of Dynamical Systems. The Role of Stability Pre-
     serving Mappings.* New York, Marcel Dekker, Inc., 2001.

Movchan, A. A.
[1] Stability of processes with respect to two measures. *Prikl. Math. Mekh.*
     24 (1960) 988-1001. [Russian].

Mukhametzyanov, I. A.
[1] Energy method for constructions of the Lyapunov function for nonstationary systems, *Vest. Ross. Univ. Druzh. Nar., Ser. Prikl. Mat. Inf.* **1** (1994) 42–44. [Russian].

Nemytskii, V. V. and Stepanov, V. V.
[1] *Qualitative Theory of Differential Equations.* Moscow-Leningrad: GITTL, 1949. [Russian].

Newman, P. K.
[1] Some notes on stability conditions. *Review of Economic Studies* **72** (1959) 1–9.

Noldus, E., Galle, A. and Josson, L.
[1] The computation of stability regions for systems with many singular points. *Int. J. Control* **17** (1973) 644–652.

Noldus, E., Vingerhoeds, R., and Loccufier, M.
[1] Stability of analogue neural classification networks. *Int. J. Syst. Sci.* **25**(1) (1994) 19–31.

Ohta, Y., Imanishi, H., Gong, L. and Haneda, H.
[1] Computer generated Lyapunov functions for a class of nonlinear systems. *IEEE Trans. Circuits Syst., I, Fundam. Theory Appl.* **40**(5) (1993) 343–354.

Olas, A.
[1] Construction of optimal Lyapunov function for systems with structured uncertainties. *IEEE Trans. Autom. Control* **39**(1) (1994) 167–171.

Pandit, S. G. and Deo, G. D.
[1] *Differential Systems Involving Impulses.* Berlin: Springer-Verlag, 1982.

Pavlidis, T.
[1] Stability of systems described by differential equations containing impulses. *IEEE Trans. Automatic Control* **AC-12** (1967) 43–45.

Peng, Xiaolin
[1] A generalized Lyapunov matrix-function and its applications. *Pure Appl. Math.* **7**(1) (1991) 119–122. [Chinese].
[2] Lipschitz stability of nonlinear differential equations. *Pure Appl. Math.* **9**(1) (1993) 51–56. [Chinese].

Persidskii, K. P.
[1] On stability of motion at first approximation. *Math. Sbor.* **40** (1933) 284–293. [Russian].

Piontkovskii, A. A. and Rutkovskaya, L. D.

[1] Investigation of certain stability theory problems by vector Lyapunov function method. *Autom. i Telemekh.* **10** (1967) 23–31. [Russian].

Porter, B.

[1] *Synthesis of Dynamical Systems.* London: Nelson, 1969.

Pota, H. R. and Moylan, P. J.

[1] A new Lyapunov function for interconnected power systems. *IEEE Trans. Autom. Control* **37**(8) pt.1 (1992) 1192–1196.

Prabhakara, F. S.; El-Abiad, A. H. and Koivo, A. J.

[1] Application of generalized Zubov's method to power system stability. *Internat. J. Control, I. Ser.* **20** (1974) 203–212.

Rodden, J. J.

[1] Numerical application of Lyapunov stability theory. *Proc. JACC* (1964) 261–268.

Rondoni, L.

[1] Autocatalytic reactions as dynamical systems on the interval. *J. Math. Phys.* **34**(11) (1993) 5238–5251.

Rosenbrock, H. H.

[1] An automatic method for finding the greatest or least value of a function. *Comp. Journal* **3** (1960) 175–184.

Rosier, L.

[1] Homogeneous Lyapunov function for homogeneous continuous vector field. *Syst. Control Lett.* **19**(6) (1992) 467–473.

Rouche, N., Habets, P. and Laloy, M.

[1] *Stability Theory by Liapunov's Direct Method.* New York: Springer-Verlag, 1977.

Samoilenko, A. M. and Perestyuk, N. A.

[1] *Impulsive Differential Equations.* Singapore: World Scientific, 1995.

Sastry, V. R.

[1] Finite regions of attraction for the problem of Lur'e. *Int. J. Control* **14** (1971) 789–790.

Schwartz, C. A. and Yan, A.

[1] Construction of Lyapunov functions for nonlinear systems using normal forms. *J. Math. Anal. Appl.* **216**(2) (1997) 521–535.

Sedaghat, H.

[1] A class of nonlinear second order difference equations from macroeconomics. *Nonlin. Anal.* **29**(5) (1997) 593–603.

Sevastyanov, B. A.
[1] Theory of branching stochastic processes. *Uspekhi Matem. Nauk* **6** (1951) 47–99. [Russian].

Shaw, M. D.
[1] Generalized stability of motion and matrix Lyapunov functions. *J. Math. Anal. Appl.* **189** (1995) 104–114.

Šiljak, D. D.
[1] *Large-Scale Dynamic Systems: Stability and Structure.* New York: North Holland, 1978.
[2] *Decentralized Control of Complex Systems.* Boston, etc.: Academic Press, Inc., 1991.
[3] *Nonlinear Systems.* New York: John Wiley & Sons, 1969.

Šiljak, D. D. and Weissenberger, S.
[1] Regions of exponential stability for the problem of Lur'e. *Regelungstechnik* **17** (1969) 27–29.

Simmons, G. F.
[1] *Differential Equations, with Applications and Historical Notes.* New York: McGraw Hill, 1972.

Simon, H. A.
[1] The architecture of complexity. *Proc. Amer. Philos. Soc.* **106** (1962) 467–482.

Simonovits, A.
[1] Chaotic dynamics of economic systems. *Szigma* **18** (1985) 267–277.

Sivasundaram, S. (Ed.)
[1] *Nonlinear Problems in Aviation and Aerospace.* Amsterdam: Gordon and Breach Science Publishers, 2000.

Sivasundaram, S. and Martynyuk, A. A. (Eds.)
[1] *Advances in Nonlinear Dynamics.* Amsterdam: Gordon and Breach Science Publishers, 1997.

Skowronski, J. M.
[1] *Nonlinear Liapunov Dynamics.* Singapore: World Scientific, 1990.

Slyn'ko. V. I.
[1] On polystability of autonomous systems. *Int. Appl. Mekh.*, (to appear).

Sree Hari Rao, V.
[1] On boundedness of impulsively perturbed systems. *Bull. Austral. Math. Soc.* **18** (1978) 237–242.

Szarski, J.
[1] *Differential Inequalities.* Warszawa: PWN, 1967.

Tchuente, M. and Tindo, G.
[1] Suites generees par une equation neuronale a memoire. *C. R. Acad. Sci., Paris, Ser. I.* **317**(6) (1993) 625–630. [French].

Vinograd, R. E.
[1] The inadequacy of the method of characteristic exponents for the study of nonlinear differential equations. *Mat. Sbornik* **41** (1957) 431–438. [Russian].

Vorotnikov, V. I.
[1] To the problems of stability in a part of variables. *Prikl. Mat. Mekh.* **63** (1999) 736–745. [Russian].

Vrkoć, I.
[1] Integral stability. *Czech. Math. J.* **9**(84) (1959) 71–128.

Walker, J. A. and McClamroch, N. H.
[1] Finite regions of attraction for the problem of Lur'e. *Int. J. Control* **6** (1967) 331–336.

Ważewski, T.
[1] Systemes des equations et des ineqalites differentielles ordinaires aux deuxiemes. *Ann. Soc. Poln. Mat.* **23** (1950) 112–166.

Weissenberger, S.
[1] Stability-boundary approximations for relay-control systems via a steepest-ascent construction of Lyapunov functions. *Trans. ASME, J.Basic Eng.* **88** (1966) 419–428.
[2] Stability Regions of Large-Scale Systems. *Automatica* **9** (1973) 653–663.

Willems, J. L.
[1] The computation of finite stability regions by means of open Liapunov surfaces. *Int. J. Control* **10** (1969) 537–544.

Yoshizawa, T.
[1] *Stability Theory by Liapunov's Second Method.* Tokyo: The Math. Soc. of Japan, 1966.

Zubov, V. I.
[1] Some sufficiently criteria of stability of nonlinear system of differential equations. *Prikl. Mat. and Mekh.* **17** (1953) 506–508.
[2] To the theory of second method of A.M.Liapunov. *Dokl. Akad. Nauk SSSR* **99** (1954) 341–344. [Russian].

[3]   *Methods of A.M.Liapunov's and its Applications.* Leningrad: Izdat. Leningrad. Gos. Universitet, 1957. [Russian].

[4]   *Mathematical Investigation of Automatic Control Systems.* Leningrad: Mashinostroeniye, 1979. [Russian].

[5]   Analytic construction of Lyapunov functions. *Russ. Acad. Sci., Dokl., Math.* **49**(2) (1994) 414–417.

SUBJECT INDEX